Japan in Space

Japan in Space

National Architecture, Policy, Legislation and Business in the 21st Century

Masataka Ogasawara and Joel Greer (Eds.)

Published, sold and distributed by Eleven International Publishing
P.O. Box 85576
2508 CG The Hague
The Netherlands
Tel.: +31 70 33 070 33
Fax: +31 70 33 070 30
e-mail: sales@elevenpub.nl
www.elevenpub.com

Sold and distributed in USA and Canada
Independent Publishers Group
814 N. Franklin Street
Chicago, IL 60610
USA
Order Placement: (800) 888-4741
Fax: (312) 337-5985
orders@ipgbook.com
www.ipgbook.com

Eleven International Publishing is an imprint of Boom uitgevers Den Haag.

ISBN 978-94-6236-203-1
ISBN 978-90-8974-528-6 (E-book)

© 2021 Masataka Ogasawara & Joel Greer | Eleven International Publishing

This publication is protected by international copyright law.
All rights reserved. No part of this publication may be reproduced, stored in a retrieval system, or transmitted in any form or by any means, electronic, mechanical, photocopying, recording or otherwise, without the prior permission of the publisher.

Table of Contents

Abbreviations	vii
Preface	ix
Chapter 1 — Brief History of Japan's Space Development: Organizations, Policies, and Laws	1
Chapter 2 — Japan's Key Space Development Organizations Today	9
Chapter 3 — Recent Legislation: 2016 Space Activities Act	17
Chapter 4 — Recent Legislation: 2016 Remote Sensing Act	29
Chapter 5 — For Foreign Investors: Other Applicable Laws In and Outside Japan	39
Chapter 6 — Japan's Space Road Map: 2020 Basic Space Plan	47
Chapter 7 — Japan's Burgeoning Space Industry	67
Select Sources	85
Annexes	93
Annex 1 – Act on Launching of Spacecraft, etc. and Control of Spacecraft (Act No. 76 of 2016)	95
Annex 2 – Review Standards and Standard Period of Time for Process Relating to Procedures under the Act on Launching of Spacecraft, etc. and Control of Spacecraft	125
Annex 3 – Application Manual for Act on Launching of Spacecraft, etc. and Control of Spacecraft	143
Annex 4 – Act on Ensuring Appropriate Handling of Satellite Remote Sensing Data (Act No. 77 of November 16, 2016)	223

Table of Contents

Annex 5 – Guidelines on Measures, etc. Under Act on Ensuring Appropriate Handling of Satellite Remote Sensing Data   247

Annex 6 – Application Manuals for Act on Ensuring Appropriate Handling of Satellite Remote Sensing Data   287

Annex 7 – Basic Space Law (Law No. 43 of 2008)   321

Annex 8 – Summary of Space Industry Vision 2030 Creating Space Use in the Fourth Industrial Revolution   331

Annex 9 – Changes to the Basic Space Plan, June 30, 2020   337

About the Authors   381

ABBREVIATIONS

AJIL	American Journal of International Law
CNT	Carbon Nanotubing
ESA	European Space Agency
FCC	Federal Communications Commission
FX Act	Foreign Exchange and Foreign Trade Act
GITAI	GITAI Japan Co., Ltd.
GPS	Global Positioning System
ISAS	Institute of Space and Aeronautical Science
ISS	International Space Station
IST	Interstellar Technologies Corporation
JASDF	Japan Air Self-Defense Force
JAXA	Japan Aerospace Exploration Agency
J-SPARC	JAXA-Space Innovation through Partnership and Co-creation
JSAT	SKY Perfect JSAT Corporation
LEO	Low Earth Orbit
LNG	Liquified Natural Gas
METI	Ministry of Economy, Trade, and Industry
MHI	Mitsubishi Heavy Industries, Ltd.
MITI	Ministry of International Trade and Industry
NAL	National Aerospace Laboratory of Japan
NASA	National Aeronautics and Space Administration
NASDA	National Space Development Agency
PDAS	PD Aerospace Co.
QZSS	Quasi-Zenith Satellite System
SSA	Space Situational Awareness
UNOPS	United Nations Office for Project Services
ZLW	Zeitschrift für Luft- und Weltraumrecht

Preface

This book offers a short survey of a large topic: Japan's space program, including its space organizations, laws, and policies, as well as a look at some of the important and interesting players in Japan's space industry today. It is meant for readers who have a general interest in Japan's space development or who (whether in or outside Japan) may be considering participation in a growing and increasingly diverse array of Japanese space-related business opportunities.

Japan was an early pioneer in space technology and, in 1970, became the fourth spacefaring nation.[1] As of the writing of this book, the Japanese satellite *Hayabusa2* has recently returned to Earth from a pioneering mission to collect surface and sub-surface samples from an asteroid. Yet Japan was relatively slow to embrace the commercial potential of space and, relatedly, was "a latecomer in the field of national space legislation," with enactment of a foundational Basic Space Law only in 2008.[2] This book addresses the reasons for this circumstance while tracing the history of Japan's space organizations, policies, and laws in Chapter 1.

The focus of this book, however, is on the present day. Chapter 2 delineates the current organizational mainstays of Japan's space program. These leading organizations are the product of a long process of consolidation, and Japan now has a coordinated centralized administration to conceive, plan, and execute its space objectives. One of these goals, as discussed, is to encourage the private sector – major established corporations, mid-sized firms, and young start-ups alike – to consider and take advantage of the wide-ranging commercial possibilities of space development and use.

Supporting new business opportunities while managing the attendant risks, especially in an area such as space development that carries not insignificant hazards and also is the subject of international treaty obligations, entails the need for more specific national legislation. Chapters 3 and 4 address in some detail the two laws Japan enacted in 2016 to regulate its space activities: the Act on Launching of Spacecraft, etc. and Control of Spacecraft, and the Act on Ensuring Appropriate Handling of Satellite Remote Sensing Data.

In addition to these two space laws, non-Japanese participants in Japan-based space activities should be aware of more general Japanese foreign investment legislation, which is introduced in Chapter 5. This chapter also discusses the possibility that such

---

[1] *See* S. Aoki, "Current Status and Recent Developments in Japan's National Space Law and its Relevance to Pacific Rim Space Law and Activities" ("Current Status and Recent Developments in Japan's National Space Law"), 35 *Journal of Space Law*, Issue 2, 2009, p. 363, at https://spacelaw.sfc.keio.ac.jp/sitedev/archive/current.pdf.

[2] *Id.*

PREFACE

participants may be subject to space-related laws of their home jurisdiction in addition to those of Japan.

Since 2008, pursuant to its Basic Space Law, the Japanese government has issued several versions of the country's Basic Plan on Space Policy. This important document sets forth broad policy goals and specific programs of Japan's space development. Chapter 6 examines the most recent version of the Basic Plan on Space Policy issued in mid-2020.

Chapter 7 describes a number Japanese companies – some large, some small – which currently are involved in various space-related businesses. The firms featured in this chapter are just a sample of Japan's private sector engaged in the commercialization of space, but they are illustrative of the sophistication and creativity being brought to bear as Japan and the world enter the 21st century's third decade.

CHAPTER 1 BRIEF HISTORY OF JAPAN'S SPACE DEVELOPMENT: ORGANIZATIONS, POLICIES, AND LAWS

I   ORIGINS AND EARLY YEARS OF JAPAN'S SPACE DEVELOPMENT PROGRAM

From the end of World War II to the 1951 Treaty of San Francisco, which officially ended hostilities between Japan and the United States and restored Japan's national sovereignty, Japan was unable to pursue aerospace development. Thus the history of Japan's space development began during the early 1950s, not long before the Soviet Union's 1957 launch of Sputnik 1.

Taking the lead was a research group at the University of Tokyo led by aeronautical engineer Dr. Hideo Itokawa, popularly known in Japan as "Dr. Rocket." Dr. Itokawa and his team initially developed the so-called "Pencil Rocket" – just 23 centimeters long, 1.8 centimeters in diameter, and weighing 200 grams – which was launched in 1955 to an altitude of 600 meters.[1] Efforts in the next several years concentrated on advancing Japanese rocket engineering with larger projectiles and creating a nascent space program infrastructure. By the early 1960s Japan's "Kappa" rockets were reaching altitudes of 200 kilometers, a Space Development Council was established in the Prime Minister's office, and the country had a launch complex called the Kagoshima Space Center.[2]

During the 1960s Japan's space development budget grew and its space program administration evolved. An early example was the Institute of Space and Aeronautical Science (ISAS), a national research organization that was founded at the University of Tokyo in 1964 and worked on developing rocket systems using solid fuel to launch

---

1   See Japan Aerospace Exploration Agency, *Shoki-Kogata-Rocket-Jidai* [Early Small Rockets Era], at https://spaceinfo.jaxa.jp/ja/early_small_rockets.html; *see also* "Pencil Rocket," at https://www.u-tokyo.ac.jp/en/whyutokyo/hongo_hi_010.html; "Pencil Rocket in Michikawa," at https://global.jaxa.jp/article/interview/sp1/episode-4_p2_e.html.
2   See Japan Aerospace Exploration Agency, Uchinoura-Uchukukan-Kansokujo [Uchinoura Space Center], at https://spaceinfo.jaxa.jp/ja/usc.html; "Japan's Space Program," July 26, 2019, at https://www.nippon.com/en/japan-data/h00501/japan%E2%80%99s-space-program.html; Brief History of Japanese Space Research, at https://www.isas.jaxa.jp/e/japan_s_history/brief.shtml; T. Wakimoto, *A Guide to Japan's Space Policy Formulation: Structures, Roles and Strategies of Ministries and Agencies for Space* (*A Guide to Japan's Space Policy Formulation*) (Pacific Forum, April 2019), p. 10, note 44, at https://pacforum.org/wp-content/uploads/2019/04/issuesinsights_Vol19WP3_0.pdf.

atmospheric and astronomical satellites.[3] In 1968, the Space Development Council was reorganized as the Space Activities Commission and charged with coordinating Japanese space development policy.[4] A year later, the National Space Development Agency (NASDA) was created and developed rockets to launch satellites using liquid fuel.[5] In 1970, Japan successfully launched its first communications satellite, *Ohsumi*, which ISAS operated.[6]

Japan began to engage internationally in space-related activities during this period. The United States and Japan started formally to cooperate on unclassified rocket technology through the 1969 *Exchange of Notes Concerning Cooperation in Space Exploration*.[7] In the early 1970s, Japan also began partnering with Europe on space science programs, and this cooperation continues with endeavors such as the BepiColombo satellite mission, which is a joint project by the European Space Agency (ESA) and the Japan Aerospace Exploration Agency (JAXA) to explore Mercury over the coming decade.[8]

From the outset, and for many years thereafter, Japan's space development was driven by the Japanese government and involved satellite delivery and tracking systems for purposes unrelated to national security. In 1967, Japan ratified the Treaty on Principles Governing the Activities of States in the Exploration and Use of Outer Space, including the Moon and Other Celestial Bodies (1967 Outer Space Treaty). Among other things, the 1967 Outer Space Treaty prohibits the placing of weapons of mass destruction in space and restricts the use of the Moon and other celestial bodies "exclusively for peaceful purposes."[9] Two years later, the language in Japan's Law Concerning the National Space Development Agency of Japan (*Law* No. 50 of June 23, *1969*) echoed that of the 1967 Outer Space Treaty in stipulating that the country's space development was "exclusively for peaceful purposes."[10] While some nations considered that the 1967 Outer Space Treaty did not proscribe space development for non-aggressive military purposes, the position of Japan, which of course has a pacifist constitution that

---

3 See S. Kozuka and M. Sato, *Uchū business notameno uchū hō nyumon* [*An Introduction to Space Law for Space Business*] (Yuhikaku, 2018), p. 20; Brief History of Japanese Space Research, at https://www.isas.jaxa.jp/e/japan_s_history/brief.shtml; "Current Status and Recent Developments in Japan's National Space Law," p. 375, notes 33 and 35.
4 See "Current Status and Recent Developments in Japan's National Space Law," p. 371.
5 See *id.*, p. 375, note 35; An Introduction on Space Law for Space Business, p. 164; NASDA History, at https://global.jaxa.jp/about/history/nasda/index_e.html.
6 See Ohsumi, at https://www.isas.jaxa.jp/en/missions/spacecraft/past/ohsumi.html.
7 See "Current Status and Recent Developments in Japan's National Space Law," pp. 364-71.
8 See *id.*, p. 364; https://sci.esa.int/web/bepicolombo.
9 Treaty on Principles Governing the Activities of States in the Exploration and Use of Outer Space, including the Moon and Other Celestial Bodies, Article IV, at https://www.unoosa.org/oosa/en/ourwork/spacelaw/treaties/outerspacetreaty.html.
10 See *Uchū kaihatsu jigyodan hō* [Law Concerning the National Space Development Agency of Japan] (Law No. 50 of June 23, 1969, as amended), Article 1, at https://www.unoosa.org/oosa/en/ourwork/spacelaw/nationalspacelaw/japan/nasda_1969E.html (unofficial English translation).

formally renounces war, was, then and until 2008, to construe "peaceful" as meaning "non-military" and to restrict its space program to activities with non-military purposes.[11]

Despite its quick accession to the 1967 Outer Space Treaty, Japan was slower to ratify other major international space treaties drafted in the late 1960s and early 1970s. Not until 1983 did Japan ratify the 1968 Agreement on the Rescue of Astronauts, the Return of Astronauts, and the Return of Objects Launched into Outer Space, the 1972 Convention on International Liability for Damage Caused by Space Objects, and the 1976 Convention on Registration of Objects Launched into Outer Space. The delay related partly to the Japanese's government's view that accession to these international treaties was not urgent, and partly to the government's desire to examine whether ratification would oblige Japan to introduce new domestic space-related legislation. Having concluded that no new domestic law was required at that time, Japan ratified each of the three treaties in 1983.[12]

Professor Setsuko Aoki of Keio University has commented that Japan was "a latecomer in the field of national space legislation."[13] She attributes this circumstance mainly to an inadequate level of space-related activity by Japanese non-governmental actors until the early 21st century, without which Japan had little need for national legislation to regulate and support a commercial space industry.[14] Professor Aoki also links this slow pace of private space development to Japan's policy of restricting its space program to activities with non-military purposes: "It is often pointed out that without a continuous governmental military program, it is difficult to establish a robust space industry under which the private business sector can flourish."[15] Thus, Professor Aoki notes, while various Japanese government ministries commissioned a number of communications, broadcasting, and meteorological satellites from Japanese companies beginning in the 1980s, a commercial satellite launch did not occur in Japan until 2008.[16]

---

11 See "Current Status and Recent Developments in Japan's National Space Law," p. 367; *A Guide to Japan's Space Policy Formulation*, p. 1.
12 See K. Uga, *Chikujo kaisetsu uchū nihō* [*Commentary on Two Space Laws*] (Kobundo 2019), pp. 2-3. Japan has not ratified the Agreement Governing the Activities of States on the Moon and Other Celestial Bodies.
13 "Current Status and Recent Developments in Japan's National Space Law," p. 363.
14 *Id.*, pp. 366-67.
15 *Id.*, p. 367.
16 *Id.*, pp. 367-69. In 1989, the Japan Communications Satellite Company (JCSC) commissioned the manufacture by Hughes Aircraft Company and launch from Kourou, French Guiana (the location the spaceport of France and the ESA) of the first Japanese commercial communications satellite, JCSAT-1. See JCSAT-1, at https://nssdc.gsfc.nasa.gov/nmc/spacecraft/display.action?id=1989-020A. In 2008, JCSC's then-successor, JSAT Corporation, operated a communications satellite that Mitsubishi Electronics manufactured and Mitsubishi Heavy Industries launched, which was the first such commercial launch in Japan. *See id.* JSAT Corporation is now SKY Perfect JSAT Corporation, discussed further in Chapter 7.

II INTERNATIONALIZATION AND MODERNIZATION OF JAPAN'S SPACE DEVELOPMENT PROGRAM

Notwithstanding the absence of domestic commercial space activity, Japan continued to develop more powerful and precise rockets, including the H-I and H-II vehicles in the 1980s and 1990s, respectively, to deliver satellites into space.[17] The H-II was some 200 times taller, and well over one million times heavier, than the "Pencil Rocket." Japan and the United States continued to cooperate on space-related commercial and scientific activities, as reflected in various documents: the 1990 *Exchange of Notes concerning the Policy and Procedure of R&D and Procurement of Artificial Satellites*;[18] the 1995 *Agreement between the United States and Japan Concerning Cross-Waiver of Liability for Cooperation in the Exploration and Use of Space for Peaceful Purposes*; and the 1998 *U.S.-Japan Joint Statement on GPS Cooperation*. During this period Japan also deepened its bilateral cooperative activities with the United States. Japanese astronauts first participated in a U.S. Space Shuttle mission in 1992, and in 1994 astronaut Chiaki Mukai became the first Japanese woman to travel to space when she joined a Space Shuttle mission (she also was a member of a second Space Shuttle spaceflight in 1998).[19]

In 1998, Japan signed the 15-nation International Space Station (ISS) intergovernmental agreement (along with a *Memorandum of Understanding between the Government of Japan and the National Aeronautics and Space Administration of the United States of America Concerning Cooperation on the Civil International Space Station*). Japan has participated actively in ISS missions, including by contributing the Japanese Experimental Module, a research facility also called *Kibō* ("hope" in Japanese), as well as the cargo spacecraft *Kounotori* ("white stork" in Japanese) and the H-IIB rocket on which *Kounotori* was launched.[20]

Within Japan reorganization of its space program administration began to proceed in earnest at the start of the new millennium. In 2001, supervision of the Space Activities Commission was assumed by the Ministry of Education, Culture, Space, Science and Technology, although around this time comprehensive strategic space planning began to be undertaken by a newly created Council for Science and Technology Policy

---

17 See About H-I Launch Vehicle, at https://global.jaxa.jp/projects/rockets/h1/index.html; About H-II Launch Vehicle, at https://global.jaxa.jp/projects/rockets/h2/index.html; *see also* An Introduction on Space Law for Space Business, p. 157.
18 According to Professor Aoki, this agreement, which opened up Japan's satellite procurement to U.S. and other foreign manufacturers, also was a significant factor in slowing the emergence of a domestic Japanese commercial satellite industry. See "Current Status and Recent Developments in Japan's National Space Law," p. 368.
19 *See Mukai Chiaki: JAXA-No-Uchūhikoshi* [Mukai Chiaki: Astronauts of JAXA], at https://iss.jaxa.jp/astro/mukai/; Mukai Chiaki, at https://www.britannica.com/biography/Mukai-Chiaki.
20 *See Kibou-To-Konotori-Ga-Hiraku-Tsugino-Stage* [Stage opened by Kibou and Konotori], at https://www.jaxa.jp/projects/iss_human/; About H-II Transfer Vehicle "KOUNOTORI," at https://global.jaxa.jp/projects/rockets/htv/.

(CSTP) in the Prime Minister's Cabinet Office (the Cabinet Office is an agency headed by the Prime Minister that handles day-to-day affairs of the Cabinet).[21] Two years later, pursuant to the Law on the Japan Aerospace Exploration Agency (2002 JAXA Law), the ISAS, NASDA, and another aerospace laboratory were consolidated into JAXA, which became responsible for developing and launching Japanese satellites into space (again, "for peaceful purposes only"), as well as other projects such as asteroid and planetary exploration.[22]

More fundamentally, the enactment in 2008 of the Basic Space Law (2008 Basic Space Law), Japan's first national space legislation, established the Strategic Headquarters for Space Development in Japan's Cabinet Office, to lead the country's space development policymaking and coordination.[23] The Strategic Headquarters for Space Development thus became the command center for Japan's space policymaking, charged with bringing together under one roof the various elements of this policy that previously had been handled by the Ministry of Education, Culture, Sports, Science and Technology and other ministries and agencies. Among other things, the Strategic Headquarters for Space Development is tasked with formulating a "Basic Space Plan" to create a framework for, and spell out measures to advance, Japan's space development. The Strategic Headquarters for Space Development delivered the first Basic Space Plan in 2009 and has since issued three updated versions of this document, most recently in 2020. This latest iteration of the Basic Space Plan is discussed in Chapter 6.

Prior to enactment of the 2008 Basic Space Law, there was concern in Japan that the country lacked a unified national strategy for the development and use of space, and that various vertically divided ministries and agencies were promoting space-related initiatives without much coordination or overarching sense of purpose. With the 2008 Basic Space Law and creation of the Strategic Headquarters for Space Development, Japan took a major step to address this concern. Japan has continued to pursue a more unified strategy to develop and use space, and to formulate a more comprehensive vision of the relationship between space-related activities and society as a whole, to this day.

---

21 *See* S. Pekannen, P. Kallender-Umezu, *In Defense of Japan: From the Market to the Military in Space Policy* (*In Defense of Japan*) (Stanford University Press, 2010), p. 35.

22 *See Dokuritsughōseihōjin uchūkōkū kenkyūkaihatsu kikōhō* [Law on the Japan Aerospace Exploration Agency] (Law No. 161 of December 13, 2002), at http://law.e-gov.go.jp/htmldata/H14/H14HO161.html; https://global.jaxa.jp/about/law/law_e.pdf (unofficial English translation); *Enkaku* [History], at https://www.jaxa.jp/about/history/index_j.html.

23 *See An Introduction on Space Law for Space Business*, p.161; *Uchū kihon hō* [Basic Space Law] (Law No. 43 of 2008), at https://www8.cao.go.jp/space/law/law.html (Japanese). An unofficial English translation of the Basic Space Law by ZeLo is included in Annex 7.

III   The 2008 Basic Space Law and Subsequent Developments

The 2008 Basic Space Law was an important milestone in Japan's space development for additional, more specific reasons. First, while acknowledging the "fundamental principle of pacifism of the Constitution of Japan," the 2008 Basic Space Law provided that the Japanese government would "implement policy measures as necessary for the promotion of Space Development and Use [defined as the "development and use of space"] to contribute to ensuring international peace and security and to the national security of our country."[24] This was the first time that Japan expressly conveyed in legislation that space development would take account of and "contribute to" national security objectives. The original impetus behind this shift had actually occurred in the prior decade, when North Korea launched a *Taepodong* missile in 1998 that passed over Japan and landed in the Pacific Ocean. North Korea had tried to put a satellite into orbit with the missile, and while this effort failed the event impressed upon Japanese policymakers the need to begin considering a revision of Japan's longstanding policy to use space exclusively for non-military purposes, a process that culminated in passage of the 2008 Basic Space Law.[25]

Second, as noted above, since the 1950s the Japanese government had led the planning, funding, and implementation of Japan's space program, albeit with design and production assistance from large Japanese and other heavy industrial manufacturers employed as subcontractors. This government-driven, public demand approach was natural as the needs of the program were technically complex and expensive, but, also as mentioned, the emergence of a commercial satellite launching sector in Japan had lagged well behind that in other major spacefaring nations. By the mid-2000s, however, the prospect of greater involvement of the private sector in satellite launches and other space-related activities had arisen, and the Japanese government took note. Thus Article 16 of the 2008 Basic Space Law, entitled "Promotion of Space Development and Use by Private Business Operators," provides:

> In view of the important role that the private sector plays in Space Development and Use [defined as the "development and use of space"], the

---

24   Basic Space Law, Article 1 and Article 14, unofficial English translation by ZeLo at Annex 7.
25   Even earlier, however, several years after the shock of the 1998 Taepodong missile launch, Japanese space policymaking documents had begun to reflect national security concerns. *See In Defense of Japan*, pp. 35-38. In addition, Japan launched several information gathering satellites beginning in 2003 that were used to monitor North Korea as well as to assist with natural disaster management. *See* http://news.bbc.co.uk/2/hi/asia-pacific/2892641.stm; http://news.bbc.co.uk/2/hi/asia-pacific/5333560.stm. These satellites operated under the jurisdiction of the civilian Cabinet Intelligence and Research Office, rather than Japan's Ministry of Defense. *See* "Current Status and Recent Developments in Japan's National Space Law," p. 380. For further discussion about Japan's move to space security *see* R. Wilson, "Japan's Gradual Shift Toward Space Security," May 2020, at https://aerospace.org/sites/default/files/2020-05/Wilson_JapansGradualShift_20200428_0.pdf.

national government shall, when engaging in its own Space Development and Use activities, not only give consideration to procuring goods and services in a systematic manner utilizing the skills of the private sector, but also to improving launch sites (facilities for launching rockets), laboratories, other equipment and facilities, etc., promoting the transfer of the results of research and development concerning Space Development and Use to the private sector, promoting the commercialization of the results of research and development concerning Space Development and Use in the private sector, and implementing tax and financial services measures to facilitate the private sector's investment in activities concerning Space Development and Use, as well as other necessary measures.[26]

Here, also for the first time in its legislation, Japan recognized the commercial potential of space and pledged to support private sector efforts to take advantage of this opportunity. The 2008 Basic Space Law further enjoined the Japanese government to enact legislation regulating space activities, "to contribute to promoting the interests of our country in international society and promoting Space Development and Use in the private sector."[27]

While not immediate, important developments ensued. For example, in 2012 three space smaller organizations that had operated under Japan's former Ministry of International Trade and Industry (MITI) were consolidated into a new entity, Japan Space System (JSS), under the supervision of MITI's successor, the powerful Ministry of Economy, Trade, and Industry (METI).[28] JSS's mandate, among other things, is to "promote commercialization, globalization and improvement of global competitiveness regarding 'space systems,'" and JSS has supported various projects, including to develop low-cost advanced small satellites systems in collaboration with Japanese electronics giant NEC Corporation and to create a database and evaluation guidelines for using commercial off-the-shelf equipment and technologies in space.[29]

In addition, since enacting the 2008 Basic Space Law Japan has taken further legislative steps to address the challenges that accompany its new approach to the

---

26 2008 Basic Space Law, Article 16, unofficial English translation by ZeLo at Annex 7.
27 *Id.*, Article 35(2).
28 *See* https://ssl.jspacesystems.or.jp/en/about/message/index.html.
29 *See* https://ssl.jspacesystems.or.jp/en/about/overview/index.html; https://ssl.jspacesystems.or.jp/en_project_asnaro/; and https://ssl.jspacesystems.or.jp/en_project_servis/. JSS and NEC are marketing the small satellite systems in emerging economies. *See* R. Dunphy, *Space Industry Business Opportunities in Japan: Analysis on the Market Potential for EU SMEs Involved in the Earth-Observation Products & Services* (*Space Industry Business Opportunities in Japan*), EU-Japan Centre for Industrial Cooperation, October 2016, p. 40, at https://www.eu-japan.eu/sites/default/files/publications/docs/2016-10-space-industry-business-opportunities-japan-ryuichi-dunphy-min.pdf.

security and commercial possibilities of space development. In 2012, the 2002 JAXA Law was amended to delete the stipulation that JAXA pursue space development "only" for peaceful purposes.[30] Japan also passed two major space-related laws in 2016: the Act on Launching of Spacecraft, etc. and Control of Spacecraft (2016 Space Activities Act); and the Act on Ensuring Appropriate Handling of Satellite Remote Sensing Data (2016 Remote Sensing Act).[31]

While each of these two laws will be discussed at greater length below, generally speaking they both serve to establish a governmental regulatory regime for private sector entities that pursue certain space-related activities in Japanese territory. The 2016 Space Activities Act sets out requirements for any "person" (*i.e.*, natural or legal person) that seeks permission to launch or operate a satellite from Japan and establishes liability rules for any damages arising from these activities. The 2016 Remote Sensing Act establishes rules for any "person" (again, natural or legal person) that wishes to use radio equipment in Japan to control certain satellite remote sensing devices and obtain data from these devices, including setting out rules on the use and provision of such data that may be sensitive from a personal privacy or national security perspective.

As such, both pieces of legislation reflect Japan's commitment to support commercial space-related activities through, in Professor Aoki's words, "the establishment of a system that clarifies matters for start-ups or other companies that develop small-scale rockets or that undertake to operate remote sensing satellites by identifying the government agency to which license applications are to be submitted and clarifying the conditions for license approval and the procedures for supervision."[32] While acknowledging the commercial promise of space development, both laws thus also exhibit an understanding that such opportunity entails risks which the Japanese government must monitor and regulate.

* * *

In sum, Japan's space program initiatives – its space organizations, policies, and laws – have evolved greatly since the early 1950s. The following chapters examine these topics in more detail.

---

30 See *An Introduction on Space Law for Space Business*, p. 163; *compare* Article 4 of 2002 version, at https://global.jaxa.jp/about/law/law_e.pdf (unofficial English translation) *with* amended version, at http://www.japaneselawtranslation.go.jp/law/detail/?id=3194&vm=04&re=02 (in Japanese and unofficial English translation).

31 See *An Introduction on Space Law for Space Business*, p. 169; *Jinkō eisei-tō no uchiage oyobi jinkō eisei no kanri ni kansuru hōritsu* [Act on Launching of Spacecraft, etc. and Control of Spacecraft] (Act No. 76 of 2016) (2016 Space Activities Act); and *Eisei rimōtosenshingu kiroku no tekiseina toriatsukai no kakuho ni kansuru hōritsu* [Act on Ensuring Appropriate Handling of Satellite Remote Sensing Data] (Act No. 77 of 2016) (2016 Remote Sensing Act).

32 S. Aoki, "New Law Aims to Expand Japan's Space Business," ("New Law Aims to Expand Japan's Space Business"), March 3, 2017, at https://www.nippon.com/en/currents/d00294/new-law-aims-to-expand-japan%E2%80%99s-space-business.html.

Chapter 2 Japan's Key Space Development Organizations Today

I Introduction

Chapter 1 sketched the historical evolution of Japan's major space development organizations. Today the organizational foundations of Japan's space program are: (1) JAXA, created pursuant to the 2002 JAXA Law; and (2) certain central government bodies that have been established since enactment of the 2008 Basic Space Law.

II JAXA

As explained previously, JAXA, Japan's premier space development organization, was created by integrating three pre-existing bodies – ISAS, NASDA, and the National Aerospace Laboratory of Japan (NAL) – in 2003. As set out in the 2002 JAXA Law, JAXA's purpose is "to facilitate the development of academic research at universities or other institutes, the enhancement of the level of space science and technology … and aeronautics science and technology, and the promotion of space development and utilization."[1] To fulfill this purpose, JAXA conducts research on space science, aeronautical science, and related technology issues, including in collaboration with universities, and manages the development, launch, operation, and tracking of satellites. These satellites have various functions, including communications, weather observation, astronomical observation, and planetary exploration.[2] In addition, JAXA operates and monitors *Kibō*, the Japanese Experimental Module that is part of the ISS, and trains Japanese astronauts who conduct ISS missions. Of course, JAXA also continues to develop new, more powerful rocket systems, including the H3 Launch Vehicle that is slated to succeed the H-II models.[3]

In 2015, JAXA's standing as an R&D institution was bolstered when its status was changed from an independent administrative agency to a "National Research and Development Agency."[4] JAXA also underwent internal restructuring and, as of 2020, JAXA's staff numbers approximately 1500 people who work mainly in a number of

---

[1] 2002 JAXA Law, Article 4, at https://global.jaxa.jp/about/law/law_e.pdf (unofficial English translation). While the JAXA Law was enacted in 2002, JAXA was not formed until October 2003.
[2] *See* Utilizing Space through Satellites, at https://global.jaxa.jp/projects/sat/index.html; *see also* 2002 JAXA Law, Article 18.
[3] *See* About H3 Launch Vehicle, at https://global.jaxa.jp/projects/rockets/h3/.
[4] *See* Introduction of JAXA, at https://global.jaxa.jp/about/jaxa/index.html.

directorates: Space Transportation Technology Directorate, two Space Technology Directorates, Human Spaceflight Technology Directorate, Research & Development Directorate, and Aerospace Technology Directorate. Besides these directorates, JAXA houses the ISAS and a Space Exploration Center (which is assessing involvement in a lunar orbiting project with ISS partners). JAXA currently has eight "field centers" across Japan that handle various program activities, including space and aeronautical research and development, rocket engine testing, satellite launch operations, and other space-related scientific activities. In addition, JAXA has offices in the United States, France, Russia, and Thailand. JAXA's annual budget has been approximately USD 1.8 billion in recent years.[5]

As part of its restructuring in 2015, JAXA also established a Space Exploration Innovation Hub Center to research space exploration technologies and spur innovation in this area, including by collaborating with companies and academic institutions. According to the Space Exploration Innovation Hub Center's Director:

> While space exploration today is supported by many research activities and government support, the role of industry will become very important in the future. If companies are not able to get into space activities and start new businesses, there will be no hope of a sustainable space program. However, it's also true that the bar to entry into space is high for the average company. Therefore, JAXA Space Exploration Innovation Hub Center's . . . policy is "Dual Utilization," which focuses on the feasibility of commercialization/ innovation on the ground, with the aim of applying it to future space exploration, and we believe that this approach will be extremely effective in bringing sustainable growth to space development as industry sequentially expands its role.[6]

Recent projects in which JAXA's Space Exploration Innovation Hub Center has partnered with Japanese firms and universities have focused on technologies in the fields of medicine and healthcare, agriculture, power, construction, automated driving, communications, and many others.[7]

---

5   For information in this paragraph, *see* JAXA's website in English at http://www.isas.jaxa.jp/en/missions/ (ISAS); http://www.exploration.jaxa.jp/e/ (Space Exploration Center); https://global.jaxa.jp/about/transition/index.html (Staff and Budget Transition); https://global.jaxa.jp/about/org/pdf/org_e.pdf (JAXA Organization Chart); https://global.jaxa.jp/about/centers/index.html (Field Centers and International Offices); *see also* in Japanese: *Soshiki* [Organizations], at https://www.jaxa.jp/about/org/index_j.html; *Jimusho-sisetsu* [Offices and Facilities], at https://www.jaxa.jp/about/centers/index_j.html; and *Yosan no suii* [Budget Changes], at https://www.jaxa.jp/about/transition/index_j.html.
6   Message from our Director, at https://www.ihub-tansa.jaxa.jp/english/; *Habucho aisatsu* [Greetings from the Hub Director], at https://www.ihub-tansa.jaxa.jp/director2.html.
7   *See* Request for Proposal Adoption Results in recent years, at https://www.ihub-tansa.jaxa.jp/english/.

More recently, in 2018, JAXA began a program called Space Innovation through Partnership and Co-creation (J-SPARC) whose goal is "to create new ideas for space related-businesses through collaboration between private companies and JAXA."[8] In this program JAXA and its partners conduct market research, business concept development, joint feasibility studies, and joint technological development and validation. J-SPARC now supports some 20 projects, including, as of September 2020, the "co-creation of world's first space robot business" with the Japanese space robotics start-up GITAI Japan Co., Ltd. (GITAI), which is developing space robotics to perform operations, maintenance, scientific experiments, and other tasks on the ISS.[9] As part of the venture, GITAI has built a mock-up of the ISS's Japanese Experimental Module *Kibō* on which GITAI has been testing its robots. From JAXA's perspective, such collaboration will help "promote the sustainable development of low Earth orbit (LEO) missions including the *Kibō* mission. Robots that will support on-orbit work of astronauts are expected to lead to promoting the private-sector participation in activities and increasing the sophistication of research on *Kibō*, as well as help enhance efficiency in astronaut operations."[10]

III   Central Government Bodies

As discussed in the previous chapter, pursuant to the 2008 Basic Space Law, overarching responsibility for leading and coordinating Japan's space development program lies with the Strategic Headquarters for Space Development, which is housed in the Japanese Cabinet Office. The Strategic Headquarters for Space Development is comprised of

---

8   See *Aratana jigyo wo souzeisuru kenkyukaihatsu puroguramu {Uchū Innovation Partnership (J-SPARC)} no kaishi ni tsuite* [Research and Development Program to Co-Create New Businesses - Launch of the Japan Space Innovation Partnership (J-SPARC)], at https://www.jaxa.jp/press/2018/05/20180511_jsparc_j.html; *see also* https://aerospacebiz.jaxa.jp/solution/j-sparc/; "GITAI and JAXA to Embark on the Co-creation of World's First Space Robot Business," September 10, 2020, at https://global.jaxa.jp/press/2020/09/20200910-1_e.html.

9   *GITAI to JAXA, sekaihatsu no {Uchū yo sagyo robot jigyo} no soshutsu ni muketa jigyo concepto kyoso katsudo wo shido* [GITAI and JAXA to Embark on the Co-creation of World's First Space Robot Business] September 10, 2020, at https://www.jaxa.jp/press/2020/09/20200910-1_j.html; https://global.jaxa.jp/press/2020/09/20200910-1_e.html.

10   *Id.*

Japan's Prime Minister, its formal head, as well as the Chief Cabinet Secretary, the Minister for Space Policy, and all other Ministers of State.[11]

One of the major functions of the Strategic Headquarters for Space Development is to formulate and produce a Basic Space Plan, which it first did in 2009 and has since revised periodically, most recently in 2020. The 2008 Basic Space Law provides that the Basic Space Plan is to address:

> (i) basic policies concerning the promotion of Space Development and Use, (ii) policy measures which the national government shall implement in a comprehensive and systematic manner, and (iii) in addition to the matters provided for in the preceding two items, matters necessary for the national government's implementation of policy measures concerning Space Development and Use.[12]

As would be expected, the initial Basic Space Plan and its subsequent iterations set out broad policy goals and specific programmatic activities for Japan's space development. The policy goals reflect the importance of space development to national security and economic growth. The programs include various plans, among other things, for developing satellites and launch vehicles, promoting space exploration initiatives, encouraging private sector involvement, and supporting endeavors to update space-related laws and boost international cooperation. In recent years, the annual budget of the Strategic Headquarters for Space Development has averaged some USD 3.35 billon.[13]

In Chapter 6 we examine in detail the fourth, updated version of the Basic Space Plan issued in 2020. At this juncture, however, it bears noting that Japan's then-Prime Minister Shinzo Abe, when commenting on the forthcoming publication of this latest

---

11 See About Strategic Headquarters for Space Development, at https://www8.cao.go.jp/space/hq/about.html (Japanese).While not addressed above in detail, a number of Japanese ministries also conduct space-related R&D, sometimes in cooperation with JAXA, including the Ministry of Foreign Affairs, Ministry of Internal Affairs and Communications (MIC), METI (discussed in Chapter 1), Ministry of Defense, Ministry of the Environment, Ministry of Land, Infrastructure, Transport and Tourism, and the Ministry of Agriculture, Forestry and Fisheries. See "Current Status and Recent Developments in Japan's National Space Law," pp. 376-78; *A Guide to Japan's Space Policy Formulation*, pp. 7-20, at https://pacforum.org/wp-content/uploads/2019/04/issuesinsights_Vol19WP3_0.pdf.
12 Basic Space Law, Article 24(2), unofficial translation by ZeLo at Annex 7.
13 See *Reiwa 2 nendo tosho yosan an oyobi reiwa gan nendo hose yosan an ni okeru uchū kankei yosan ni tsuite* [Budget for space utilization in the proposed initial budget for fiscal 2020 and the proposed supplementary budget for fiscal 2020], at https://www8.cao.go.jp/space/budget/r02/fy02yosan_01hosei.pdf; *Heisei 30 nendo hosei yosan an oyobi heisei 31 nendo tosho yosan an ni okeru uchū kankei yosan ni tsuite* [Space-related budget in the supplementary budget plan for FY 2019 and the initial budget plan for FY 2018], at https://www8.cao.go.jp/space/budget/h31/fy30hsei_31yosan.pdf; *Heisei 29 nendo hosei yosan an oyobi heisei 30 nendo tosho yosan an ni okeru uchū kankei yosan ni tsuite* [Space-related budget in the supplementary budget plan for FY 2018 and the initial budget plan for FY 2017], at https://www8.cao.go.jp/space/budget/h30/fy30yosan.pdf.

iteration of the Basic Space Plan at the end of 2019, chose to highlight the commercial potential of space:

> In Japan, we are seeing an increasing number of start-ups and other initiatives taking on the challenge of developing new space businesses. Efforts are even being made towards the commercialization of suborbital flights in Japan within the early 2020s. I ask you, in particular the Cabinet Office and the Ministry of Land, Infrastructure, Transport and Tourism, to advance the creation of an environment that fosters new space businesses, including accelerating the development of the system necessary to materialize that. I also ask you to implement this newly revised Plan steadily, such as through utilization of satellites and other technologies in disaster response, as well as initiatives to ensure space security.[14]

Within the Strategic Headquarters for Space Development are two important bodies that were created pursuant to the Act for the Establishment of the Cabinet Office.[15] The first is the Committee on National Space Policy, which is tasked with investigating and deliberating on Japan's major space development and utilization policies, including space-related safety and security initiatives, in consultation with the Prime Minister.[16] The Committee on National Space Policy is assisted in its work by a number of subcommittees that are described briefly below.[17]

- *The National Institute of Aerospace Research and Development Subcommittee.* The National Institute of Aerospace Research and Development Subcommittee supervises administration of R&D activities conducted by JAXA in consultation with the Prime Minister.[18]
- *Basic Policy Subcommittee.* The Basic Policy Subcommittee was established in 2019 to help formulate the Basic Space Plan issued in 2020 and to consider a broad range of approaches to space policy.[19]

---

14 *Uchū kaihatsu senryaku honbu* [Strategic Headquarters for Space Development], http://www.kantei.go.jp/jp/98_abe/actions/201912/13uchu.html; Prime Minister in Action, December 13, 2019, at http://japan.kantei.go.jp/98_abe/actions/201912/_00025.html.
15 See *Naikaku hu setchi hō* [Act for the Establishment of the Cabinet Office] (Act No. 89 of July 16, 1999).
16 See *Uchū seisaku iinkai setchi konkyo* [Grounds for Establishing Space Policy Committee], at https://www8.cao.go.jp/space/comittee/konkyo.html.
17 See *Kaisai jōkyō* [Events Status], at https://www8.cao.go.jp/space/comittee/kaisai.html.
18 See *Uchū seisaku iinkai kokuritsu kenkyu kaihatsu hojin uchū koku kenkyu kaihatsu kiko bunkakai no unei ni tsuite* [Space Policy Committee, Operation of the Subcommittee of the National Aerospace Laboratory of Japan], at https://www8.cao.go.jp/space/comittee/bunkakai/bunkakai-dai13/sankou3.pdf; *Uchū seisaku iinkai rei* [Space Policy Committee Order], Article 5(2), at https://elaws.e-gov.go.jp/document?lawid=424CO0000000186.
19 See *Uchū seisaku bukai no setchi ni tsuite* [Space Policy Committee has committed to establishment of the Basic Policy Subcommittee], June 24, 2019, at https://www8.cao.go.jp/space/comittee/01-kihon/kihon-dai1/siryou2.pdf.

- *Space Security Subcommittee.* The Space Security Subcommittee considers initiatives to ensure the stable use of outer space, efforts to strengthen Japan's security capabilities utilizing outer space, and ways to strengthen the Japan-U.S. alliance through space cooperation.[20]
- *Civil Use of Space Subcommittee.* The Civil Use of Space Subcommittee considers initiatives aimed at addressing ways to use space to enhance the safety, security, and prosperity of society, as well as initiatives aimed at creating new space-related industries.[21]
- *Space Industry Science and Technology Infrastructure Subcommittee.* The Space Industry Science and Technology Infrastructure Subcommittee is responsible for programs to maintain and strengthen the space industry's scientific and technological foundations.[22] Two other sub-subcommittees have been set up within the Space Industry Science and Technology Infrastructure Subcommittee. The first is the Sub-subcommittee for Standards and Safety of the Space Activity Law, which examines matters related to ensuring the safety of launches of satellites and their launch vehicles, as well as for protecting the space environment.[23] The second is the Space Legislation Sub-subcommittee, which considers governmental compensation for any mishaps involving orbital satellite activities (debris removal, refueling, satellite repair, etc.).[24] In addition, a third entity, the Space Science and Exploration Sub-subcommittee has been established under the joint authority of the Space Industry Science and Technology Infrastructure Subcommittee and the Civil Use of Space Subcommittee to examine initiatives in the field of solar system scientific exploration.[25]

In addition to the Committee on National Space Policy, a second body established under the Cabinet Office Establishment Law works under the Strategic Headquarters for Space Development: the Space Development Strategy Promotion Secretariat. This Secretariat oversees the comprehensive promotion of Japan's space development and use policy and related matters, as well as the coordination of administrative agencies regarding the development and use of space. Besides these broad mandates, the Space Development

---

20 See *Kongo no uchū seisaku iinkai no kento taisei ni tsuite* [Future study structure of the Space Policy Committee], February 2, 2017, at https://www8.cao.go.jp/space/comittee/dai36/siryou1.pdf.
21 See id.
22 See id.
23 See *Uchū katsudo hō siko ni muketa aratana shoiinkai no setchi (kaiso) ni tsuite (an)* [Establishment (Reorganization) of a New Subcommittee for the Enforcement of the Space Activities Law (Draft)], October 30, 2018, at https://www8.cao.go.jp/space/comittee/dai73/siryou4.pdf.
24 See *Kongo no uchū sesaku iinkai no kentou taisei ni tsuite* [Future study structure of the Space Policy Committee], August 31, 2018, at https://www8.cao.go.jp/space/comittee/27-minsei/minsei-dai21/sankou3-1.pdf.
25 See *Kongo no uchū seisaku iinkai no kento taisei ni tsuite* [Future study structure of the Space Policy Committee], February 2, 2017, at https://www8.cao.go.jp/space/comittee/dai36/siryou1.pdf.

Strategy Promotion Secretariat also is responsible for maintaining and managing navigation satellites that provide geo-spatial positioning for governmental and public uses.[26]

---

26 See *Naikaku hu setchi hō*[Act for the Establishment of the Cabinet Office] (Act No. 89 of July 16, 1999), Article 40-4(1), at https://www8.cao.go.jp/space/cao/jurisdiction.html.

Chapter 3 Recent Legislation: 2016 Space Activities Act

I Introduction

The 2016 Space Activities Act was enacted to regulate and facilitate the growing commercialization of Japan's space industry and the growing ambition of Japanese private sector actors to enter the field of space development and use.[1] As such, the law aims to provide a clear legal regime on which both these actors and the Japanese government may rely. As Professor Aoki noted in 2017, "[a]longside technological development and financing, the design of the legal and regulatory system is a key determinant of success or failure in space business. The new Space Activities Act is sure to give a major boost to this business in Japan, which has both technological strength and great potential."[2]

Broadly speaking, the 2016 Space Activities Act sets out government requirements for any "person"[3] that seeks permission to launch or operate a satellite from Japan and establishes liability rules for damages arising from these activities. In so doing, the 2016 Space Activities Act reflects Japan's duty under the 1967 Outer Space Treaty:

> to bear international responsibility for national activities in outer space, including the moon and other celestial bodies, whether such activities are carried on by government agencies or non-governmental entities, and for assuring that national activities are carried out in conformity with the provisions set forth in the present Treaty. The activities of non-governmental entities in outer space, including the moon and other celestial bodies, shall require authorization and continuing supervision by the appropriate State party to the Treaty.[4]

---

1. See *Jinkō eisei-tō no uchiage oyobi jinkō eisei no kanri ni kansuru hōritsu* [Act on Launching of Spacecraft, etc. and Control of Spacecraft] (Act No. 76 of 2016) (2016 Space Activities Act). The text of the 2016 Space Activities Act in Japanese and English may be found at https://www8.cao.go.jp/space/english/activity/documents/space_activity_act.pdf. The English version is included in Annex 1.
2. New Law Aims to Expand Japan's Space Business, at https://www.nippon.com/en/currents/d00294/new-law-aims-to-expand-japan%E2%80%99s-space-business.html.
3. When used in a legal context such as in the 2016 Space Activities Act or other legislation, the Japanese word for "person" (*mono*) means natural persons and legal persons (*e.g.*, corporations).
4. Treaty on Principles Governing the Activities of States in the Exploration and Use of Outer Space, including the Moon and Other Celestial Bodies, Article VI, at https://www.unoosa.org/oosa/en/ourwork/spacelaw/treaties/outerspacetreaty.html.

Further, as space-related activities are dangerous and can pose serious risks to people and property, the 2016 Space Activities Act accords with Japan's international treaty obligation to compensate for harm caused by such activities in its territory:

> Each State Party to the Treaty that launches or procures the launching of an object into outer space, including the moon and other celestial bodies, and each State Party from whose territory or facility an object is launched, is internationally liable for damage to another State Party to the Treaty or to its natural or juridical persons by such object or its component parts on the Earth, in air or in outer space, including the moon and other celestial bodies.[5]

The first article of the 2016 Space Activities Act captures the various elements noted above:

> The purpose of this Act is to ensure the accurate and smooth implementation of conventions concerning the development and use of outer space, ensure public safety and protect people affected by relevant damage by establishing a system for permission and license related to the launching of spacecraft, etc. and the control of spacecraft in Japan, as well as a system for compensation for damage caused by a fall, etc. of a spacecraft, etc. in Japan . . . thereby to contribute to the improvement of the lives of the citizens as well as the development of the economy and society.[6]

In addition, Article 3 states:

> For the enforcement of this Act, the national government is to pay due consideration for the strengthening of technical competence and international competitiveness of Japanese industries related to the launching of spacecraft, etc., and the control of spacecraft, as a part of policy measures for the promotion of the development and use of outer space by private businesses as provided in Article 16 of the Space Basic Act.[7]

Concretely, the 2016 Space Activities Act addresses duties in three principal areas: (i) obtaining permission to launch spacecraft (with "spacecraft" defined as "an artificial object which is used by putting it into Earth orbit or beyond, or placed on a celestial body other than the Earth," and "spacecraft, etc." defined as "spacecraft and a vehicle for

---

5   Id., Article VII. See also New Law Aims to Expand Japan's Space Business, at https://www.nippon.com/en/currents/d00294/.
6   2016 Space Activities Act, Art. 1.
7   Id., Article 3. For Article 16 of Japan's Basic Space Law, see Chapter 1.

CHAPTER 3   RECENT LEGISLATION: 2016 SPACE ACTIVITIES ACT

launching the spacecraft"[8]); (ii) obtaining permission to control spacecraft; and (iii) providing compensation for "launch vehicle fall damage" and "spacecraft fall damage," as those terms are defined in the 2016 Space Activities Act. Each of these three sets of obligations is discussed in greater detail below.

II   Permission to Launch Spacecraft

Article 4(1) of the 2016 Space Activities Act provides: "A person who intends to implement the launching of a spacecraft, etc. using a launch site located in Japan or onboard a ship or aircraft with Japanese nationality must obtain permission from the Prime Minister for each instance of launching."[9] Thus, an individual or firm that launches a rocket with a satellite or satellites is subject to jurisdiction under the 2016 Space Activities Act only if done so from a facility that either is located in Japanese territory or on a Japanese ship or aircraft. Individuals or companies that do so outside Japan (or not on Japanese ships or aircraft) are not required to obtain permission pursuant to the 2016 Space Activities Act, even if those individuals or firms are Japanese. Professor Aoki has commented that this strict territorial limitation of jurisdiction, which she says is exceptional among spacefaring nations:

> will help to avoid double-licensing requirements for Japanese companies, given that a foreign government will often authorize and supervise the launch and operation of space objects in and from its territory irrespective of the nationality of those involved. The SAA [2016 Space Activities Act] also limits uncertainty regarding the actual exercise of enforcement jurisdiction over the activities of Japanese entities overseas. Thus, the law facilitates space business.[10]

Professor Aoki also notes, however, that "Japanese actors might encounter regulatory voids in seeking to conduct space activities abroad, given the nonexistence of national acts and other governmental administration in many states[,]" and that this may create gaps in the regime of the 1967 Outer Space Treaty that hinders Japan from meeting its

---

8   Id., Article 2(ii) and (iii). Thus, "spacecraft" do not include sub-orbital vehicles. Nonetheless, Professor Aoki observes that "spacecraft" is broadly defined and while typically it would be a satellite, "in theory other types of objects are also included. A Moon rover, for example, would likely count." S. Aoki, "Symposium on the New Space Race – Domestic Legal Conditions for Space Activities in Asia" ("Domestic Legal Conditions for Space Activities in Asia"), *AJIL Unbound*, Vol. 113, April 1, 2019, pp. 106, 107, at https://www.cambridge.org/core/services/aop-cambridge-core/content/view/093CF942F1D5A2F04AC3C2A9B7FC3D93/S239877231900014Xa.pdf/domestic_legal_conditions_for_space_activities_in_asia.pdf.
9   2016 Space Activities Act, Article 4(1).
10   Id.

obligations under this convention.[11] She recommends that more countries enact national space legislation to address this potential issue.

Practically, individual or companies that wish to obtain permission to launch vehicles with satellites under the 2016 Space Activities Act must submit to the Cabinet Office's National Space Policy Secretariat various information, including about themselves, the launch vehicle design or any certification of the launch vehicle, the location, design, and equipment of the launch site, the launch schedule and methods, the launch trajectory, and the purpose and use of the satellite(s) being launched.[12] While Article 4(1) states that the permission to launch is to be granted "for each instance of launching," the 2016 Space Activities Act provides for the possibility of obtaining "type certification" for a given launch vehicle design so that an applicant, if it receives such certification, does not need to submit the same information repeatedly for detailed review.[13] Permission will be granted to the designated "launch operator"[14] only if the launch vehicle, site, and plan are deemed to meet requisite safety standards, including as to the radio equipment for communicating with the satellite, and so long as the satellite's purpose and use "are not likely to cause any adverse effect on the accurate and smooth implementation of the conventions on development and use of outer space and ensuring public safety."[15]

In addition to meeting the requirements above, a launch operator must take "security measures for compensation for damages," that is, steps to guarantee compensation for damage caused by a launch, including through liability insurance and an indemnification arrangement with the government.[16] This topic will be discussed in section IV below.

As a rule, it takes an applicant four to six months to obtain a decision or, if the applicant has received "type certification" for its launch vehicle, one to three months.[17] In this regard, and to facilitate the completion and processing of applications (as well as

---

11 *Id.*
12 2016 Space Activities Act, Article 4(2)(i)-(v). The Cabinet Office's National Space Policy Secretariat, to which applications are sent, provides a manual to assist applicants. For the English version, *see* https://www8.cao.go.jp/space/english/activity/documents/apmnl.pdf, included in Annex 3: *see also* https://www8.cao.go.jp/space/english/activity/application.html.
13 Similarly, applicants may obtain "compliance certification" for a given launch site. *See id.*, Articles 6(i) and 13(1), and H. Yotsumoto & D. Ishikawa, "Japan," ("Japan") *The Space Law Review*, Edition 1, December 2019, at https://thelawreviews.co.uk/edition/the-space-law-review-edition-1/1211969/japan.
14 2016 Space Activities Act, Article 7(1) (defining "launch operator" as "a person who obtained permission under Article 4, paragraph (1)"). Pursuant to Article 10(1), if a launch operator wishes to transfer "the business with respect to the launching of the spacecraft, etc." permitted under Article 4(1), the transferring launch operator and the transferee need to obtain advance government authorization for the transfer.
15 *Id.*, Article 6(i)-(iv).
16 *Id.*, Article 9(1).
17 *See Jinkō eisei-tō no uchiage oyobi jinkō eisei no kanri ni kansuru hōritsu ni motozuku shinsa kijun & hyōjun shorikikan* [Review Standards and Standard Period of Time for Process Relating to Procedures under the Act on Launching of Spacecraft, etc. and Control of Spacecraft], November 15, 2017, p. 3 (English version), at https://www8.cao.go.jp/space/english/activity/documents/reviewstand.pdf. The English version of this document is included in Annex 2.

increase the likelihood of approval), it is recommended that applicants coordinate in advance with the National Space Policy Secretariat.

III   PERMISSION TO CONTROL SPACECRAFT

Article 20(1) of the 2016 Space Activities Act stipulates: "A person who intends to implement the control of a spacecraft using a spacecraft control facility located in Japan must obtain a license from the Prime Minister for each of the spacecraft."[18] Unpacking this provision, "control of spacecraft" means "to detect the position, attitude and condition of a spacecraft and to control these using a spacecraft control facility."[19] In turn, "spacecraft control facility" means:

> radio equipment equipped with functions to detect signals indicating the position, attitude and condition of a spacecraft transmitted by spacecraft-borne radio equipment . . . either directly or by receiving it via other radio equipment using electromagnetic waves, or to detect the position of the spacecraft by transmitting signals to the spacecraft either directly or via other radio equipment and then receiving the reflected signals from the spacecraft directly or via other radio equipment, or by other means, and to transmit signals to control the position, attitude and condition of the spacecraft to the spacecraft-borne radio equipment directly or via other radio equipment using electromagnetic waves.[20]

Again, jurisdiction is limited to activities at such a facility that is located in Japanese territory. Further, and similar to the requirements for an application to launch spacecraft, individuals and companies that wish to obtain permission to control spacecraft are obliged to submit information about themselves, the control facility location, and the methods of controlling the spacecraft, as well as the configuration, purpose, and orbit (if relevant) of the spacecraft.[21]

---

18 2016 Space Activities Act, Article 20(1). It bears noting that, in both the 2016 Space Activities Act and the 2016 Remote Sensing Act, the Japanese word *kyoka* is used and is translated variously as "permission" and "license" in the English versions of these statutes on the Japanese government websites. In English, *kyoka* means "permission," "license," "authorization," or "approval." In other words, *kyoka* means "license" in the sense of being given permission to do something, and not an official document that gives one permission to do something. Accordingly, notwithstanding that these statutes (and some commentators) sometimes utilize the word "license" in this context, the word "permission" is used above except when directly quoting the English version of the statutory language from the Japanese government websites.
19 *Id.*, Article 2(vii).
20 *Id.*, Article 2(vi).
21 *Id.*, Article 20(2)(i)-(vii).

Additional requirements in the 2016 Space Activities Act regarding permission to control spacecraft are noteworthy as they go beyond measures to ensure public safety and establish duties to mitigate space debris. Under Article 22, such permission also is contingent upon meeting the following conditions:

> the configuration of the spacecraft, a mechanism for the prevention of dispersion of its components and parts has been implemented, or that the configuration of the spacecraft otherwise complies with the standard specified by Cabinet Office Order as being those that are not likely to cause an adverse effect on the prevention of the harmful contamination of outer space including the Moon and other celestial bodies and the prevention of potentially harmful interference with activities of other countries in the peaceful exploration and use of outer space provided in Article 9 of the Outer Space Treaty[;]

and

> the control plan requires the implementation of measures to avoid collision with other spacecraft or other measures specified by Cabinet Office Order which are necessary for the prevention of harmful contamination of outer space, etc. as well as termination measures, and the applicant (in the case of an individual, including the representative in case of death) has sufficient ability to execute the control plan[.][22]

---

22 *Id.*, Article 22(ii) and (iii). Article 9 of the 1967 Outer Space Treaty states: "In the exploration and use of outer space, including the moon and other celestial bodies, States Parties to the Treaty shall be guided by the principle of co-operation and mutual assistance and shall conduct all their activities in outer space, including the moon and other celestial bodies, with due regard to the corresponding interests of all other States Parties to the Treaty. States Parties to the Treaty shall pursue studies of outer space, including the moon and other celestial bodies, and conduct exploration of them so as to avoid their harmful contamination and also adverse changes in the environment of the Earth resulting from the introduction of extraterrestrial matter and, where necessary, shall adopt appropriate measures for this purpose. If a State Party to the Treaty has reason to believe that an activity or experiment planned by it or its nationals in outer space, including the moon and other celestial bodies, would cause potentially harmful interference with activities of other States Parties in the peaceful exploration and use of outer space, including the moon and other celestial bodies, it shall undertake appropriate international consultations before proceeding with any such activity or experiment. A State Party to the Treaty which has reason to believe that an activity or experiment planned by another State Party in outer space, including the moon and other celestial bodies, would cause potentially harmful interference with activities in the peaceful exploration and use of outer space, including the moon and other celestial bodies, may request consultation concerning the activity or experiment." At https://www.unoosa.org/oosa/en/ourwork/spacelaw/treaties/outerspacetreaty.html.

According to Professor Aoki, the Cabinet Office standards track the space debris mitigation guidelines of the UN Committee on the Peaceful Uses of Outer Space and the Inter-Agency Space Debris Coordination Committee.[23]

Further, the 2016 Space Activities Act sets forth expressly and specifically the space debris mitigation criteria. Subclauses of Article 22 stipulate that an application for permission to control spacecraft must contain termination measures for spacecraft: (i) to control the spacecraft's re-entry and burn in the atmosphere, or to guide and retrieve the spacecraft or parts thereof that do not burn but descend and land on Earth, while ensuring public safety;[24] (ii) to control the spacecraft to put it into Earth's orbit from which its altitude will not decrease over time, without risk of adversely affecting other spacecraft;[25] (iii) to control the spacecraft so as to put it into orbit around a celestial body other than Earth or guide it to fall to that body, without "risk of significantly deteriorating the environment of the celestial body;"[26] or, if the above measures are not possible, (iv) steps "to suspend the control of the spacecraft after taking measures to prevent the unexpected activation and explosion or other measures that are necessary for the prevention of harmful contamination of outer space, etc. as specified by Cabinet Office Order, and notifying the Prime Minister of the position, attitude and condition of the spacecraft."[27]

Article 28(1) further provides that when a "spacecraft control operator" (defined in Article 23(1) as the "person who obtained the license under Article 20, paragraph 1") intends to terminate control of its spacecraft and take appropriate termination measures, it must inform the Japanese government "in advance."[28] Professor Aoki observes that the 2016 Space Activities Act "establishes a precedent in that it sets forth detailed space debris mitigation measures. No other act has criteria and regulations that are as clear, detailed, or strict."[29]

---

23 *See* Domestic Legal Conditions for Space Activities in Asia, p. 107; *see also* United Nations Office for Outer Space Affairs, Space Debris Mitigation Guidelines of the Committee on the Peaceful Uses of Outer Space (2010), at https://www.unoosa.org/pdf/publications/st_space_49E.pdf, and Inter-Agency Space Debris Coordination Committee, IADC Space Debris Mitigation Guidelines (Sept. 2007), at https://orbitaldebris.jsc.nasa.gov/library/iadc_mitigation_guidelines_rev_1_sep07.pdf.
24 2016 Space Activities Act, Article 22(iv)(a).
25 *Id.*, Article 22(iv)(b).
26 *Id.*, Article 22(iv)(c).
27 *Id.*, Article 22(iv)(d).
28 *Id.*, Article 28(1). In addition, Article 25 requires that, if the spacecraft control operator loses control of the spacecraft due to collision with another spacecraft or another accident and there is no possibility of recovery, it must "promptly" notify the Japanese government of this circumstance and provide information to assist in identifying the spacecraft's post-accident location, and the permission granted under Article 20 ceases to be effective.
29 *See* Domestic Legal Conditions for Space Activities in Asia, p. 107.

IV Compensation for "Launch Vehicle Fall Damage" and "Spacecraft Fall Damage"

The 2016 Space Activities Act establishes a compensation regime to address third-party damages arising from space-related activities. This regime entitles third parties to compensation for harm they incur due to "launch vehicle fall damage" and "spacecraft fall damage," terms that, along with others, are explained below.

1 Launch Vehicle Fall Damage

As noted above, pursuant to Article 9(1) and (2) of the 2016 Space Activities Act, a launch operator may not launch any spacecraft "unless it has taken security measures for compensation for damages," in particular, by executing "a launch vehicle fall damage liability insurance contract" and "a launch vehicle fall damage liability indemnification contract."[30] This language needs unpacking.

Preliminarily, "launch vehicle fall damage" is defined as:

> damage caused to human life or body, or property on the ground surface or water surface, or an aircraft in flight or other flying objects caused by the fall, collision or explosion of a spacecraft, etc. in whole or part, that has not been successfully separated from the launch vehicle after the lift-off, or the launch vehicle after all of the spacecraft have been successfully separated; provided, however, that damages suffered by workers of the person implementing the launching of the spacecraft, etc. or other persons specified by Cabinet Office Order as those having a close business relationship with the person implementing the launching of the spacecraft, etc. in the course of their businesses are excluded.[31]

Article 35 of the 2016 Space Activities Act stipulates: "A person who implements the launching of spacecraft, etc. using a launch site located in Japan or onboard a ship or aircraft with Japanese nationality is liable to compensate for any launch vehicle fall damage caused by the person in relation to the launching of the spacecraft, etc."[32] This liability is strict and therefore the injured party does not need to prove negligence.[33]

Moreover, Article 36(1) channels this liability exclusively to the launch operator: "In the case referred to in the preceding Article, no person other than the person

---

30 2016 Space Activities Act, Article 9(1) and (2).
31 Id., Article 2(viii).
32 Id., Article 35.
33 See id.; Domestic Legal Conditions for Space Activities in Asia, p. 107.

implementing the launching of the spacecraft, etc., who is to be held liable to compensate for the damage pursuant to that Article, is liable to compensate for the damage."[34] Thus, for example, pursuant to the 2016 Space Activities Act liability does not apply to launch vehicle manufacturers. Professor Aoki has noted that "[t]his channeling of liability would seem to be disadvantageous to launch operators, but it can be expected to enhance the competitive position of the Japanese companies providing this service, because it reassures customers around the world who are seeking to have their satellites put into orbit."[35]

Article 38(1) does permit a launch operator that provides compensation under Article 35 to seek reimbursement from "any other person [who] is to be held liable for the cause of the damage."[36] However:

> if the person who is to be held liable was the supplier of materials or other goods or services used for the launching of the spacecraft, etc. (excluding the manager and operator of the launch site used for the launching of the spacecraft, etc.), the person who compensated for the damage has a right to reimbursement from the supplier only if the damage was caused by the intentional conduct of the supplier or its workers.[37] (Emphasis added.)

In addition to this narrow exception to the application of strict liability for "launch vehicle fall damage," Article 37 allows a court, in its discretion, to consider "if a natural disaster or other force majeure also contributed to the occurrence of the launch vehicle fall damage" when determining liability and damages.[38]

As noted above, the launch operator must execute "a launch vehicle fall damage liability insurance contract," which is:

> a contract wherein an insurer . . . promises to compensate for the loss suffered by a person implementing the launching of a spacecraft, etc. as a result of accruing for a liability to compensate for launch vehicle fall damage (excluding launch vehicle fall damage caused by the fall, collision or explosion of a spacecraft, etc. primarily caused by an act of terrorism or any other events specified by Cabinet Office Order as those for which the calculation of the reasonable amount of insurance premiums would be

---

34 2016 Space Activities Act, Article 36(1).
35 New Law Aims to Expand Japan's Space Business, at https://www.nippon.com/en/currents/d00294/.
36 2016 Space Activities Act, Article 38(1).
37 Id.
38 See id., Article 37.

difficult . . .) and making the compensation therefor, and the policyholder promises to pay insurance premiums to the insurer.[39]

Pursuant to Article 9(2), the Cabinet Office determines the appropriate amount of insurance coverage a launch operator is to obtain, as well as a supplemental amount of compensation coverage in a separate indemnification contract (discussed below), "in light of the protection of victims of the launch vehicle fall damage, considering the design of the launch vehicle, the location of the launch site, or other situations[.]"[40] As of July 2019, this amount of insurance coverage was JPY 20 billion (approximately USD 192 million) for all launch vehicle types.[41] Under Article 39(1), "[v]ictims of launch vehicle fall damage are entitled to receive the payment from insurance proceeds under the launch vehicle fall damage liability insurance contract, in relation to their respective claims for damages, prior to other creditors."[42]

As indicated above, besides executing a contract with insurance providers, and before conducting its spacecraft launch, a launch operator is required to enter into "a launch vehicle fall damage liability indemnification contract" with the Japanese government, which is:

> a contract wherein the government promises to indemnify the person implementing the launching of spacecraft, etc. against losses caused by making a compensation for the launch vehicle fall damage for which the person became liable that cannot be covered by the launch vehicle fall damage liability insurance contract or any other means for compensation for launch vehicle fall damage.[43]

Pursuant to the 2016 Space Activities Act, after assessing the coverage afforded under the launch vehicle fall damage liability insurance contract, the Japanese government may execute a contract with the launch operator to provide indemnification up to a certain ceiling for third-party damages that are not covered by the liability insurance.[44] More specifically, Article 40(2) states that the Japanese government may: (i) supplement through an indemnification contract the insurance coverage obtained by a launch

---

39　*Id.*, Article 2(ix).
40　*Id.*, Article 9(2).
41　See *Jinkō eisei-tō no uchiage oyobi jinkō eisei no kanri ni kansuru hōritsu sekōkisoku* [Regulation for Enforcement of the Act on Launching of Spacecraft, etc. and Control of Spacecraft] (Cabinet Office Order No. 50 of 2017), Article 9-2; *see also* Japan, note 23, referencing Schedule to Art. 9-2 of the Regulation for Enforcement of the Act on Launching Artificial Satellites and Managing Satellites (Cabinet Ordinance No. 50 of 2017), at https://thelawreviews.co.uk/edition/the-space-law-review-edition-1/1211969/japan.
42　2016 Space Activities Act, Article 39(1).
43　*Id.*, Article 2(x).
44　See *id.*, Article 40(1) and (2).

operator for launch vehicle fall damage "that is not covered by a launch vehicle fall damage liability insurance contract"; and (ii) supplement such insurance coverage through additional indemnification "in an amount . . . which is appropriate from the standpoint of the reinforcement of international competitiveness of Japanese industries related to the launching of spacecraft, etc.[.]"[45] While not set out in the 2016 Space Activities Act, the combined maximum of insurance and indemnification is JPY 350 billion (approximately USD 3.3 billion).[46]

2 Spacecraft Fall Damage

The 2016 Space Activities Act defines "spacecraft fall damage" as:

> damage to human life or body, or property on the ground surface or water surface, or an aircraft in flight or other flying object caused by the fall or explosion of a spacecraft successfully separated from the launch vehicle; provided, however, that damages suffered by workers of the person implementing the control of the spacecraft or other persons specified by Cabinet Office Order as those having a close business relationship with the person implementing the control of the spacecraft in the course of their business are excluded.[47]

Similar to Article 35, Article 53 provides: "A person who implements the control of a spacecraft using a spacecraft control facility located in Japan is liable to compensate for spacecraft fall damage caused by the person in relation to the control of the spacecraft."[48] The liability for spacecraft control operators also is strict, but like the rules for launch operators a court may, in its discretion, consider "if a natural disaster or other force majeure also contributed to the occurrence of the spacecraft fall damage" when determining spacecraft control operators' liability and damages.[49]

Unlike the regime for launch operators, however, liability is not channeled only to spacecraft control operators under the 2016 Space Activities Act. Thus, third parties that have suffered spacecraft fall damage may pursue compensation from any other

---

45 Id., Article 40(2); see also Article 2(ix) ("excluding launch vehicle fall damage caused by an act of terrorism or any other events specified by Cabinet Office Order" in scope of launch vehicle fall damage liability insurance contract coverage).
46 See Jinkō eisei-tō no uchiage oyobi jinkō eisei no kanri ni kansuru hōritsu sekōkisoku [Regulation for Enforcement of the Act on Launching of Spacecraft, etc. and Control of Spacecraft] (Cabinet Office Order No. 50 of 2017), Article 32-2; see also Japan, at https://thelawreviews.co.uk/edition/the-space-law-review-edition-1/1211969/japan.
47 2016 Space Activities Act, Article 2(xi).
48 Id., Article 53.
49 Id., Article 54.

parties to which such damage may be attributable, and there is no limitation on reimbursement claims by spacecraft control operators from such other parties.

In addition, and also unlike the regime for launch operators, the 2016 Space Activities Act does not oblige spacecraft control operators to take "security measures for compensation for damages," such as obtaining liability insurance. The rationale for this difference was that the risk of damage from falling satellites to people and property on Earth is deemed to be relatively low. Further, there was concern that, as a requirement to obtain this insurance is not an international norm, imposing such an obligation on Japan-based operators might negatively affect their international competitiveness.[50]

---

50 *See Jinkō eisei-tō no uchiage oyobi jinkō eisei no kanri ni kansuru hōritsu ni motozuku daisansha songaibaishōseido ni kansuru gaidorain* [Guideline on the Compensation for Third Parties under the Act on Launching of Spacecraft, etc. and Control of Spacecraft] (2019/5/31), p. 15, at https://www8.cao.go.jp/space/application/space_activity/documents/guideline5.pdf.

Chapter 4 Recent Legislation: 2016 Remote Sensing Act

I   Introduction

The 2016 Remote Sensing Act regulates satellite remote sensing activities in Japan, including the use and handling of satellite remote sensing data, that is, data in the form of electromagnetic energy (*e.g.*, radio, microwave, infrared, ultraviolet, and x-ray waves) from Earth that satellite-based sensors detect.[1] Satellite remote sensing has commercial and other applications in various fields, such as agriculture and natural resource management (*e.g.*, monitoring land use/erosion, vegetation, soil moisture, ocean and coastal resources, and hydrocarbons), meteorology (*e.g.*, tracking weather systems including hurricanes), natural disaster response (*e.g.*, monitoring landslides, volcanoes, post-earthquake damage), and topographic mapmaking.[2]

While it is a space-related technology with many potential commercial functions, the collection and use of satellite remote sensing data raises concerns regarding individual privacy and national security interests. The 2016 Remote Sensing Act seeks to address such issues and clarify the obligations of private sector actors so as to facilitate their entry into satellite remote sensing business ventures. It covers two main areas: (i) permission to use satellite remote sensing instruments; and (ii) regulations on handling satellite remote sensing data, including certification to handle such data. The key elements of these two areas are discussed below, along with several miscellaneous provisions.

---

1   *See Eisei rimōtosenshingu kiroku no tekiseina toriatsukai no kakuho ni kansuru hōritsu* [Act on Ensuring Appropriate Handling of Satellite Remote Sensing Records] (Act No. 77 of November 16, 2016) ("2016 Remote Sensing Act"). The text of the 2016 Remote Sensing Act in Japanese and English may be found at https://www8.cao.go.jp/space/english/rs/rs_act.pdf. The English version is included Annex 4.

2   For further detailed discussion about the fields of application for satellite remote sensing in Japan and potential opportunities in this area, as well as Japanese companies involved in activities related to satellite remote sensing, *see Space Industry Business Opportunities in Japan*, pp. 60-161, at https://www.eu-japan.eu/sites/default/files/publications/docs/2016-10-space-industry-business-opportunities-japan-ryuichi-dunphy-min.pdf.

II   Permission to Use Satellite Remote Sensing Instruments

Article 4(1) of the 2016 Remote Sensing Act provides: "A person who intends to conduct the Use of Satellite Remote Sensing Instruments by use of a Ground Radio Station for Command and Control located in Japan (excluding Specified User Organization) must obtain a license from the Prime Minister per [*i.e.*, for each of its] Satellite Remote Sensing Instruments."[3] This provision and the defined terms it contains need unpacking.

First, "Satellite Remote Sensing Instruments" means:

> equipment onboard a satellite which is used by [being] launched into the Earth orbit (hereinafter referred to as "Earth Orbiting Satellite"), which detects electromagnetic waves emitted or reflected from objects existing on the ground surface or water surface (including underground or underwater which is near the surface) or in the air (hereinafter referred to as "Ground Emitted Electromagnetic Waves, etc."), and which records information concerning intensity, frequency and phase of Ground Emitted Electromagnetic Waves, etc. as well as information on the position and other status of the Earth Orbiting Satellite when that information was detected . . . in electronic or magnetic records (meaning records created electronically, magnetically and any other form which cannot be recognized through human senses and which is used for information processing in computers; the same applies hereinafter), and has the function of sending such information to the ground; when this function is operated under appropriate conditions, the strength of those electronic or magnetic records received on the ground and made recognizable through visualization on a computer screen (hereinafter referred to as "distinguishing accuracy of target" in this Article and Article 21, paragraph (1)) fulfills the criteria specified by Cabinet Office Order which determines the thresholds as capable of discerning the movement of vehicles, ships, aircraft and other moving facilities; and which has a ground radio station capable of sending or receiving a signal necessary to activate or deactivate these functions and those electronic or magnetic records to other ground radio

---

[3]   2016 Remote Sensing Act, Article 4(1). Again, when used in a legal context such as in the 2016 Remote Sensing Act or other legislation, the Japanese word for "person" (*mono*) means natural persons and legal persons (*e.g.*, corporations). Also, as noted in the previous chapter, in both the 2016 Remote Sensing Act and the 2016 Space Activities Act, the Japanese word *kyoka* is used and is translated variously as "permission" and "license" in the English versions of these statutes on the Japanese government websites. In English, *kyoka* means "permission," "license," "authorization," or "approval." In other words, *kyoka* means "license" in the sense of being given permission to do something, and not an official document that gives one permission to do something. Accordingly, notwithstanding that these statutes (and some commentators) sometimes utilize the word "license" in this context, the word "permission" is used above except when directly quoting the English version of the statutory language from the Japanese government websites.

CHAPTER 4   RECENT LEGISLATION: 2016 REMOTE SENSING ACT

stations (meaning electrical equipment for sending or receiving codes using electromagnetic waves, and a computer connected to such equipment via telecommunication line; the same applies hereinafter) by the use of electromagnetic waves.[4]

The Cabinet Office Order standards referenced above ("criteria specified by Cabinet Office Order which determines the thresholds as capable of discerning the movement of vehicles, ships, aircraft and other moving facilities") relate to a satellite remote sensing device's designated level of resolution for object discernment that triggers national security concerns (*i.e.*, devices with high levels of resolution). As such, the 2016 Remote Sensing Act's regulatory regime applies only to satellite remote sensing devices with resolution levels designated by the Japanese government.[5]

Next, "Ground Radio Station for Command and Control" is:

a ground radio station which has the function of transmitting signals necessary for the operation of Satellite Remote Sensing Instruments to those Satellite Remote Sensing Instruments, directly or via other ground radio stations using electromagnetic waves; and for which signals include the time at which to activate the function of detecting Ground Emitted Electromagnetic Waves, etc. of the Satellite Remote Sensing Instruments, the time at which to send electronic or magnetic records recording detected information (hereinafter referred to as "Electromagnetic Data of Detected Information") to the ground, communication method used for sending such information, and the determination and change of distinguishing accuracy of target.[6]

Lastly, "Use of Satellite Remote Sensing Instruments" means:

to operate Satellite Remote Sensing Instruments onboard an Earth Orbiting Satellite by the use of Ground Radio Station for Command and Control operating on its own or managed by another person, and to send

---

4   *Id.*, Article 2(ii). "Satellite" here is defined in almost the same manner as "spacecraft" in the 2016 Space Activities Act: "an artificial object which is used by [being] launched into Earth orbit or beyond, or placed on a celestial body other than the Earth." *Id.*, Article 2(i).

5   Designated levels of resolution for certain data include, for example: (i) for optical sensors, two meters or less; (ii) for Synthetic Aperture Radar (SAR) sensors, three meters or less; (iii) for hyper-spectral sensors, ten meters or less, and the range of identifiable bands is more than 49; and (iv) for thermal infrared sensor, five meters or less. *See Eisei rimōtosenshingu kiroku no tekiseina toriatsukai no kakuho ni kansuru hōritsu shikōkisoku* [Regulation for Enforcement of the Act on Ensuring Appropriate Handling of Satellite Remote Sensing Data] (Cabinet Office Order No. 41 of August 9, 2017), Article 3(1), at https://www8.cao.go.jp/space/english/rs/rs_cabinetofficeorder.pdf.

6   2016 Remote Sensing Act, Article 2(iii).

Electromagnetic Data of Detected Information to the ground after setting methods for sending necessary signals to that Satellite Remote Sensing Instruments from that Ground Radio Station for Command and Control.[7]

It may be recalled that under Article 4(1) that the 2016 Remote Sensing Act applies only to the operation of Satellite Remote Sensing Instruments from Ground Radio Stations for Command and Control that are "located in Japan." Accordingly, permission is not required for satellite remote sensing devices that are controlled from such facilities outside Japanese territory.[8]

As a practical matter, applicants seeking to obtain permission as a satellite remote sensing device user under the 2016 Remote Sensing Act must submit to the Cabinet Office's National Space Policy Secretariat various information, including about: themselves; the type, structure and capability of the satellite remote sensing devices; the orbit of the satellite equipped with the devices; the location, structure and capability both of the ground radio stations that are to be used to transmit signals to operate the satellite remote sensing devices and of those stations that are to receive the electromagnetic data transmitted from the devices; and the methods of managing this electromagnetic data.[9]

Before granting permission, the Japanese government must be satisfied that the application meets certain requirements to ensure that unauthorized individuals and entities may not use the satellite remote sensing devices and that the data from these devices is not improperly disclosed or lost. Specifically, the applicant must show that: (i) "the necessary and appropriate measures have been taken to prevent persons other than the applicant from the Use of Satellite Remote Sensing Instruments;"[10] (ii) "measures for prevention of divulgence, loss or damage of Satellite Remote Sensing Data and any other necessary and appropriate measures to be specified by Cabinet Office Order for the safety management of the Satellite Remote Sensing Data have been taken";[11] and (iii) "the applicant . . . has the ability to properly implement measures to prevent persons other than the applicant provided in item (i) from the Use of Satellite Remote Sensing Instruments and measures for safety management of the Satellite Remote Sensing Data provided in the preceding item."[12] The Japanese government will also review the application to ensure, among other things, that the applicant is not a threat

---

7   *Id.*, Article 2(iv).
8   *See id.*, Article 4(1); Japan, at https://thelawreviews.co.uk/edition/the-space-law-review-edition-1/1211969/japan. Permission pursuant to the 2016 Remote Sensing Act also is not required for "Specified User Organizations," which are defined as national and local governmental entities. *See id.*, Article 2(v).
9   *Id.*, Article 4(2)(i)-(vi). The Cabinet Office's National Space Policy Secretariat, to which applications are sent, provides a manual to assist applicants. For the English version, *see* https://www8.cao.go.jp/space/english/rs/rs_manual.pdf; *see also* https://www8.cao.go.jp/space/english/rs/application.html. The English version is included in Annex 6.
10 *Id.*, Article 6(i).
11 *Id.*, Article 6(ii).
12 *Id.*, Article 6(iii).

CHAPTER 4   RECENT LEGISLATION: 2016 REMOTE SENSING ACT

(*e.g.*, a terrorist) and that the applicant and Use of Satellite Remote Sensing Instruments pose minimal risk to national and international security.[13]

On average it takes about two months for an applicant to receive a decision from the Japanese government regarding permission for Use of Satellite Remote Sensing Instruments.[14] Similar to the permitting procedure under the 2016 Space Activities Act, applicants are encouraged to consult in advance with the National Space Policy Secretariat to facilitate the completion and processing of applications, as well as increase the likelihood of approval.

In addition, once an applicant has received permission, thereby becoming a "Satellite Remote Sensing Instruments User" under the 2016 Remote Sensing Act,[15] it is obliged to encrypt both the communication signals used to operate the satellite remote sensing devices and the electromagnetic data transmitted from these devices.[16] Further, the Satellite Remote Sensing Instruments User is proscribed from disclosing the encryption codes to any other person (except such other persons who manage the designated Ground Radio Stations for Command and Control and Receiving Stations), and must "take measures for the prevention of divulgence, loss or damage" of these codes.[17] The Satellite Remote Sensing Instruments User also must keep a log recording the status of its Use of Satellite Remote Sensing Instruments.[18]

If, after obtaining permission under Article 4(1), a Satellite Remote Sensing Instruments User wishes to make any changes (other than minor changes) to the information provided in the application to the Japanese government regarding the satellite remote sensing device (*e.g.*, structure, capability, orbit), radio stations, or methods of managing the remote sensing data, the user must obtain permission.[19] The 2016 Remote Sensing Act also stipulates that, if the satellite equipped with the permitted remote sensing device deviates from the orbit for which the user permission was granted

---

13   *Id.*, Article 5(i) and (iii), 6(iv). The particular phrase used is "Ensuring Peace of the International Community, etc." and is defined earlier in the 2016 Remote Sensing Act as "ensuring the peace and security of the international community and the national security of Japan provided in Article 14 the Basic Space Act[.]" *Id.*, Article 2(vi). Article 14 of the 2008 Basic Space Law in turn provides: "The national government shall implement policy measures as necessary for the promotion of Space Development and Use to contribute to ensuring international peace and security and to the national security of our country." 2008 Basic Space Law, unofficial ZeLo translation at Annex 7.
14   See Applying for a License pertaining to Use of Satellite Remote Sensing Instruments and a Certification of Persons Handling Satellite Remote Sensing Data, at https://www8.cao.go.jp/space/english/rs/application.html.
15   See 2016 Remote Sensing Act, Article 7(1).
16   See *id.*, Articles 8(1) and (2). *See also* English version of Guidelines on Measures, etc. Under Act on Ensuring Appropriate Handling of Satellite Remote Sensing Data, December 29, 2017, at https://www8.cao.go.jp/space/english/rs/rs_guideline.pdf. This version is included in Annex 5.
17   See *id.*, Articles 8(3), (4), and (5).
18   See *id.*, Article 12(1).
19   See *id.*, Article 7(1).

under Article 4(1), the Satellite Remote Sensing Instruments User must "immediately" suspend operation of the device until the satellite returns to its prescribed orbit.[20]

The 2016 Remote Sensing Act allows a Satellite Remote Sensing Instruments User to terminate its Use of Satellite Remote Sensing Instruments "at any time."[21] When the remote sensing device user does so, it must inform the Japanese government and must take appropriate measures to send a signal to the device to stop it from functioning or suspend its functioning, at which point the permission granted under Article 4(1) to that user ceases to be effective.[22]

III REGULATIONS ON HANDLING SATELLITE REMOTE SENSING DATA, INCLUDING CERTIFICATION TO HANDLE SUCH DATA

Article 2(vi) of the 2016 Remote Sensing Act defines "Satellite Remote Sensing Data" as:

> Electromagnetic Data of Detected Information which is sent to the ground by a person other than Specified User Organization by Use of Satellite Remote Sensing Instruments through the Ground Radio Station for Command and Control located in Japan, and electronic or magnetic records processed from that Electromagnetic Data of Detected Information; and when such information and processed data fall under the criteria of information specified by Cabinet Office Order where the use of such information is likely to cause adverse effect on ensuring the peace and security of the international community and the national security of Japan provided in Article 14 of the Basic Space Act (hereinafter referred to as "Ensuring Peace of the International Community, etc.") in view of their distinguishing accuracy of target, the extent and degree of modification of Electromagnetic Data of Detected Information through processing, the elapsed time since that Electromagnetic Data of Detected Information was recorded, and other circumstances; and copies of that information on electromagnetic recording medium (means recording medium of electronic or magnetic records).[23]

---

20 *See id.*, Article 9.
21 *Id.*, Article 15(1).
22 *See id.*, Article 15(2)(i) and (ii). Article 15(2)(ii) provides that a Satellite Remote Sensing Instruments User may send a signal "to the Satellite Remote Sensing Instruments to stop its function until a restart signal (meaning a signal necessary to recover the function of detecting Ground Emitted Electromagnetic Waves, etc. if the function has been suspended . . .) is received[.]" Because the permission granted to this user thereby ceases to be effective, Article 15(4) stipulates that such former user may provide information on the restart signal only to a person who has obtained permission under Article 4(1) to operate the satellite remote sensing device. *See id.*, Article 15(4).
23 *Id.*, Article 2(vi). As noted, "Specified User Organization" means national and local governmental entities. *See id.*, Article 2(v).

CHAPTER 4   RECENT LEGISLATION: 2016 REMOTE SENSING ACT

An individual or company in possession of Satellite Remote Sensing Data is defined as a "Satellite Remote Sensing Data Holder" and is required to take measures to prevent the unauthorized disclosure, loss, or damage of such data.[24]

Pursuant to Article 18, a Satellite Remote Sensing Data Holder may provide Satellite Remote Sensing Data only to a Satellite Remote Sensing Instruments User or to a person that has obtained certification from the Japanese government to handle such data.[25] As to the latter, Article 21(1) states:

> A person handling Satellite Remote Sensing Data (excluding Specified Data Handling Organization) may, upon application, obtain a certification from the Prime Minister that states that person is found to be capable of properly handling Satellite Remote Sensing Data, according to the categories of Satellite Remote Sensing Data specified by Cabinet Office Order having regard to circumstances such as Distinguishing Accuracy of Target, the scope and degree of information changed as a result of processing of Electromagnetic Data of Detected Information, or the time elapsed since the relevant Electromagnetic Data of Detected Information was recorded.[26]

As indicated earlier, "Distinguishing Accuracy of Target" means the strength of "electronic or magnetic records received on the ground and made recognizable through visualization on a computer screen" in relation to a satellite remote sensing device's government-designated level of resolution for object discernment that triggers national security concerns.[27] The "categories of Satellite Remote Sensing Data specified by Cabinet Office Order" are "raw data" and "standard data," with the former comprising information from optical, hyper-spectral, thermal infrared, and certain SAR sensors which is not subject to radiometric or geometric processing, and the latter consisting of raw data that is subject to corrections through radiometric or geometric processing.[28]

An applicant for this certification must submit information about itself and the category and methods of using and managing the Satellite Remote Sensing Data, as well as about the ground station where the data is received (if the applicant receives the data at

---

24 See id., Article 2(viii). This definition excludes governmental organizations that handle Satellite Remote Sensing Data. See also id., Article 20.
25 Id., Article 18(1) and (2). Article 18(3) provides certain exceptions in this respect, including "in an urgent situation when measures must be taken to rescue human life, for disaster relief or for other emergencies."
26 Id., Article 21(1).
27 Id., Article 2(ii).
28 See Eisei rimōtosenshingu kiroku no tekiseina toriatsukai no kakuho ni kansuru hōritsu sekoukisoku [Regulation for Enforcement of the Act on Ensuring Appropriate Handling of Satellite Remote Sensing Data] (Cabinet Office Order No. 41 of August 9, 2017), Article 3(1), at https://www8.cao.go.jp/space/english/rs/rs_cabinetofficeorder.pdf. The levels of object discrimination accuracy designated by the Japanese government vary depending on the type of sensor and whether the data is classified as raw or standard.

a ground station).²⁹ The Japanese government will grant the certification unless, among other things, the applicant is deemed a threat (*e.g.*, a terrorist), and so long as the government concludes there is minimal risk to national and international security "by taking into consideration the purpose and methods of use of Satellite Remote Sensing Data by the applicant, the capability to carry out analysis or processing of Satellite Remote Sensing Data, measures to ensure the safety management of Satellite Remote Sensing Data and other circumstances."³⁰ Individuals or companies that obtain such certification must keep a log recording the status of their handling of the Satellite Remote Sensing Data.³¹

As noted above, Article 18(1) of the 2016 Remote Sensing Act also states that a Satellite Remote Sensing Data Holder may provide remote sensing data to an individual or company that has obtained permission as a Satellite Remote Sensing Instruments User under Article 4(1). Both with respect to such a device user and an individual or company that has obtained certification under Article 21, the Satellite Remote Sensing Data Holder must clearly specify the category of the data being provided (raw or standard) and must "provide the information using cryptography or any other method of transmission whereby it is not easy to restore the contents thereof or any other method specified by Cabinet Office Order as necessary and appropriate for prevention of acquisition and use of Satellite Remote Sensing Data by any person other than the recipient of the Satellite Remote Sensing Data[.]"³² If the Japanese government considers that use of Satellite Remote Sensing Data likely poses a risk to national or international security, the government may issue an order to the Satellite Remote Sensing Data Holder to prohibit provision of the data "designating the scope and time period."³³ Such a prohibition "must be limited to the minimum extent required" to ensure national and international security.³⁴

---

29 2016 Remote Sensing Act, Article 21(2).
30 *Id.*, Article 21(3)(ii).
31 *Id.*, Article 23. An individual or company that has obtained certification under Article 21 and wishes to change the information about the purpose and methods of using and managing the satellite remote sensing data, or about the ground station receiving such data, must obtain government certification. *See id.*, Article 22(1).
32 *Id.*, Article 18(1) and (2).
33 *Id.*, Article 19(1). This order does not apply to "a natural person who has neither domicile nor residence in Japan or a corporation or any other organization which does not have a principal office in Japan that handles Satellite Remote Sensing Data in a foreign country[.]" *Id.* Instead, for such a "Foreign Handler," the Japanese government would "request" it "not to provide" the Satellite Remote Sensing Data. *Id.*, Article 19(3).
34 *Id.*, Article 19(2).

IV MISCELLANEOUS PROVISIONS

In addition to the foregoing, the 2016 Remote Sensing Act provides: "[t]he Prime Minister may give necessary guidance, advice and recommendations to Satellite Remote Sensing Instruments Users or Satellite Remote Sensing Data Holders in order to ensure the proper handling of Satellite Remote Sensing Data in Japan."[35] Further, the Prime Minister may, "to the extent necessary for the enforcement of this Act, request a Satellite Remote Sensing Instruments User or Satellite Remote Sensing Data Holder . . . to provide necessary reports or have Cabinet Office officials enter its office or any other places of business to inspect books, documents or other items or to question relevant persons[.]"[36]

---

35 *Id.*, Article 28.
36 *Id.*, Article 27(1).

Chapter 5 For Foreign Investors: Other Applicable Laws In and Outside Japan

As noted above, the 2016 Space Activities Act and Remote Sensing Act apply to activities located in Japan (or in certain instances "onboard a ship or aircraft with Japanese nationality"). Thus, non-Japanese participants conducting relevant space-related activities in Japan would be subject to this legislation. In addition, overseas companies and businesspeople interested in establishing a space-related corporation or joint venture company in Japan should be aware of other applicable laws and regulations both in and outside Japan. This chapter provides an introductory overview of this topic.

I Japanese Corporate Structures

A non-Japanese company or individual that wishes to set up a corporation in Japan would likely do so using one of the two following structures, both of which are governed by Japan's Companies Act.[1]

1 Kabushiki Kaisha (KK)

The most common corporate entity in Japan is the joint stock corporation (*kabushiki kaisha*) (KK). KKs can be established with minimal capital (even just one Japanese Yen) and a sole director. A KK's ownership interests are represented as shares (*kabushiki*) and its holders are shareholders (*kabunushi*). Certain types of shares, such as preferred shares, may be issued in addition to common shares.

In terms of internal corporate structure, the minimum for a KK is a sole director (*torishimariyaku*) and no board of directors. There is no corporate secretary system under Japanese corporate law, but if a board of directors (*torishimariyaku-kai*) is established a statutory auditor (*kansayaku*) will be appointed to audit the directors' conduct. Three or more directors must be appointed to establish a board of directors. Directors and statutory auditors are appointed by shareholders through a resolution at the general meeting of shareholders (*kabunushi sokai*). Directors have a right to represent

---

1 See *Kaisha hō* [Companies Act] (Act No. 86 of 2005), Japanese and unofficial English translation at http://www.japaneselawtranslation.go.jp/law/detail/?id=3206&vm=04&re=02 and http://www.japaneselawtranslation.go.jp/law/detail/?re=02&vm=02&al[]=C&ky=%E6%A0%AA%E5%BC%8F&page=18.

the company, but the board of directors also may designate a representative director(s) (*daihyou torishimariyaku*) to be the company's official representative. KKs are required to have an internal constitutional document, Articles of Incorporation (*teikan*), which sets out the company's structure and shareholdings.

KKs are registered in a public company registration system that includes the name and address of any representative directors, as well as the names of directors and statutory auditors. Names of shareholders and their shareholding ratio, however, are not registered in this registration system.

There are no residence requirements to be a shareholder, director, or statutory auditor of a KK. Japan maintains a system of seal registration (*inkan toroku*) for individuals and companies, and some government offices request a certificate of a registered seal and require this seal to be stamped on corporate filing or application documents. Japanese residents only may make this seal registration. Thus, non-resident companies or individuals may need to take additional steps, such as preparing an affidavit or other certification materials, which can be used instead of the certificate of a registered seal. The filing and application documents, as well as court documents, must be in Japanese or have a Japanese translation. There are, however, no language requirements for contracts to which a KK or other Japanese company is a party.

2   Godo Kaisha (GK)

Another possible Japanese corporate structure for an overseas investor is a limited liability company (*godo kaisha*) (GK), the ownership and management of which are more closely connected than in a KK. A GK's interests are owned by members and a representative member that makes the decisions about the GK's daily business. This representative member may be a legal entity, but if so an individual executive officer (*shokumu shikkousha*) who makes such daily business decisions would be appointed.[2]

II   JOINT VENTURE COMPANY

If an overseas entity wishes to set up a joint venture (JV) with a Japanese firm, the first choice for structuring the JV would be a KK. As mentioned, a KK could be established with a board of directors of three or more directors and one statutory auditor, and each JV partner would be able to designate its nominated directors on the board. If there are two JV partners, each partner could own 51% and 49% of the shares, respectively. The

---

[2] More information in English regarding the establishment of KK and setting up it as a business may be found on the website of the Japan External Trade Organization, at https://www.jetro.go.jp/en/invest/setting_up/.

parties would need to draft and execute a shareholders (or JV) agreement that details their respective rights and obligations.[3]

III JAPANESE FOREIGN INVESTMENT REGULATIONS

In general, foreign investment in Japan by overseas parties, whether establishing a new company or setting up a JV, is not strictly regulated. There are *post-facto* reporting requirements for inbound investment to authorities, such as the Bank of Japan, under the Japanese Foreign Exchange and Foreign Trade Act (FX Act).[4] These requirements are mainly for statistical record-keeping purposes and are not especially burdensome.

On the other hand, investment by overseas investors in certain sectors and especially "core sectors" that relate to Japan's national security is subject to prior notice obligations and needs approval from the Japanese government. Manufacture of satellites, rockets, and their launching and navigation systems typically falls under these regulated sectors, and inbound investment in this area would be required to give prior notification and obtain government approval. There are certain exemptions from such prior notice obligations under the FX Act, but limitations in this respect apply to the "core sectors," including space-related manufacturing. Thus, for example, a non-Japanese investor aiming to acquire 10% or more voting shares in a Japanese space-related company would not be exempt from the prior notification requirement. Nor would this investor be exempt if it planned to become a director or statutory auditor of the Japanese company, or if sought access to certain confidential technical information. Investors that are foreign government entities are subject to stricter restrictions in this regard.

Accordingly, an overseas firm or individual seeking to invest in Japan's space business would be prudent to assume that its investment will require prior notice to and approval from the Japanese government. Typically, 30-days prior notification is required for a non-Japanese investor that plans to invest in Japan, and a resident agent must make the relevant filing on behalf of the investor.

---

3 Non-Japanese investors in Japanese space activities also may be reassured to know that in recent years Japan has made strides to become a jurisdiction that is friendly to international arbitration, which is the preferred dispute resolution mechanism for a high-technology sector such as space. Japan is a signatory to the 1958 Convention on the Recognition and Enforcement of Foreign Arbitral Awards (commonly known as the New York Convention), has a modern national arbitration law, and its courts have developed a solid track record in enforcing international arbitration awards. In addition to the availability of arbitral rules of the Japan Commercial Arbitration Association, the Japan International Dispute Resolution Center opened facilities in Osaka and Tokyo in 2018 and 2020, respectively, which are available to parties whose arbitral proceedings are administered under the rules of other, non-Japanese arbitration institutions. *See* ZeLo, Japan International Dispute Resolution Center: Advanced Facilities in Osaka and Tokyo for Virtual Arbitration Hearings, August 24, 2020, at https://zelojapan.com/en/4982.

4 *See Gaikoku kawase oyobi gaikoku bōeki hō* [Foreign Exchange and Foreign Trade Act] (Act No. 228 of December 1, 1949), Japanese and unofficial English translation at http://www.japaneselawtranslation.go.jp/law/detail/?printID=&id=21&re=02&vm=03.

Japan's Ministry of Finance has published a list of factors that the government uses to screen such inbound investments.[5] Screening factors for prospective overseas investors include information concerning the investor, such as "capital structure, beneficial ownership and business relationships, and the foreign investor's plan and track record of its behaviors relating to the investment," as well as about the investment's impact on the Japanese entity, including "the number and share of shares, equities, voting rights, subscription certificates or corporate bonds that have been acquired or are to be acquired by the foreign investor."[6]

A number of factors also specifically concern national security, for example: (i) the extent of the inbound investment's impact on maintaining bases of "production and technologies in the business sectors that relate to protection of national security, maintenance of public order, or safeguard of public safety," as well as on the supply and quality of "goods or services that relate to protection of national security, maintenance of public order, or safeguard of public safety, in ordinary and emergency situations";[7] (ii) potential "leakage of technologies or information that relate to protection of national security, maintenance of public order, and safeguard of public safety; or use of these technologies or information against the objectives of protection of national security, maintenance of public order, or safeguard of public safety;"[8] and (iii) the impact on national security of the overseas investor "(1) attending, regarding business activities in core sectors, the investee company's executive board or committees that have the authority to make important decisions for the business activities (2) making proposals, regarding business activities in core sectors, in a written or electronic form, to the investee company's executive board, committees that have the authority to make important decisions, or members of these executive board/committees, requiring their responses and/or actions by certain deadlines."[9]

Separately, it bears noting that Japanese businesses which are engaged in space-related activities outside Japan also are subject to the FX Act, and these companies need to be aware of provisions in this legislation that regulate the export of rockets and satellites overseas.[10] For example, launching a rocket assembled in Japan at an overseas launch site is considered export and is subject to the FX Act. The FX Act further regulates the export of information regarding rocket and satellite technologies, and Japanese

---

5   *See* "Factors to be considered in authorities' screening of prior-notification for Inward Direct Investment and Specified Acquisition under the Foreign Exchange and Foreign Trade Act," Japan Ministry of Foreign Affairs, May 8, 2020, at https://www.mof.go.jp/english/international_policy/fdi/gaitamehou_20200508.htm.
6   *See id.*, items 5 and 6.
7   *See id.*, items 1 and 3.
8   *See id.*, item 2.
9   *See id.*, item 10.
10 *See* Article 48 of *Gaikoku kawase oyobi gaikoku bōeki hō* [Foreign Exchange and Foreign Trade Act] (Act No. 228 of December 1, 1949), Export Trade Control Order Article 1, paragraph (1), Appended Table 1, 1., (xvii), 4., etc., Japanese and unofficial English translation at http://www.japaneselawtranslation.go.jp/law/detail/?printID=&id=21&re=02&vm=03.

businesses must comply with the relevant requirements when working with partners outside Japan.[11]

IV Applicable Laws Outside Japan

In addition to adhering to Japan's space and foreign investment legislation, non-Japanese participants in Japan-based space activities will need to comply with any applicable laws of their own jurisdictions. Some two dozen nations regulate space activities, and a number of countries require their nationals (legal and natural persons) to obtain a license or other permission for space activities conducted from outside as well as in their territories.[12] An Australian individual or company, for example, is required to obtain an overseas launch certificate to launch a space object from outside Australia.[13] The United States requires its citizens to obtain a license before carrying out space vehicle launch or reentry activities in the United States or elsewhere in the world.[14] France and Norway each require their nationals to obtain authorization before conducting space launches from territory outside, as well as within, their respective jurisdictions.[15] Under United Kingdom (U.K.) space legislation, a U.K. national or subject must obtain a license before launching or operating a space object, or engaging in any space activity, whether conducted in or from outside the United Kingdom.[16]

Accordingly, individuals and companies from these and other such jurisdictions likely would be subject to double-licensing requirements if they aim to participate in Japan-based space activities, and they should bear this responsibility in mind while seeking appropriate legal guidance. As one commentator has noted, national space laws:

---

11 *See id.*, Article 25(1), Foreign Exchange Order Article 17, paragraph (1), Appended Table 1, 4., etc., Japanese and unofficial English translation at http://www.japaneselawtranslation.go.jp/law/detail/?printID=&id=21&re=02&vm=03. *See also* Center for Information on Security Trade Control, *Overview of Japan's Export Controls* (4th ed., 2015), pp. 1-15, 51-53, at https://www.cistec.or.jp/english/export/Overview4th.pdf.

12 *See* P.S. Dempsey, "National Laws Governing Space Activities: Legislation, Regulation, & Enforcement" (National Laws Governing Space Activities), *Northwest. J. Int. Law Bus.*, v. 36, no. 1, 2016, pp. 115, 119. For a comprehensive compendium of national space legislation, *see* vols. 5 and 6 of *Space Law Basic Legal Documents*, eds. M. Benkö & K.H. Böcksteigel (Eleven International Publishing, 2019).

13 *See id.*, p. 120, and F.G. von der Dunk, "Launching from 'Down Under': The New Australian Space Activities Act of 1998," 2000, p. 136, Space, Cyber, and Telecommunications Law Program Faculty Publications, University of Nebraska, at https://digitalcommons.unl.edu/spacelaw/37/.

14 *See* National Laws Governing Space Activities, pp. 125-26, and M. Smith, "United States" (United States), *The Space Law Review*, ed. 2, December 2020, at https://thelawreviews.co.uk/edition/the-space-law-review-edition-2/1235331/united-states.

15 *See* National Laws Governing Space Activities, p. 121, 124.

16 *See id.*, pp. 121-22, and T.I. Mainura, "The United Kingdom's Outer Space Act 1986," *Int. J. of Social Sciences and Humanities Invention*, v. 5, no. 5, May 30, 2018, at https://valleyinternational.net/index.php/theijsshi/article/view/1278.

differ when it comes to the details. This is particularly true for licensing procedures. The duration of issuing a license varies from a few weeks to six months or longer. Some states issue licenses for free while others impose licensing fees. Some states require the operator to hold third-party liability insurance, other states do not. The measures established to supervise activities differ a lot, including fines and penalties for not complying with terms of the license. Finally, some (but not all) states impose on the license holder the obligation to apply for the transfer of ownership, if the license holder decides to sell the space object – with which the activity in outer space is undertaken – to another owner.[17]

Non-Japanese participants in Japan-based space activities also should be sure to comply with applicable space-related laws and regulations other than their jurisdiction's national space legislation. For instance, United States entities involved in launching and operating communications satellites (including small satellites) from Japan likely need to obtain a license from the Federal Communications Commission (FCC), which, among other things, oversees procedures and requirements for the licensing and operation of satellite communications facilities, including satellites and ground stations. In 2018, California-based start-up firm Swarm Technologies agreed to pay a fine of USD 900,000 for launching several small satellites aboard an Indian rocket and using ground stations in the United States without FCC authorization.[18] As a result of this incident, the FCC issued a second enforcement advisory regarding compliance with satellite communications licensing requirements.[19] In addition, United States entities will want to carefully heed security-related U.S. international export controls on rocket and other

---

17 M. Gerhard and K. Gungaphul-Brocard, "The Impact of National Space Legislation on Space Industry Contracts," in *Contracting for Space – Contract Practice in the European Space Sector* (Ashgate Publishing, 2016), p. 66.
18 *See* L. Grush, "FCC slams spaceflight company with $900,000 fine over illegal satellite launch," December 20, 2018, at https://www.theverge.com/2018/12/20/18150684/swarm-technologies-illegal-satellite-launch-fcc-settlement-fine. For more information regarding FCC oversight of satellite communications, *see* United States.
19 *See* FCC Enforcement Advisory on Satellite Communications Licensing, April 12, 2018, at https://www.fcc.gov/document/enforcement-advisory-satellite-communications-licensing.

space-related technologies, violations of which may lead to substantial civil and criminal penalties.[20]

---

20 *See* Dr. G. Robinson, "Impact of the U.S. International Traffic in Arms Regulations (ITAR) on International Collaboration Involving Space Research, Exploration, and Commercialization," *ZLW 58. Jg.* 3, 2009, pp. 426-27. In 1998, Boeing Co. agreed to pay a USD 10 million civil penalty for the unauthorized transfer of certain information to Russia and Ukraine in connection with a communication satellite launch project in violation of United States defense-related export control legislation. *See* "Boeing to Pay $10-Million Sea Launch Fine," *Los Angeles Times*, October 2, 1998, at https://www.latimes.com/archives/la-xpm-1998-oct-02-fi-28500-story.html, and W. Pincus, "Boeing Fined $10 Million for Data Transfer to Ukraine, Russia," *Washington Post*, October 3, 1998, at https://www.washingtonpost.com/archive/politics/1998/10/03/boeing-fined-10-million-for-data-transfer-to-ukraine-russia/307b7709-655a-4572-84f9-0202ce5e5352/.

Chapter 6 Japan's Space Road Map: 2020 Basic Space Plan

I   Introduction

This chapter explains the goals and measures set out in the fourth, most recent version of the Basic Plan on Space Policy issued by Japan's Strategic Headquarters for Space Development on June 30, 2020 (2020 Basic Space Plan or 2020 Plan).[1] As noted earlier, when announcing this Plan's forthcoming publication, then-Prime Minister Abe chose to begin by mentioning the commercial potential of space and "the increasing number of start-ups and other initiatives taking on the challenge of developing new space businesses" before referencing the importance of initiatives to enhance space security.[2] The sequence of this emphasis was repeated in the 2020 Basic Space Plan's Preamble, the first paragraph of which reads:

> Today, the role of space systems is becoming more important than ever in the context of our national security, economy, and society. This trend is likely to further develop in the years to come. Under these circumstances, we are entering an era in which space activities are driven by public-private collaborations instead of conventional government-led initiatives. As a result, wide-ranging fields of industries are being invigorated through space utilization. Moreover, progress in space exploration expands the sphere of human activities beyond the earth's orbit to reach the lunar surface, and further beyond into deep space. **Space becomes more important as a frontier of science and technology, and as a driving force for economic growth. It has also potential to be a major driver for the economic growth of Japan.**[3] (Emphasis in original.)

---

1   See *Uchū kihon keikaku* [Basic Plan on Space Policy] approved by the Cabinet office on June 30, 2020 (2020 Basic Space Plan), in Japanese at https://www8.cao.go.jp/space/plan/kaitei_fy02/fy02.pdf. An unofficial English translation of the 2020 Basic Space Plan by ZeLo is included in Annex 9 (page references below, however, are to the Japanese version).
2   See The Prime Minister in Action, December 13, 2019, at http://japan.kantei.go.jp/98_abe/actions/201912/_00025.html.
3   2020 Basic Space Plan, Preamble, p. 3.

The Preamble's second paragraph then addresses space as a security issue:

> On the other hand, space security is a matter of urgency, as an increasing number of countries, including the United States, **regard outer space as a "Warfighting Domain" or an "Operational Domain"**, while there is suggestion of the increasing threats in outer space, including the development of various means of disruption, not limited to destruction of satellites by missile strikes or other ways as has been discussed. In addition, a growing number of small and micro-satellite constellations is growing as a "game changer" in the space industry. Japan's space equipment industry is, however, lagging behind these trends. While space-related technologies are undergoing a rapid evolution, **the reinforcement of the industrial, scientific and technological basis is an immediate challenge to maintain the autonomy of space activities that Japan has built up since the end of the Second World War.**[4] (Emphasis in original.)

The remainder of the 2020 Plan is divided into four sections: (1) Current Situation Surrounding Japan's Space Policy; (2) Goals of Japan's Space Policy; (3) Basic Stances in Promoting Japan's Space Policy; and (4) Specific Approaches to Japan's Space Policy. Each of these sections is discussed below.

II Current Situation Surrounding Japan's Space Policy

The first section of the fourth Basic Space Plan, Current Situation Surrounding Space Policy, addresses how Japan's government perceives the external environment for its space policy as of June 2020. The circumstances and concerns underlying Japan's space policy may be found in this chapter.

1 Growing Importance of Outer Space in Security

Although the 2020 Plan's Preamble mentioned the significant economic aspects of Japan's space policy before noting security issues, the first section begins by stating that the importance of outer space to security has "grown significantly" and explaining that Japan is advancing various initiatives for security-related uses of space.[5] One such

---

4 Id.
5 See 2020 Basic Space Plan, p. 4.

initiative was the formation in mid-2020 of a unit, the Space Operations Squadron, in Japan's Air Self-Defense Force (JASDF).[6] Among other things, the new Space Operations Squadron is tasked with enhancing Japan's "space situational awareness" and monitoring debris and other possible sources of interference to Japanese satellites.[7] In doing so, the 2020 Plan suggests, Japan is keeping in step with the United States and its European allies. In December 2019, the United States created a U.S. Space Force (following the establishment of a U.S. Space Command for military operations in space), and in November 2019 the North Atlantic Treaty Organization formally declared space to be an "operational area."[8]

2   *Increasing Dependence on Society's Space Systems*

The 2020 Basic Space Plan acknowledges that space-based navigation and communications systems are firmly established in modern economic and social activities, and echoes Japan's "Fifth Science and Technology Basic Plan" (approved by Japanese Cabinet in 2016) regarding the significance of space to realizing "Society 5.0" towards which the country is to strive.[9] In this scheme, Society 5.0 represents a future "society that balances economic advancement with the resolution of social problems by a system that highly integrates cyberspace and physical space," and follows the hunting society (Society 1.0), agricultural society (Society 2.0), industrial society (Society 3.0), and information society (Society 4.0).[10]

More concretely, the aim is to enhance the acquisition and distribution of information obtained from "physical space," including geo-positioning and remote sensing data from outer space, for processing and analysis by ever-more sophisticated cloud service databases and artificial intelligence (AI) in order to address various social and economic issues. Information about areas and communities affected by natural disasters or about long-term climate change trends, for example, can be obtained from satellites and assessed with advanced computer analytics to assist in disaster-response efforts or to address global environmental challenges.[11] Of course, satellites also will be essential to

---

6   *See* Launch of the Space Operations Squadron, July 2020, at https://www.mod.go.jp/e/jdf/no125/specialfeature.html; *The National Defense Program Guidelines for FY2019 and beyond*, p. 24, at https://www.cas.go.jp/jp/siryou/pdf/h31boueikeikaku.pdf.

7   *See* 2020 Basic Space Plan, p. 4; Launch of the Space Operations Squadron, July 2020, at https://www.mod.go.jp/e/jdf/no125/specialfeature.html.

8   *See* "The US now has a Space Force and a Space Command. It's not clear what each will do, but commanders are excited," December 27, 2019, at https://www.businessinsider.com/us-establishes-space-force-to-support-space-command-operations-2019-12; "Foreign Ministers take decisions to adapt NATO, recognize space as an operational domain," November 20, 2019, at https://www.nato.int/cps/en/natohq/news_171028.htm.

9   *See* "Fifth Science and Technology Basic Plan" approved by Cabinet Office on January 22, 2016, Chapter 2 (2), p. 10, at https://www.cas.go.jp/jp/siryou/pdf/h31boueikeikaku.pdf.

10 *See* Society 5.0, at https://www8.cao.go.jp/cstp/english/society5_0/index.html.

11 *See* 2020 Basic Space Plan, p. 5.

enabling communications systems to connect seamlessly in fifth generation (5G) mobile networks that Japan is helping to develop and bring to fruition.

3   Worsening of Risks Hindering the Sustainable and Stable Use of Outer Space

The 2020 Basic Space Plan raises concerns about congestion in outer space due both to space debris and larger numbers of satellites in orbit amid lower satellite production and launch costs. The 2020 Plan proposes actions to address these risks, such as international rulemaking.[12]

4   Expanded Space Activities in Other Countries

The 2020 Basic Space Plan notes that the former U.S.-Soviet bipolar order has been transformed into a multipolar order in respect of the global competition for space development and use. In this regard, the 2020 Plan highlights China, which in January 2019 became the first country to successfully land a spacecraft on the far side of the Moon and which also launched the world's largest number of satellites in 2019, as well as India, which has had its own satellite geo-positioning system since 2018.[13]

5   Expanded Space Activities in the Private Sector and the Rise of New Business Models

The 2020 Basic Space Plan observes that the private sector entities, including venture companies, are becoming increasingly active in space-related commercial activities around the world, resulting in decreasing costs of rocket launch services and the creation of new business models involving constellations of small and micro communication and observation satellites. The Japanese government is concerned that Japan is beginning to lag behind the United States and Europe in space industry technology (having once been a global leader in this field), and the 2020 Plan recommends improving the environment to encourage venture companies to pursue space-related business opportunities.[14]

---

12   See id.
13   See 2020 Basic Space Plan, p. 6.
14   See 2020 Basic Space Plan, p. 7.

6   Emergence of New Space Activities

The 2020 Basic Space Plan identifies new space activities that are different from conventional space activities (*e.g.*, geo-positioning, communications, and observation). Examples of new space activities include sub-orbital flights (flights reaching an altitude of around 100 kilometers), removal of space debris, satellite repair, and processing of satellite data with cutting-edge AI technologies.[15]

7   Rapid Evolution of Science and Technology

The 2020 Basic Space Plan observes that technological evolution is rapidly bringing about innovative changes in space activities, including design and manufacture of small and micro satellites, satellite constellations, space optical communications, quantum key distribution, AI, and robotics. The 2020 Plan further notes that the United States and European countries have for many years used scientific and technological advances made for national security purposes to strengthen the competitiveness of the space and related industries. Japan, however, had not formally recognized the relationship between national security and space development and use until the 2008 Basic Space Law, and thus the pipeline of cutting-edge science and technology that has long existed in the U.S. and Europe is less developed in Japan.[16]

III   GOALS OF JAPAN'S SPACE POLICY

The second section of the 2020 Basic Space Plan sets out two overarching goals for Japan's space policy: 1. contributing to diverse national interests; and 2. strengthening the comprehensive foundations that support Japan's space activities, including the industrial, scientific, and technological bases of these activities.[17]

1   Contributing to Diverse National Interests

As used in the 2020 Basic Space Plan, the term "diverse" means a truly wide range of national interests, and this section is divided into four discrete (albeit still broad) objectives.

---

15   *See id.*
16   *See id.*
17   *See* 2020 Basic Space Plan, p. 9.

A   **Ensuring Space Security**

As discussed earlier, the Japanese government has recognized the increasing importance of outer space to national security, and it now places great importance on security-related initiatives related to space. Accordingly, the 2020 Basic Space Plan identifies three components of the objective to ensure security in space.[18]

- To ensure the sustainable and stable use of outer space by enhancing the ability to obtain information from outer space and strengthening mission assurance, and by playing a larger role in the creation of international rules.
- To further improve various capabilities such as data collection, telecommunications, and geo-positioning using space, and to strengthen the capacity for ensuring superiority of space use in both times of peace and armed contingencies, including to guarantee the functioning of these capabilities and to block an adversary's command and control telecommunications.
- As part of the effort to reinforce the Japan-U.S. alliance, to comprehensively strengthen Japan-U.S. space-related cooperation in the security area, including the division of roles with the United States in maintaining space security, and also to pursue wide-ranging collaboration and cooperation in space with partners other than the United States. In particular, Japan will bolster its efforts in the Indo-Pacific region with a view to contributing to the maintenance and promotion of a "Free and Open Indo-Pacific."

Japan's *National Defense Program Guidelines for FY2019 and beyond*, an important defense policy document, had stated that Japan will promote efforts to secure the use of space in peacetime and armed contingencies.[19] Item (ii) above reiterates this goal. In addition, it is noteworthy that item (iii) mentions the pursuit of cooperation in security-related space endeavors among partners other than the United States. This suggests the Japanese government considers that, despite the central importance of the Japan-U.S. alliance, it is beneficial to collaborate with additional partners to ensure the security of outer space.

B   **Contributing to Disaster Management and Infrastructure Maintenance, and Resolving Global Issues**

Japan often suffers from various natural disasters, with several powerful typhoons hitting the country's mainland each year, severe flooding annually, and major earthquakes and tsunamis striking periodically. Thus, for many decades Japan has been pursuing multifaceted efforts in disaster management. In this regard, the 2020 Basic Space Plan includes the objective of enhancing the development and use of advanced satellite-based

---

18   *See id.*
19   *See National Defense Program Guidelines for FY2019 and beyond*, p. 4, at https://www.mod.go.jp/e/publ/w_paper/wp2019/pdf/DOJ2019_Special_1.pdf.

geo-positioning, communications and broadcasting, and Earth observation systems to help deal with major natural disasters and also to address the maintenance and management of aging infrastructure in Japan.

With regard to global issues such as environmental challenges, the 2020 Plan sets forth the objective that, "in cooperation with the international community, Japan will demonstrate its leadership and contribute to the resolution of global issues that are becoming more serious, such as the global energy situation, climate change, the environment, food, public health, large-scale natural disasters, etc., thereby contributing to the achievement of SDGs [sustainable development goals]."[20]

C Creation of New Knowledge through Space Science and Exploration

This space policy objective concerns the field of research activities. The 2020 Basic Space Plan notes that Japan is well regarded internationally for its space science and exploration, and the aim continues to be to strengthen efforts in these areas to foster new knowledge and skills.[21]

D Realizing Economic Growth and Innovation with Space as the Driving Force

The 2020 Basic Space Plan also identifies a space policy objective in respect of economic and industrial fields. The aim in this respect is to enhance innovation and economic growth in Japan by strengthening and expanding the use of space science and technologies.

2 *Strengthening the Comprehensive Foundations that Support Japan's Space Activities, including the Industrial, Scientific, and Technological Foundations*

To help achieve the objectives set forth in the first goal discussed above, the 2020 Basic Space Plan sets forth a second goal that emphasizes the need to strengthen the industrial, scientific, and technological foundations of space development and use activities. Indeed, the 2020 Plan states that Japan will "revamp" the ecosystem for its space industry by encouraging this sector to become self-sustaining and by promoting greater international cooperation.[22]

Initiatives to bolster the industrial, scientific, and technological bases of Japan's space activities are to include: upgrading space transportation systems; developing and employing satellites that incorporate cutting-edge technologies such as space optical communications, quantum key distribution, AI and robotics, micro satellites, and

---

20 *See* 2020 Basic Space Plan, p. 10.
21 *See id.*
22 *See id.*

satellite constellations; working on the regulatory environment for collecting and analyzing geospatial information, including satellite data; cultivating and mobilizing human resources able to develop space technologies, including by promoting exchanges among space industry and non-space industry professionals; and striving to advance international rulemaking and cooperation in the area of space development and use.[23]

IV Basic Stances in Promoting Japan's Space Policy

The third section of the 2020 Basic Space Policy addresses four basic stances that Japan has adopted in its space policy promotion efforts.

1 Goal-Oriented Space Policy Based on Needs for Security, Industrial Use, etc.

As explained previously, before the enactment of the 2008 Basic Space Law Japan considered that it lacked a unified national strategy for promoting its space policy. In addition, there was the perception that Japan's space policy had been pursued without a clear and comprehensive vision of how to use the opportunities and challenges of space development in and for society as a whole. Aware of this issue, the Japanese government sets out in the 2020 Basic Space Plan the importance of "a goal-oriented space policy based on needs for security, industrial use, etc." as the first basic stance in promoting the development and use of space.[24]

In particular, the 2020 Plan highlights the significance of considering fully the needs of the wide range of stakeholders that use or benefit from space development, including in connection with security, disaster management, infrastructure maintenance, and industrial activity. Relevant ministries and agencies are to collaborate and, based on an appropriate division of roles between the public and private sectors, ensure that necessary resources will be invested for the strategic goals of Japan's space policy, that this policy will be implemented effectively, and that there will be close coordination with related strategic policy making for economic growth, science and technology (including the promotion of Society 5.0), maritime issues, and national security.[25]

Additionally, the 2020 Plan provides that Japan's Self-Defense Forces are to analyze global trends in space development and use and take on the challenge of creating and demonstrating the efficacy of innovative space technologies that serve Japan's strategic goals.[26]

---

23 See id.
24 See 2020 Basic Space Plan, p. 11.
25 See id.
26 See id.

2   *Space Policy that Provides Predictability of Investment so as Maximize the Vitality of the Private Sector*

The 2020 Basic Space Plan recognizes that historically the public sector, specifically Japan's government, has accounted for the large majority of the "demand" for the space-related goods and services provided by the country's industrial and technological corporations. Further, given the complexity and cost of space projects, long-term planning and heavy advance investment by these companies are needed. As such investment will not be forthcoming if public sector demand is uncertain, the 2020 Plan stresses that predictability of public sector demand is a key factor to stimulate private sector investment in Japan's space program.[27]

Therefore, Japan's government pledged to publish a "Basic Space Plan Process Chart" in order to ensure the predictability of public sector demand. This process chart is to set out concrete space-related measures that Japan intends implement over the next decade, with the aim of providing sufficient predictability as to demand that will spur private sector investment in Japan's space industry.[28]

3   *Space Policy to Effectively and Efficiently Use Human Resources, Funds, and Intellectual Property*

Japan's financial condition continues to be characterized by large budget deficits where government spending significantly exceeds tax revenues. Consequent budgetary constraints affect space policy as they do other areas of policy making. Accordingly, the 2020 Basic Space Plan highlights the need to use funds for space policy as effectively and efficiently as possible. This posture also applies to the effective and efficient use of human resources and intellectual property rights. Thus, the 2020 Plan states that clear policy targets will be set in advance, and evaluations of achievement will be conducted subsequently.[29]

In addition, the 2020 Plan identifies the need to facilitate the transfer of cutting-edge technologies developed for national security purposes to the space industry, as well as the importance of encouraging personnel exchanges between the space industry and other industries.[30]

---

27   *See id.*
28   *See id.*
29   *See id.*
30   *See* 2020 Basic Space Plan, p. 12.

4   Space Policy in Strategic Collaboration with Allies and Partners

The 2020 Basic Space Plan explains that Japan's government has said it will proactively engage in international rulemaking on space policy in cooperation with its allies and partners. Particularly in this regard, the Japanese government has pledged to strengthen its cooperative efforts in the "Free and Open Indo-Pacific" region, a commitment that, besides being meant to counter China's growing influence, reflects the level of space-related development activity in recent years by countries in the Indo-Pacific region, including India, Australia, Bangladesh, Indonesia, Malaysia, New Zealand, North Korea, the Philippines, Singapore, Korea, Thailand, and Vietnam. Meanwhile, the Plan notes, China is cementing its space program's position as the world's third largest after those in the United States and Russia.[31]

Reiterating that space development is highly important to national security, the 2020 Plan highlights an urgent need to establish a cooperative framework for space development in the Indo-Pacific region. In particular, Japan and India are fortifying bilateral cooperation; the first Japan-India Space Dialogue was held on March 8, 2019, in which Japan and India's space ministries and agencies agreed to promote mutual understanding and cooperation in space development.[32] It seems Japan will continue its efforts to build a cooperative framework for space development with India and other Indo-Pacific partners.

V   SPECIFIC APPROACHES TO SPACE POLICY

The fourth and final section of the 2020 Basic Space Plan presents specific approaches that Japan is to take in four areas, which by now should sound familiar: (1) ensuring space security; (2) contributing to disaster response, national infrastructure maintenance, and solutions to global issues; (3) creating new knowledge through space science and exploration; and (4) realizing space-driven economic growth and innovation. A number of key specific approaches are discussed in more detail below.

1   Specific Approaches to Ensuring Space Security

A   Quasi-Zenith Satellite System (QZSS)

Japan's QZSS program began with the launch of the first quasi-zenith satellite, *Michibiki* ("guidance" in Japanese), in 2010. Currently the program consists of four satellites, one of which is geostationary while the other three travel in elliptical, geosynchronous orbits, and at nearly all times at least one of these satellites is traveling over Japan. The QZSS can

---

31   *See id.*
32   *See id.*

be used for security needs and other purposes, such as disaster management, and is meant to support and enhance the United States-operated Global Positioning System (GPS). The efficacy of GPS in Japan has been hindered by the country's many mountains (about 75% of Japan is mountainous) and also large city skyscrapers, which block GPS signals and can lead to positioning errors. Operating in conjunction with GPS, the QZSS expands the area and duration of positioning capacity, including in Japan's mountainous and urban areas, and the QZSS provides greater positioning accuracy (to a meter or less) than GPS as well.[33] Concerns about potential malfunction of (or attack on) GPS also have led Japan to move to augment its QZSS program.[34] Thus, according to the 2020 Basic Space Plan, by 2023 there will be a total of seven QZSS satellites in operation that will be able to function fully without GPS.[35]

B   X-Band Defense Satellite Communications

X-band satellite communications are satellite-based communication networks that operate at frequency bands used by governments, including military forces, and have features allowing for stable communications even in inclement weather. As noted earlier, until the 2008 Basic Space Law was enacted there were restrictions on the security-related use of space technology by Japan's military, the Self-Defense Forces, which were not allowed to own communications satellites (they leased privately-owned satellites for internal communications). After passage of the 2008 Basic Space Law, however, Japan's Ministry of Defense worked with Japanese companies to develop and launch its own X-band communication satellites, *Kirameki*-1 and *Kirameki*-2 (*kirameki* means "spark" or "glitter" in English), in 2017 and 2018. According to the 2020 Basic Space Plan, Japan plans to launch a third X-band communication satellite in 2022.[36]

C   Information-Gathering Satellites

The Japanese government initiated an information-gathering satellite program after North Korea launched a missile that flew over Japan in 1998. As explained earlier, the first information-gathering satellites were launched in 2003, before enactment of the 2008 Basic Space Law that formally reversed Japan's longstanding policy to use space exclusively for non-military purposes. While some might consider these satellites to have been a possible exception to this policy, at the time Japanese officials said that they were not directly related to the situation in North Korea and also were to be used

---

33 *See* What is the Quasi-Zenith Satellite System (QZSS)?, at https://qzss.go.jp/en/overview/services/sv02_why.html.
34 *See* "Japan Prepares for GPS Failure with Quasi-Zenith Satellites," at https://spacewatch.global/2019/01/japan-prepares-for-gps-failure-with-quasi-zenith-satellites/.
35 *See* 2020 Basic Space Plan, p. 13; *see also id.*; "Japan mulls seven-satellite QZSS system as a GPS backup," May 15, 2017, at https://spacenews.com/japan-mulls-seven-satellite-qzss-system-as-a-gps-backup/.
36 *See* 2020 Basic Space Plan, p. 13.

to help respond to natural disasters (in addition, they were operated by the civilian Cabinet Intelligence and Research Office, not Japan's Ministry of Defense).[37]

There are three types of information-gathering satellites: optical satellites; radar satellites; and information-relay satellites. Optical satellites collect images from the ground during the day, radar satellites collect images at night and in bad weather, and information-relay satellites receive data from optical and radar satellites and transmit the images to Earth. Currently, according to the 2020 Basic Space Plan, Japan operates four optical and radar satellites.[38] Further, the 2020 Plan says, Japan aims to put 10 information-gathering satellites into orbit.[39]

D   Space Situational Awareness (SSA)

Space situational awareness (SSA) systems monitor orbiting artificial objects, typically old satellites and rockets or their fragments that are classified as "space debris," track these objects' orbits and their potentially hazardous proximity to operational satellites and other spacecraft and forecast their re-entry to earth. Up to this point, JAXA has played a key role in Japan's SSA program, conducting surveillance with optical telescopes and radars at the Bisei and Kamisaibara Space Guard Centers in Okayama Prefecture.[40]

According to the 2020 Basic Space Plan, given the increasing importance of space in terms of national security Japan's Ministry of Defense is aiming by 2022 to build a more comprehensive SSA system, including an optical telescope and radar as well as analytic capabilities, in cooperation with JAXA and the United States.[41] Ministry of Defense officials have participated in Global Sentinel, a multilateral SSA conference that the U.S. military has sponsored annually since 2016, to learn more about SSA and strengthen Japan's cooperation with the United States and other partners in this field.[42]

---

37 *See, e.g.,* "Japan launches new spy satellite," September 11, 2006, http://news.bbc.co.uk/2/hi/asia-pacific/5333560.stm; "Current Status and Recent Developments in Japan's National Space Law," p. 380. *See also* Space Development and Utilization, Information Gathering Satellite (IGS), at https://www.mext.go.jp/a_menu/kaihatu/space/kaihatsushi/detail/1299560.htm; Article 4-2(2)(i) of the *Naikakubanbō soshikirei* [Cabinet Order on the Organization of Cabinet Secretariat], Cabinet Order No. 219 of July 31, 1957, as amended (defining information gathering satellites as "artificial satellites in order to collect imaging information useful for assuring the security of Japan, addressing large-scale disasters, and other important policies of the Cabinet"), at https://elaws.e-gov.go.jp/document?lawid=332CO0000000219.
38 *See* 2020 Basic Space Plan, p. 13.
39 *See id.,* p. 14.
40 *See* Bisei Space Guard Center (https://global.jaxa.jp/about/centers/bsgc/index.html, Kamisaibara Space Guard Center, at https://global.jaxa.jp/about/centers/ksgc/index.html.
41 *See* 2020 Basic Space Plan, p. 14.
42 *See* 2020 Defense White Paper, Section 3 Response in Areas of Space, Cyber, and Electromagnetic Waves, at https://www.mod.go.jp/j/publication/wp/wp2020/html/n31301000.html.

2   *Specific Approaches to Contributing to Disaster Management, Infrastructure Maintenance, and Solutions to Global Issues*

As noted above, certain programs that help ensure Japan's space security, including the QZSS and information-gathering satellites, also can assist in areas such as disaster management and thus fall under this second category of specific approaches. In addition to these programs are several others discussed below.

A   **Meteorological Satellites**

The Japan Meteorological Agency has for decades operated the *Himawari* ("sunflower" in Japanese) series of weather satellites, which play an important role in weather observation and forecasting, not least regarding typhoons that strike Japan every year. Currently *Himawari*-8 is in operation, and the 2020 Basic Space Plan says its successor *Himawari*-9 will begin operating in 2022.[43] Future successors are expected to start operation by 2029.[44]

B   **Greenhouse Gas Observation Satellites**

To help address the threat of global warming, JAXA co-developed and launched the satellite *Ibuki* ("breath" in Japanese) in 2009 and its improved successor *Ibuki*-2 in 2018 to monitor the major greenhouse gases: carbon dioxide and methane.[45] *Ibuki* and *Ibuki*-2 are the only such satellites that track greenhouse gas emissions. Not merely can they identify various particular emission sources with a high degree of precision, given their orbital perspective they also help provide a comprehensive picture of greenhouse gas distribution around the world.[46] JAXA has noted that "[t]he usability and roles of satellite observation data as a verification tool of greenhouse gasses inventories were stated for the first time in '2019 Refinement to the 2006 IPCC Guidelines on National Greenhouse Gas Inventories,' which was agreed in Intergovernmental Panel on Climate Change, IPCC, held in Kyoto in May 2019."[47]

According to the 2020 Basic Space Plan, *Ibuki*-3 will be launched in 2023 to further enhance understanding of greenhouse gas emission sources and the accuracy of emission estimates.[48]

---

43   *See* 2020 Basic Space Plan, p. 16.
44   *See id.*
45   JAXA's partners were Japan's Ministry of the Environment and the National Institute for Environmental Studies.
46   *See* Ibuki-2 (GOSAT-2), at https://www.satnavi.jaxa.jp/e/project/gosat2/.
47   *Id.*
48   *See* 2020 Basic Space Plan, p. 16.

3   Specific Approaches to Creating New Knowledge through Space Science and Exploration

A   **Japan's Space Science and Exploration Activities to Date**

Japan has an impressive track record in space science and exploration. ISAS,[49] for instance, along with the United States National Aeronautics and Space Administration (NASA) and ESA, is a major player in international space science initiatives. ISAS has done major research in the fields of cosmic-ray physics, infrared astronomy, high-energy astrophysics, and astrobiology. ISAS also has conducted research on the upper layer of the atmosphere and space plasma physics, performed engineering experiments on attitude control systems, re-entry technology, and navigation technology, and helped develop Japan's launch vehicles using solid fuel rocket technology. In addition, ISAS has spearheaded dozens of satellite missions engaged in solar, asteroid, comet, lunar, and planetary exploration.[50]

For example, Japan has received widespread recognition for its pioneering *Hayabusa* ("Peregrine falcon" in Japanese) asteroid exploration missions. The first satellite mission, *Hayabusa*, was launched in 2003 to the asteroid 25143 Itokawa (named after Japan's "Dr. Rocket," Dr. Hideo Itokawa), which belongs to the near-Earth Apollo Group. Upon reaching 25143 Itokawa in 2005, *Hayabusa* conducted orbital surveys before landing on the asteroid and collecting surface samples. When *Hayabusa* returned these materials to Earth in 2010, it became the first mission to have obtained surface samples from an astronomical body outside Earth's gravity. Following this success, *Hayabusa2* launched in 2014 to another near-Earth asteroid, 162173 Ryugu. Upon reaching its destination in 2018, *Hayabusa2* conducted orbital surveys and collected surface and also sub-surface samples from the asteroid, which the spacecraft successfully returned to Earth for analysis in December 2020.[51]

B   **Continued Participation in International Space Exploration**

ISAS and Japan's other space organizations will be involved in a number of additional lunar, planetary, and astronomical satellite missions over the next decade, including in cooperation with American and European partners. Indeed, according to the 2020 Basic Space Plan, Japan intends to build on its past achievements by continuing to participate in international space science and exploration projects.[52]

For example, JAXA is one of the signatory partners of the October 2020 Artemis Accords, a multilateral cooperation agreement on space exploration that is led by NASA and also involves ESA and the space agencies from Australia, the United

---

49 As may be recalled, ISAS is the Institute of Space and Aeronautical Science and is now part of JAXA.
50 See Research Area, at https://www.isas.jaxa.jp/en/about/research_area/.
51 See *Mainichi Japan*, "Q&A: What has Japan's *Hayabusa2* mission accomplished?," December 8, 2020, at https://mainichi.jp/english/articles/20201207/p2g/00m/0na/115000c.
52 See 2020 Basic Space Plan, p. 14.

Kingdom, Canada, Italy, and the United Arab Emirates. The Artemis Accords envision landing men and the first women on the Moon by 2024, and the first astronauts on Mars by 2033.

As an initial step in connection with the Artemis Accords, Japan will contribute to the "Gateway," a manned space station that will orbit the Moon and serve as a platform for servicing lunar spacecraft, crew habitation, and research. Even before agreeing to the Artemis Accords, in July 2020, Japan's Ministry of Education, Culture, Sports, Science and Technology had signed a joint declaration with NASA to collaborate, among other things, on the Gateway.[53] Pursuant to this declaration, Japan is planning to provide power components to the Gateway's Habitation and Logistics Outpost by 2022, and thereafter to provide batteries and thermal control pumps to the Gateway's International Habitation Module. In addition, the declaration provides that Japan will continue to pursue development of cargo transportation, resupply, and docking capabilities for service missions both to the Gateway and ISS. The declaration also says that Japan and NASA will discuss future cooperative efforts regarding exploration missions to the Moon and Mars.[54]

4 *Specific Approaches to Realizing Economic Growth and Innovation Driven by Space*

A **Expanded Use of Satellite Data (Satellite Remote Sensing and Geospatial Information)**

According to the 2020 Basic Space Plan, Japan aims to double the size of the space industry (approximately USD 11.5 billion in 2017) by the early 2030s. The 2020 Plan forecasts not just continued development of the aerospace sector, but also economic growth driven by new uses of space technology, for example, by utilizing satellite data in fields such as agriculture, forestry, disaster prevention, and transport and logistics, including in collaboration with efforts to revitalize Japan's regional economies.[55]

For instance, the 2020 Plan notes that Japan's central and regional governments have established a Task Force on the Use of Satellite Remote Sensing Data in order to improve the sophistication and efficiency of operations that use satellite data, including by examining both how different national ministries might use satellite remote sensing data and how local governments can be involved in such projects. In this regard, the 2020 Plan promotes more active utilization of Japan's G-Space Information Center, a

---

53 *See* Joint Exploration Declaration of Intent for Lunar Cooperation (JEDI) between the Ministry of Education, Culture, Sports, Science, and Technology of Japan and the National Aeronautics and Space Administration of the United States of America, July 10, 2020, at https://www.mext.go.jp/content/20200714-mxt_uchukai02-000008680_1.pdf.
54 *See id.*
55 *See* 2020 Basic Space Plan, p. 22.

facility set up in 2016 by a government-industry-academic coalition to provide a single place where users from the public and private sectors can distribute and access geospatial data (*i.e.*, "G-space information").[56] Among the many kinds of data available on the G-Space Information Center's website[57] is population location information (nationally, and by prefecture and municipality) obtained from smartphone application logs by Agoop Co., Ltd., and a microtopographic map of Nagano Prefecture's forests created by the Nagano Prefectural Forestry Research Center. The 2020 Plan encourages national and local government bodies to actively take advantage of G-space information, including particularly to develop disaster prevention and mitigation programs that use such information.[58]

B   Government Satellite Data Platform (Tellus)

As part of its aim to promote greater private sector use of satellite data, the Japanese government (specifically, METI) began operating the country's first platform that can analyze satellite and terrestrial data on the cloud, called Tellus, in 2019. Previously in Japan, much satellite data was not accessible to the general public and, in addition, the ability of companies and others to use such data was limited by the need for highly specialized personnel and expensive software to analyze the data. Tellus seeks to address this challenge by making available to private and public sector users the ability both to access and analyze substantial amounts of government satellite information, including, for example, about climate indicators, soil fertility and agricultural productivity, automobile stockpiles, public infrastructure, housing and construction, maritime transportation, location of fish, status of forests, and pollution. Tellus's express goal in this regard is to spur development of new markets and businesses.[59]

The 2020 Basic Space Plan highlights the need going forward to expand the variety of satellite and other data available on Tellus and to improve Tellus's analytic capabilities in order to help catalyze commercial innovation.[60]

C   Expanded Procurement from Private Sector, including Start-Up Companies, in Government Projects

The 2020 Basic Space Plan seeks to expand procurement by JAXA and other government space agencies from the private sector, including from start-up companies. Indeed, the Plan suggests, boosting government demand for space-related goods and services is an

---

56   *See id.*
57   *See* G Spatial Information Center, at https://www.geospatial.jp/gp_front/.
58   *See* 2020 Basic Space Plan, p. 22.
59   *See* About Tellus, at https://www.tellusxdp.com/en-us/about/; Satellite Data Business, at https://www.tellusxdp.com/en-us/about/satellite_data.html.
60   *See* 2020 Basic Space Plan, p. 23.

important way to grow the space industry, including the creation of more start-up firms.[61]

In this regard, the 2020 Plan urges Japan's space agencies to use the Small Business Innovation Research (SBIR) program, which began under the 1999 Act for Facilitating New Business Activities of Small and Medium-sized Enterprises (Act No. 18 of 1999) that was enacted to promote small business R&D and commercial activities based on new technologies developed by small and mid-sized businesses.[62] Among other things, the SBIR program aims to enhance opportunities for giving targeted subsidies to small and medium-sized businesses, as well as support the commercialization of productive results of the subsidized R&D. In addition, the program seeks to facilitate greater disclosure by the governmental agencies of their required specifications for technologies and services. The SBIR program also backs the use of flexible contractual arrangements such as those that provide for project milestone payments with shorter payment periods than contracts stipulating payment upon project completion, which may be more feasible for start-ups and other small firms that typically have fewer available cash reserves than larger companies.

D S-Matching and S-Booster Initiatives

The 2020 Basic Space Plan pledges to continue initiatives that the Japanese government has implemented in recent years to encourage start-ups and other businesses to enter into the space industry.[63]

One of these initiatives, "S-Matching," is an online platform launched in 2018 by the Cabinet Office and METI (with support from JAXA) to advance space business innovation by helping to connect entrepreneurs and investors.[64] S-Matching strives "to provide smooth business-matching opportunities between space business entrepreneurs and space business investors, aiming to create and foster venture businesses involving space-related initiatives, including utilization of data on space."[65] Specifically, S-Matching offers a website where individuals, start-ups, and other entrepreneurs can post their space-related ideas and ventures. Investors and larger entities can browse the website and, if interested, contact the entrepreneurs. As of December 2020, 389 "space business entrepreneurs" and 61 "space business investors" were registered on S-Matching's website.[66]

---

61 *See id.*
62 *See* What is SBIR?, at https://sbir.smrj.go.jp/about/index.html.
63 *See* 2020 Basic Space Plan, p. 23.
64 *See* Outer Space Entrepreneurs: Business Matching Platform, "S-Matching," Launched for Encouraging Investment in Space Industry, at https://www.meti.go.jp/english/press/2018/0531_004.html.
65 *Id.*
66 *See* New Space Businesses Gather Here, at https://s-matching.jp/.

Since 2017, Japan's Cabinet Office, JAXA, and other agencies also have held a space business contest called "S-Booster" to help spur innovation in the space industry.[67] S-Booster uses the contest format to solicit a wide range of space-related ideas from individuals and start-ups that, if successful, can receive financial support in the form of "prize" money, assistance from experts to help refine their ideas, and business matching opportunities. In 2019, S-Booster expanded eligibility to include contestants not just from Japan but from the entire Asia-Oceania region.

5 Specific Approaches to Enhancing the Comprehensive Foundations of Japan's Space Activities

A Mainstay Rockets

Mainstay rockets are defined as "transport systems that are indispensable for maintaining domestic operations and ensuring the autonomy of transport systems in order to achieve the missions of the government such as security."[68] The Japanese government owns these core rocket systems, which currently consist of the H-IIA and H-IIB launch vehicles, as well as the Epsilon solid-fuel rocket, which all have been used to deploy various satellites.

The 2020 Basic Space Plan states Japan has been developing the H3 launch vehicle as a successor to the H-IIA/B rockets; the H3's maiden voyage likely will take place in the first half of 2021.[69] According to JAXA, the H3 will enable Japan to "maintain its autonomous access to space to launch satellites and probes including important missions for the government. We are eager to launch commercial satellites every year as well."[70]

B Measures to Manage Space Debris

Given the hazard that space debris poses to satellites, the development of methods to remove or mitigate space debris has become increasingly important to ensure the sustainable and stable use of outer space. To address this issue, the Japanese government established a Space Debris Task Force in 2019, with the ministers of relevant ministries and agencies as members.[71] The 2020 Basic Space Plan states that multiple initiatives will be taken to combat space debris, including development of technologies to monitor and remove space debris.[72] The 2020 Plan also promotes

---

67 *See* S-Booster Space-based Business Idea Contest, at https://s-booster.jp/en/.
68 *Chūkan torimatome (an) Heisei 25-nen 5 tsuki 17-nichi uchū yusō shisutemu bukai* [Interim Report (Draft), Space Transport System Subcommittee], May 17, 2013, at https://www8.cao.go.jp/space/comittee/yusou-dai5/siryou1.pdf.
69 *See* 2020 Basic Space Plan, p. 27; "First H3 launch slips to 2021," September 11, 2020, at https://spacenews.com/first-h3-launch-slips-to-2021/.
70 About H3 Launch Vehicle, at https://global.jaxa.jp/projects/rockets/h3/.
71 *See* Task Force Ministers' Meeting on Space Debris, at https://www8.cao.go.jp/space/taskforce/debris/kaisai.html.
72 *See* 2020 Basic Space Plan, p. 30.

efforts to address space debris in other ways, such as establishing a rating system to evaluate the effectiveness of businesses that use satellites to reduce space debris and collaborating in international rulemaking on space debris.

C   Space Solar Power Generation R&D

Space solar power seeks to harness solar-generated power in space and send it to Earth. It is expected to help address issues such as global energy needs, climate change, and environmental problems, and has the potential to generate energy on Earth more efficiently than older power generation systems. According to the 2020 Basic Space Plan, Japan will strive to conduct experiments in order to implement space solar power generation.[73]

D   Strengthening Space Industry Supply Chains

The 2020 Basic Space Plan states that Japan will begin investigation to identify technologies that are critical to the development of space-related manufacturing.[74] Japan also will support the advancement of such technologies.

---

73   *See id.*, p. 30.
74   *See id.*, p. 32.

Chapter 7 Japan's Burgeoning Space Industry

I   Introduction

This chapter highlights the diversity of Japan's space industry by introducing a cross section of Japanese companies that are involved in different kinds of space-related businesses. According to the "Space Industry Vision 2030," a document issued by the Cabinet Office in 2017, the scale of Japan's space industry was approximately USD 11.5 billion, with the stated goal of doubling this figure by the early 2030s.[1] Shortly after releasing "Space Industry Vision 2030," the Japanese government announced that a fund consisting of nearly USD 1 billion of venture capital would be provided by the Development Bank of Japan and other organizations to help Japanese start-ups in the space sector.[2] As this chapter shows, new Japanese companies have emerged that are embracing the challenges and opportunities of space in the 21st century. Of course, Japan's space industry also includes a number of large, well-established companies (several of which are discussed below), which in recent years have begun to offer new space-related services beyond their traditional fields of business.

II   Rocket Development and Launch Services

1   Mitsubishi Heavy Industries, Ltd.

Established in 1950, Mitsubishi Heavy Industries, Ltd. (MHI) is Japan's largest heavy-industry corporation and the principal manufacturer of Japan's mainstay H-IIA and H-IIB rockets (as well as their predecessor, the H-II). Since 2001 there have been 42 H-IIA launches, with just one failure. There have been nine H-IIB launches since 2009, with no failures.

---

1   See *Uchū sangyō bijon 2030 ni tsuite* [About Space Industry Vision 2030], at https://www8.cao.go.jp/space/vision/vision.html; *Uchū sangyō bijon 2030* [Space Industry Vision 2030], May 29, 2017, at https://www8.cao.go.jp/space/vision/mbrlistsitu.pdf; *Uchū sangyō bijon 2030 yōyaku* [Summary of the Space Industry Vision 2030], May 29, 2017, at https://www8.cao.go.jp/space/vision/summary.pdf. An unofficial English translation of the Summary of the Space Industry Vision 2030 by ZeLo is included in Annex 8.
2   See "New fund to boost Japanese space startups," March 21, 2018, at https://spacenews.com/new-fund-to-boost-japanese-space-startups/. For discussion regarding the need for greater investment in Japanese space start-ups by larger corporations in and outside the space sector, see M. Uchino, "Can Japan Launch Itself into Becoming a Leader in Global Space Business with its New Space Legislation?," 69th International Astronautical Congress (IAC), Bremen, Germany, 1-5 October 2018 (International Astronautical Federation, 2018), pp. 11-15.

In addition, since 2007 and 2013, respectively, MHI has handled various launch services, including launch transportation services, for the H-IIA and H-IIB rockets (MHI took over these services from JAXA).[3] MHI is one of the few companies in the world with such an offering, which provides integrated services from rocket manufacture to coordination between rockets and satellites, program management, and launch transportation services (including coordination with related space manufacturers, insurance companies, JAXA and related ministries and agencies), as well as support for satellite preparation at launch sites and the launches themselves. According to MHI:

> The comprehensive services include rocket manufacture, program management, and execution of launch campaigns. MHI's deep experience in launch vehicle development and launch operations has given Japan's launch services a competitive edge in the global market . . . . MHI has earned a solid reputation for maintaining launch schedules. Aided by extensive expertise in managing large projects with multiple partners, MHI has a record of punctual launches – launches that aren't rescheduled after the start of launch operations – except for a small number postponed by weather.[4]

MHI currently is developing the H3 launch vehicle with JAXA as the successor to H-IIA/B rockets.[5]

2   IHI Aerospace Co., Ltd.

IHI Aerospace Co., Ltd. (IHI Aerospace) is a subsidiary of IHI Co., Ltd., another major Japanese heavy industry manufacturer. IHI Aerospace is the company primarily responsible for the manufacture of Japan's mainstay Epsilon rocket system, which has had four successful launches since 2013, as well as a secondary contractor for components of the H-IIA, H-IIB, and H3 rockets.[6] Among other things, IHI Aerospace also has contributed to the development of ISS's *Kibō* facility and the *Hayabusa2* asteroid probe.

In June 2020, IHI Aerospace and JAXA signed an agreement to produce a successor to the Epsilon rocket, the Epsilon S, which also envisions that IHI Aerospace will begin to

---

3   *See* MHI Launch Services, at https://www.mhi.com/products/space/launch_service.html; MHI https://www.mhi.com/products/space/launch_service.html. MHI also provides launch services for satellites from national space agencies outside Japan. These services do not include launch safety management operations (*e.g.*, ground safety assurance operations, flight safety assurance operations, and comprehensive command operations during count-down). *See also* "Current Status and Recent Developments in Japan's National Space Law," pp. 369, 399-400.
4   *See* Launch Services: Bringing the Future to Outer Space, at https://www.mhi.com/expertise/showcase/column_0002.html.
5   *See* About H3 Launch Vehicle, at https://global.jaxa.jp/projects/rockets/h3/.
6   *See* Space Development, at https://www.ihi.co.jp/ia/en/products/space/index.html.

offer launch transportation business services for the Epsilon S in the future.⁷ According to IHI Aerospace, its agreement with JAXA:

> lays down fundamental matters such as the roles of JAXA and [IHI Aerospace] at the development and operational steps with an eye to establishing a framework to independently expand launch service business with the Epsilon S to commercial market, as well as to transforming the current space transportation system into an independent and sustainable business structure by maintaining and advancing the industrial basis.⁸

Further, IHI Aerospace explains:

> In recent years, as satellites getting smaller and the manufacturing cost getting lowered, satellite applications businesses are expanding. In response to this trend, we are expecting a bright forecast of increasing demands of satellite launches. [IHI Aerospace] will focus on development and launch service business activities worldwide aiming to expand the field of the space transportation.⁹

Epsilon S's first demonstration flight is scheduled for 2023.

3 Interstellar Technologies Corporation

Interstellar Technologies Corporation (IST) is a Hokkaido-based venture firm established in 2003 to develop, manufacture, and launch liquid-fuel rockets. Its stated mission is to reduce the cost of access to space by producing and launching low-cost rockets into space.¹⁰ In 2017, IST attempted to launch a sounding rocket dubbed MOMO into space, although this effort failed. After a second failure, IST successfully sent MOMO3 to an altitude of 113.4 kilometers in 2019, thereby becoming the first private Japanese company to have manufactured and launched a rocket into space.¹¹ IST has used JAXA's Taiki Aerospace Research Field in Taiki, Hokkaido for its launches.

---

7   *See* Signing of the "Basic Agreement" concerning the Epsilon S Launch Vehicle - IHI AEROSPACE satellite launch business," June 12, 2020, at https://www.ihi.co.jp/en/all_news/2020/aeroengine_space_defense/2020-6-12/index.html.
8   *See id.*
9   *Id.*
10 *See* Mission, at http://www.istellartech.com/mission.
11 *See* History, http://www.istellartech.com/history_en.

IST has received funding from Japan's Ministry of Economy, Trade and Industry to research commercially viable micro satellite launch systems.[12] IST currently is developing a new rocket, ZERO, to put such small satellites into orbit, and is aiming to conduct its first launch of a ZERO model in 2023.[13] IST plans to use methane-containing liquefied natural gas (LNG) as a "next-generation" propellant for the ZERO rocket, and notes that livestock manure in rural Hokkaido is a rich source of methane.[14] IST has been developing its LNG propulsion system in collaboration with JAXA and the Muroran Institute of Technology, a Hokkaido university.[15]

IST's efforts also have attracted support from large Japanese companies, including Marubeni Corporation, one of Japan's largest trading firms. In 2019, the two companies formed a capital partnership via Marubeni's investment in IST. A joint press release stated:

> While the demand for using small satellites for Earth observation and communication is increasing, there is currently a shortage of rockets for transporting small satellites to space. Existing large rockets do not always meet the demand for transportation of small satellites to space in terms of launch frequency, price, and launch orbit flexibility. IST develops a rocket that make it possible to launch small satellites at high frequency and low price, and aims to create a launch service for small satellites that meets the needs of users.
>
> Marubeni started a business collaboration with IST in 2016 and has been supporting the company's sales and marketing efforts in Japan and overseas ever since. With this new investment, Marubeni and IST will further strengthen their partnership and contribute to the development of the space industry not only in Japan but also throughout the rest of the world by providing high-frequency and low-priced launch services.[16]

Despite such backing, IST has confronted financial challenges due to the COVID-19 pandemic, like many small companies. Facing delays to its planned launch of the

---

12 *See* Bid result / contract result (April 2020 to May 2020), at https://www.meti.go.jp/information_2/publicoffer/R_00_bid_news_list.html.
13 *See* "Japanese space startup developing LNG-powered rocket," May 27, 2020, at https://www.offshore-energy.biz/japanese-space-startup-developing-lng-powered-rocket/.
14 *See* "LNG, which is attracting attention as a next-generation fuel, is selected as a propellant for the orbital rocket ZERO, May 3,2020, at http://www.istellartech.com/archives/2262.
15 *See* "Japanese space startup developing LNG-powered rocket," May 27, 2020, at https://www.offshore-energy.biz/japanese-space-startup-developing-lng-powered-rocket/.
16 *See* "Capital Partnership between Marubeni and Interstellar Technologies," November 11, 2019, at https://www.marubeni.com/en/news/2019/release/201911112E.pdf.

MOMO5 rocket in 2020, for example, IST launched a crowdfunding campaign for support and raised some USD 370,000.[17]

III   SPACE TRAVEL AND TRANSPORTATION SERVICES

1   *PD Aerospace Co., Ltd.*

Established in 2007, PD Aerospace Co., Ltd. (PDAS) is a Nagoya-based company that is striving to develop a reusable, suborbital "spaceplane" with a hybrid jet-rocket engine system that can reach an altitude of 100 kilometers and conduct both manned and unmanned missions for purposes of scientific experiments (*e.g.*, Earth observation and microgravity research) and space tourism.[18] According to PDAS, "[t]o improve our understanding of our planet and to promote the utilization of near-Earth space, we aim to establish a low cost and practical space transportation infrastructure."[19]

PDAS has partnered with a number of companies, government organizations, and universities to research and develop its spaceplane project. In 2016, major Japanese travel agency H.I.S. took a 10.3% share in PDAS while ANA Holdings Inc., the parent company of ANA (All Nippon Airways) airline, took a 7% share.[20] In 2018, PDAS continued to add shareholders, including a venture capital company funded by Tohoku University, and also signed a memorandum of understanding with JAXA to collaborate within the scope of the J-SPARC program "to achieve successful business incubation of the company's reusable suborbital spaceplane through sharing of JAXA's technical capabilities, know-how, and equal cooperation between both parties to obtain new knowledge and technology breakthrough."[21] PDAS announced another capital expansion in April 2020, with the Japanese trading company Sojitz Corporation taking a stake.[22]

Most recently, in September 2020, PDAS revealed that Okinawa Prefecture had accepted PDAS's offer to develop Shimojishima Airport as a spaceport for its spaceplanes, which PDAS bills as "the first manned space tourism hub in Asia."[23]

---

17 *See* "Japanese space startup developing LNG-powered rocket," May 27, 2020, at https://www.offshore-energy.biz/japanese-space-startup-developing-lng-powered-rocket/.
18 *See* Spaceplane Project, at https://pdas.co.jp/en/business01.html.
19 *See id.*
20 *See* "Japanese heavy-hitters invest lightly in PD Aerospace's space tourism effort," December 2, 2016, at https://www.geekwire.com/2016/japan-pd-aerospace-space-tourism/.
21 *See* "Capital Expansion and Organization Enhancement for Spaceflight in 2019," December 25, 2018, at https://pdas.co.jp/documents/Press_181225_En.pdf.
22 *See* "A Total of $6.8 million Raised during Round of Series A Funding," April 15, 2020, at https://pdas.co.jp/documents/Press_200415_En.pdf.
23 *See* "Spaceport business development at island in Okinawa 'Shimoji-shima the Gate for space,'" September 11, 2020, at https://pdas.co.jp/documents/Press_200911_En.pdf.

PDAS plans to use the airport's 3000-meter runway and nearby training flight facilities "to develop the following three businesses in addition to spaceplane flight testing: 1. Tenant business (Hangar rentals for spaceplane operators) 2. Training program business for space tourists (Zero G, High G, medical check) 3. Tourism business (Entertainment in spaceport)."[24]

2   SPACE WALKER Co., Ltd.

SPACE WALKER Co., Ltd. (SPACE WALKER) is a start-up company established in December 2017 that is closely affiliated with the Tokyo University of Science, Faculty of Science and Technology Department of Mechanical Engineering Space Systems Laboratory. Like PDAS, SPACE WALKER is developing a reusable, suborbital spaceplane with a hybrid jet-rocket engine system, which uses LNG modified liquid oxygen and methane as a propellant. SPACE WALKER aims to launch the first such vehicle in 2024 for scientific research purposes.[25] Other SPACE WALKER spaceplanes are slated to be developed for small satellite launches and space tourism and put into operation in 2026 and 2029, respectively.[26] SPACE WALKER is considering using Taiki, Hokkaido, which is the site of a JAXA research field and also the Taiki Multipurpose Aviation Park, as the location of a spaceport for its spaceplanes.[27]

Given its ties with the Tokyo University of Science (with which it has a joint research agreement), SPACE WALKER possesses considerable research capabilities, and it has signed collaboration agreements with various major corporations to help it realize its plans, including Air Water Inc., Air Water Hokkaido Ind., IHI Corporation, Toray Industries, I-NET Corporation, Kawasaki Heavy Industries Ltd., and IHI Aerospace

---

24 *Id.*
25 *See* "Sub-orbital spaceplane launch schedule changes and component sales," October 2, 2020, at https://www.space-walker.co.jp/en/news/press-release/sub-orbital-spaceplane-launch-schedule-changes-and-component-sales.html.
26 *See* Spaceplane, at https://www.space-walker.co.jp/en/service.
27 *See* "Concluding a basic agreement with Air Water Inc. and Air Water Hokkaido Inc.," October 8, 2020, at https://www.space-walker.co.jp/en/news/press-release/concluding-a-basic-agreement-with-air-water-inc-and-air-water-hokkaido-inc.html.

Corporation.[28] In addition, SPACE WALKER has signed a memorandum of understanding with JAXA to collaborate within the scope of the J-SPARC program.[29]

By November 2020, SPACE WALKER had raised more than USD 5.7 million in angel pre-seed investment funding.[30]

3   *Obayashi Corporation*

In 2012, Obayashi Corporation (Obayashi), Japan's largest construction company, announced plans to complete a "space elevator" by 2050. In Obayashi's words, the space elevator:

> will be constructed to replace rockets as a way to transport large amounts of people and materials into space economically. A terminal station would be located 36,000 kilometers above the Earth, about one-tenth the distance from Earth to the moon, and a spaceport located offshore on Earth, connected by a cable 96,000 kilometers long to operate elevators. This will broaden possibilities in space solar power generation, investigating and utilizing space resources, and space travel.[31]

While still at a very early stage, as indicated above the goal of the space elevator is to provide a means of transporting people and cargo in an elevator via a cable connecting an Earth port and an orbiting space station, a method Obayashi says will be less expensive than using rockets.

Obayashi faces various challenges in bringing its space elevator to fruition, including developing cables that are light but strong enough to resist high-energy cosmic rays and

---

28 *See id.*, "Conclusion of basic transaction contract with IHI Corporation," September 29, 2020, at https://www.space-walker.co.jp/en/news/press-release/conclusion-of-basic-transaction-contract-with-ihi-corporation-2.html, "Concluded a contract with Toray Magic Carbon Co., Ld.," February 3, 2020, at https://www.space-walker.co.jp/en/news/press-release/concluded-a-contract-with-toray-carbon-magic-co-ltd.html, "Notice of business alliance with I-NET Corporation," August 22, 2019, at https://www.space-walker.co.jp/en/news/press-release/notice-of-business-alliance-with-i-net-corporation.html, "Consignment contract with Kawasaki Heavy Industries, Ltd.," April 26, 2019, at https://www.space-walker.co.jp/en/news/press-release/2019-04-26-press-release-consignment-contract-with-kawasaki-heavy-industries-ltd.html, and "Mou about commercialization of winged resusable suborbital Spaceplane with IHI/IHI Aerospace," August 1, 2018 https://www.space-walker.co.jp/en/news/press-release/mou-about-commercialization-of-winged-reusable-suborbital-spaceplane-with-ihi-ihi-aerospace.html.

29 *See* "Mou about co-creation of business concept with JAXA through J-SPARC," https://www.space-walker.co.jp/en/news/press-release/mou-about-co-creation-of-business-concept-with-jaxa-through-j-spac.html.

30 *See* "About financing," November 13, 2020, at https://www.space-walker.co.jp/en/news/press-release/about-financing-2.html.

31 *See Obayashi Corporate Report 2012*, p. 4, at https://www.obayashi.co.jp/en/ir/upload/img/ir2012en.pdf.

protecting the elevators from collisions with space debris and meteorites.[32] Regarding the first challenge, Obayashi has been working with Shizuoka University on carbon nanotubing (CNT), which is a lightweight, very strong material with good thermal conductivity properties, as a potential candidate for the cables.[33] Researchers have been conducting experiments to expose CNT materials to space conditions, including on the ISS's *Kibō* module. According to Obayashi, "Going forward, we will do detailed analysis with measurements of radiated light and investigate damage mechanisms at the atomic level. From there, we will develop technology to increase CNT durability in space."[34] Additionally, in 2018 Obayashi and Shizuoka University began tests of elevator movement in space by placing in orbit two small satellites connected by a 10-meter steel cable, and then observing how a motorized box travels between the satellites along the cable.[35]

IV   Space Resource Exploration Business

1   ispace inc.

ispace inc. (ispace) is a start-up company founded in 2010 with offices in Japan, Europe, and the United States that "aims to be a vehicle for companies on Earth to access new business opportunities on the Moon and ultimately incorporate the Moon into Earth's economic and living sphere."[36] ispace's origins lie in the Google Lunar XPRIZE competition, a multi-year contest that offered a prize to the privately funded team which first landed a robotic spacecraft on the Moon that could travel 500 meters and transmit video and images to Earth. While no team won the competition by the 2018 deadline, five finalists were chosen, including ispace's "Team HAKUTO" lunar rover.[37]

Since then ispace has focused on a new commercial Moon exploration program, HAKUTO-R. According to ispace, HAKUTO-R will consist of three lunar missions: the first, scheduled for 2022, to put a lunar lander on the Moon (which if successful would be the first private Japanese group to do so); the second, slated for 2023, to land and deploy a lunar rover capable of exploration and data collection (for this and the first mission ispace plans to use SpaceX's Falcon 9 rocket); and a third mission, as yet unscheduled, to make more lunar landings and send multiple rovers on the Moon that can collect data and also

---

32 *See* "Japan to conduct first test as part of space elevator project," September 3, 2018, at https://www.cnet.com/news/japan-to-conduct-first-test-as-part-of-space-elevator-project/.
33 *See Obayashi Corporate Report 2018*, p. 26, at https://www.obayashi.co.jp/en/ir/upload/img/ir2018en.pdf.
34 Id.
35 *See* "Japan is about to start testing the feasibility of a space elevator — starting with a small model in orbit," at https://www.businessinsider.com/japan-testing-space-elevator-model-2018-9.
36 *See* Project, at https://ispace-inc.com/project/.
37 Id.

deliver payloads from customers on Earth.[38] ispace's partners in the HAKUTO-R project include Japan Airlines, Suzuki Motors, Citizen Watch, Mitsui Sumitomo Insurance, NGK Spark Plug, Takasago Thermal Engineering, and Sumitomo Corporation.[39]

In November 2020, ispace announced it was opening an office in Denver, Colorado, a move ispace says was driven by its aim to have a U.S. presence from which to partner with NASA to advance its lunar exploration objectives, such as through the Commercial Lunar Payload Services initiative, in the spirit of the Artemis Accords.[40]

To date, ispace has raised a cumulative total investment of approximately USD 125 million, including some USD 28 million in Series B funding in 2020 from IF SPV 1st Investment Partnership (managed by Incubate Fund) as the lead investor, Space Frontier Fund (managed by Sparks Innovation for Future, Inc.), Takasago Thermal Engineering Co., Ltd., and Mitsui Sumitomo Insurance Co., Ltd.[41]

V Space Debris Removal Business

1 Astroscale Holdings Co., Ltd.

Founded by a Japanese IT entrepreneur in Singapore in 2013, Astroscale Holdings Co., Ltd. (Astroscale) is now headquartered in Tokyo and also has offices in the United States, Israel, and the United Kingdom.[42] In its own words:

> Astroscale is developing innovative and scalable solutions across the spectrum of on-orbit servicing missions, including Life Extension, In-situ Space Situational Awareness, End of Life services, and Active Debris Removal, to create sustainable space systems and mitigate the growing and hazardous buildup of debris in space. Astroscale is also defining business cases and working with government and commercial stakeholders to develop norms, regulations, and incentives for the responsible use of space.[43]

38　Id.; "Completion of final design of Mission 1 Lander (Lunar Module) for private lunar exploration program 'HAKUTO-R,'" July 30, 2020, at https://ispace-inc.com/jpn/news/?p=1636.
39　See "ispace Raises $28 Million in Series B Funding; Announces Lunar Data Business," August 20, 2020, at https://ispace-inc.com/news/?p=1681.
40　See "ispace Opening New U.S. Office to Support Artemis, Hires Program Director," November 9, 2020, at https://ispace-inc.com/news/?p=1741.
41　See id.
42　See "Astroscale to Relocate Japan HQ to Accommodate Business Expansion," November 4, 2020, at https://astroscale.com/astroscale-to-relocate-japan-hq-to-accommodate-business-expansion/.
43　See About Astroscale, at https://astroscale.com/about-astroscale/about/.

Astroscale estimates that currently orbiting Earth are 34,000 objects larger than 10 centimeters, 900,000 objects between one and 10 centimeters, and 130 million objects less than one centimeter.[44] These objects threaten satellites, most of which operate in low Earth orbit (LEO, an orbit with an altitude of 200-2000 kilometers), where, Astroscale says, the majority of space debris is located. Hence Astroscale's goal of developing satellite end-of-life and space debris removal services.

Astroscale's end-of-life service, currently in the testing phase, involves removing space debris by putting small satellites in orbit that have a magnetic docking mechanism which can capture other failing or already defunct satellites and then burn them by re-entering the Earth's atmosphere.[45] According to Astroscale, the main targets of this service are the "thousands of satellites being launched in emerging large constellations by commercial telecoms operators . . . . A small percentage of satellites will inevitably fail and, should this happen, internal means for de-orbit will be inoperable. These satellites will need a 'semi-cooperative' solution that allows for simple capture by an outside party."[46] Astroscale further proposes that all satellites destined for mid- to high LEO "be pre-engineered with a light-weight docking plate that will facilitate eventual removal from orbit, should the satellite go defunct."[47]

In addition to the satellite end-of-life service, Astroscale is working to develop the means to remove the "thousands of pieces of existing large debris, consisting of defunct satellites and spent upper stage rocket bodies, that are currently in orbit."[48] According to Astroscale, this space debris removal service needs a "'non-cooperative' grappling solution as this debris was not prepared for de-orbit before launch."[49] Astroscale says it intends to partner with national space agencies and international organizations to address this issue, and in February 2020 it announced that JAXA had selected Astroscale as the commercial partner for the initial phase of its first space debris removal project, the Commercial Removal of Debris Demonstration (CRD2) project.[50] This initial phase involves a demonstration of debris data acquisition, including by video, and is scheduled to take place at the end of 2022. Astroscale:

> will be responsible for the manufacturing, launch and operations of the satellite that will characterize the rocket body, acquiring and delivering movement

---

44 See Space Debris, at https://astroscale.com/space-debris/.
45 See End of Life (EOL), at https://astroscale.com/services/end-of-life-eol/.
46 See Space Debris, https://astroscale.com/space-debris/.
47 Id.
48 Id.
49 Id.
50 See "Astroscale Selected as Commercial Partner for JAXA's Commercial Removal of Debris Demonstration Project," February 12, 2020, at https://astroscale.com/astroscale-selected-as-commercial-partner-for-jaxas-commercial-removal-of-debris-demonstration-project/; "Regarding the conclusion of a partnership-type contract for JAXA commercial debris removal demonstration," March 23, 2020, at https://www.jaxa.jp/press/2020/03/20200323-1_j.html.

observational data to better understand the debris environment. The CRD2 project will further cement Japan's leadership in developing the technology and policies that will drive this growing market.[51]

A future second phase will aim to demonstrate the removal of large-scale debris that has been identified during phase one.

In October 2020, Astroscale said it had completed a Series E round of funding to obtain an additional USD 51 million from a group of investors led by aSTART Co., Ltd. Cumulatively, Astroscale has raised some USD 191 million, which it says makes it "the most funded on-orbit services and logistics company globally and most funded space venture in Japan."[52]

2   SKY Perfect JSAT Corporation

SKY Perfect JSAT Corporation's predecessor, Japan Communications Satellite Company, was a pioneering satellite operator that commissioned the first launch, from the spaceport of France and the ESA *in Kourou*, French Guiana, of a Japanese commercial communications satellite (JCSAT-1) in 1989. SKY Perfect JSAT Corporation (JSAT) currently owns 19 satellites, making it the largest private satellite operator in Asia. Lately, JSAT has sought to expand its space-related business activities. In 2019, the company announced it was entering into a partnership agreement with PASCO Corporation, a Japanese geospatial information collection and processing firm. The two companies say they will work together to create additional services and markets in the satellite and geospatial data businesses.[53]

More recently, in June 2020, JSAT announced it would begin collaborating with JAXA, RIKEN (a Japanese government-affiliated research institute), and Nagoya and Kyushu universities to develop the world's first satellite to remove space debris with laser ablation.[54] JSAT aims to use satellites equipped with high-energy lasers to irradiate the surface of space debris to cause the debris to emit vapors that propel it

---

51 *Id.*
52 *See* "Astroscale Brings Total Capital Raised to U.S. $191 Million, Closing Series E Funding Round," October 13, 2020, at https://astroscale.com/astroscale-brings-total-capital-raised-to-u-s-191-million-closing-series-e-funding-round/.
53 *See* "Partnership in Space Business -Seeking New Business Creation in the Joint Utilization of Satellite Ground Stations and Satellite Data," March 28, 2019, at https://www.jsat.net/common/pdf/news/news_2019_0328_en.pdf.
54 *See* Designing and Developing the World's First Satellite for Removing Space Debris with a Laser," June 11, 2020, at https://www.skyperfectjsat.space/en/news/files/pdf/34d770bf12e8d85e94a3c76d2e222be6.pdf.

into the atmosphere, where the debris will burn up.[55] According to JSAT, which plans to launch this service in 2026, the laser ablation method has advantages of safety, because there is no contact between the satellite and the debris material, and economic efficiency, because the debris will generate its own propellant which eliminates the need for a separate source of fuel to move the debris.[56]

JSAT's president acknowledges that his company's self-interest is part of the motivation to expand into the space debris removal business. "We can't avoid the problem of space debris, which will become a threat to our business in the future," he notes.[57] More broadly, however, JSAT explains that it has disposed of its own satellites in accordance with the Inter-Agency Space Debris Coordination Committee's Mitigation Guidelines,[58] and further:

> The problem of space debris is an environmental problem similar to CO2 and marine plastics. Therefore, JSAT will continue to contribute to the maintenance of a sustainable space environment, aiming to solve the problem of space debris through this project as space SDGs [Sustainable Development Goals].[59]

VI Space Entertainment

1 ALE Co., Ltd.

Established in 2011, ALE Co., Ltd. (ALE) is a start-up company in Tokyo that has been working to research and develop "Sky Canvas," an innovative technology and business that aims to use satellites which can carry and discharge artificial meteoroids that become "shooting stars" for entertainment purposes.[60] The goal is to launch satellites, each with several hundred meteoroids one centimeter in diameter, to an orbiting altitude of about

---

55 *See id.*; "Japan Firm to Develop Satellite to Remove Space Debris," June 11, 2020, at https://www.nippon.com/en/news/yjj2020061100899/; "Remove space debris without contact." SKY Perfect JSAT's challenge goes beyond movies," June 15, 2020, at https://forbesjapan.com/articles/detail/35007.
56 *See* "Company developing laser satellite to clear away space junk," July 2, 2020, at http://www.asahi.com/ajw/articles/13465274.
57 *See* "Japan Firm to Develop Satellite to Remove Space Debris," June 11, 2020, at https://www.nippon.com/en/news/yjj2020061100899/.
58 *See* Inter-Agency Space Debris Coordination Committee's Mitigation Guidelines (2010), at https://www.unoosa.org/pdf/publications/st_space_49E.pdf.
59 *See* Designing and Developing the World's First Satellite for Removing Space Debris with a Laser," June 11, 2020, at https://www.skyperfectjsat.space/en/news/files/pdf/34d770bf12e8d85e94a3c76d2e222be6.pdf.
60 *See* Sky Canvas, https://star-ale.com/skycanvas/.

500 kilometers, where upon customer demand the satellites will discharge the meteoroids to create vivid, multihued displays at night similar to fireworks.[61]

ALE's scientists and engineers have had to do a diverse amount of work to perfect this concept, from enhancing knowledge about the behavior of natural meteors to developing the material to create the artificial meteoroids of different colors and sufficient luminosity to crafting the method of discharging the meteoroids from satellites.[62] ALE has collaborated with Tohoku University on its R&D, and JAXA has supported ALE's efforts, including by providing an Epsilon rocket to launch an ALE test satellite.[63] ALE ran into technical challenges with one of its satellites in 2019 and has rescheduled the debut of Sky Canvas to 2023.[64]

In 2020, ALE received the Minister of Economy, Technology, and Industry Award for Sky Canvas.[65] ALE also took the opportunity in late 2020 to refresh its brand with a colorful and playful corporate design in the image of a pinball machine.[66] At the same time, ALE issued a statement to express its objectives:

> ALE, which has entered its 10th year since its establishment, is engaged in corporate activities such as the world's "first entertainment" that fosters the curiosity of people, including artificial shooting stars, and "accumulation and use of mid-level atmosphere data" that contributes to the elucidation of climate change. We aim to contribute to the sustainable development of humankind through the two approaches of "entertainment" and "science".... In the future ... we will widely convey the beauty of the universe and the fun of science to as many people as possible in order to further achieve our mission.[67]

---

61 See "Fireworks of the future? Japan to create fake shooting stars," October 24, 2016, at https://edition.cnn.com/style/article/on-japan-artificial-meteors/index.html.
62 Id.
63 See Our Satellites, https://star-ale.com/technology/#sec_satellites; "ALE artificial shooting star won't be realized by the end of the year due to satellite malfunction postponed to 2023," April 20, 2020, at https://www.nikkei.com/article/DGXMZO58274360Q0A420C2XY0000; About the additional selection results of 'Innovative satellite technology demonstration No. 1 theme' open call for participants," (July 2017), at https://www.kenkai.jaxa.jp/pickup/kakushin_3.html.
64 See "ALE artificial shooting star won't be realized by the end of the year due to satellite malfunction postponed to 2023," April 20, 2020, at https://www.nikkei.com/article/DGXMZO58274360Q0A420 C2XY0000; "Notice of Postponement of Realization of Artificial Shooting Star (2020 → 2023) Due to Malfunction of Artificial Satellite No. 2," April 20, 2020, at https://star-ale.com/news/2020/04/20/000161.html.
65 See "Received the 'Minister of Economy, Technology, and Industry Award' at the '34th Advanced Technology Awards for Creating Originality" by Fuji Sankei Business, June 5, 2020, at https://star-ale.com/news/2020/06/05/000163.html.
66 See "ALE Co., Ltd. Rebrands New Corporate Statements and Visuals to Arouse Interest in Space and Science for a Wider Audience," September 30, 2020, at https://star-ale.com/news/2020/09/30/000165.html.
67 See id.

In 2019, ALE completed a round of funding from investors such as Horizon Ventures, Sparks Group Co., Ltd., Shinsei Corporate Investment Ltd., and QB Capital LLC, raising nearly USD 1 million. This brought cumulative total amount invested in ALE to USD 27 million.[68]

VII Satellite Ground Station Sharing

1 Infostellar

Tokyo-based Infostellar emerged in 2016 to take advantage of the opportunity of satellite ground station downtime, that is, the significant amount of time each day that individual satellite operators' expensive ground stations are not communicating with their own orbiting satellites due to the latter being out of range. This downtime can be up to 23.5 hours a day in places such as Japan.[69]

Thus, in 2017 Infostellar developed a cloud-based platform, StellarStation, to connect satellite operators that have idle ground station capacity with other operators that wish to use the available stations to communicate with their satellites.[70] Infostellar has 13 ground stations in its network that include all the frequencies for LEO satellites, and it is aiming to add up to 12 more stations to provide service coverage around the world.[71] The company has been dubbed the "Airbnb for satellite antennas" for its innovative Ground Station-as-a-Service (GSaaS) business model.[72]

Currently, Infostellar is working to lock in fee-based contracts with satellite operators to obtain consistent revenue. It also is tackling rulemaking challenges by seeking to amend regulations in countries that require satellite operators to own a ground station; this requirement is at odds with Infostellar's flexible GSaaS business model wherein

---

68 *See* "Raised a Total of 1.2 Billion Yen in Series A Round from Horizon Ventures, etc. – Accelerate Technological Development and Commercialization of Space Entertainment Business 'Sky Canvas,'" September 5, 2019, at https://star-ale.com/news/2019/09/05/000071.html.

69 *See* Japan's Airbnb for Satellites – Infostellar, interview transcript, https://www.disruptingjapan.com/japans-airbnb-satellites-infostellar/.

70 *See* StellarStation - The Ground Station Platform for Modern Space Business, at https://www.stellarstation.com/.

71 *See* "Infostellar raises a total of $3.5 million with lead investor Airbus Ventures, Sony Innovation Fund, and New Investors Daiwa Energy & Infrastructure, Mitsubishi UFJ Capital, and Mitsubishi UFJ Lease & Finance," April 30, 2020, at https://infostellar.net/cb-finance; "Infostellar Raises Additional Capital in Attempt to Disrupt the Ground Station Market," April 30, 2020, at https://www.satellitetoday.com/business/2020/04/30/infostellar-raises-additional-capital-in-attempt-to-disrupt-the-ground-station-market/.

72 *See* "Infostellar raises $7.3M for its 'Airbnb for satellite antenna' rental services," September 14, 2017, at https://techcrunch.com/2017/09/13/infostellar-raises-7-3m/.

companies interested in entering the satellite business would not need their own expensive ground station.[73]

Infostellar raised USD 7.3 million in 2017 from investors such as Airbus Ventures and the Sony Innovation Fund.[74] In 2020, Infostellar raised an additional USD 3.5 million from its existing investors plus Daiwa Energy Infrastructure, Mitsubishi UFJ Capital, and Mitsubishi UFJ Lease & Finance.[75] In Infostellar's words:

> The purpose of this [latest] finance round is to accelerate Infostellar's business activities, specifically to hire business development and sales personnel and to strengthen Infostellar management. We will continue to develop our business activities throughout Europe and Asia, and provide satellite operators with a more flexible and expandable ground station network for full global deployment.[76]

VIII Agricultural Hybrid Remote Sensing

1 Sagri Co., Ltd.

Sagri Co., Ltd. (Sagri) was founded in 2018 and is a pioneering "smart agriculture" start-up that uses a hybrid of satellite, drone, and ground data to help farmers in Japan and elsewhere improve soil fertility and increase harvest yields.[77] In particular, Sagri uses images and other data obtained from satellites, drones, and ground-based devices to measure the Normalized Difference Vegetation Index of farmland, that is, the distribution of vegetative mass across a given agricultural region, in order to identify areas that have higher amounts of vegetation and areas that have less vegetation and may therefore be less productive.[78] Such an assessment enables farmers to ascertain which areas may need agricultural improvement and also consider how to optimize

---

73 See "Infostellar Raises Additional Capital in Attempt to Disrupt the Ground Station Market," April 30, 2020, at https://www.satellitetoday.com/business/2020/04/30/infostellar-raises-additional-capital-in-attempt-to-disrupt-the-ground-station-market/.
74 Id.
75 See "Infostellar raises a total of $3.5 million with lead investor Airbus Ventures, Sony Innovation Fund, and New Investors Daiwa Energy & Infrastructure, Mitsubishi UFJ Capital, and Mitsubishi UFJ Lease & Finance," April 30, 2020, at https://infostellar.net/cb-finance; "Infostellar raises less than 400 million yen expands adoption such as business development," May 7, 2020, at https://www.nikkei.com/article/DGXMZO58799090X00C20A5XY0000.
76 See "Infostellar raises a total of $3.5 million with lead investor Airbus Ventures, Sony Innovation Fund, and New Investors Daiwa Energy & Infrastructure, Mitsubishi UFJ Capital, and Mitsubishi UFJ Lease & Finance," April 30, 2020, at https://infostellar.net/cb-finance.
77 See Technology, at https://sagri.tokyo/technology/.
78 More specifically, Normalized Difference Vegetation Index is a graphic indicator of the greenness, representing the relative density and health of vegetation, in a satellite image.

crop rotation. In addition, Sagri is developing an AI-driven system called "ACTABA" that uses satellite and ground data to detect abandoned cultivated land, which can assist local governments with aging farming communities – an important issue in graying Japan – to evaluate ways to boost agricultural land productivity.[79]

Sagri is working to help farmers outside Japan as well. It was one of a number of entities selected by METI and the Japan External Trade Organization in 2020 to receive funding for an "Asia Digital Transformation" project.[80] Sagri will team up with a firm in Thailand to help the Thai government create an agricultural database using satellite data and AI that will be used to establish a farmland registry and collect information (cultivation/crop details, yield amounts, etc.) that individual farmers input from smartphones through an application developed by Sagri's Thai collaborator. This information will be used to help the Thai government and farmers improve agricultural productivity and better prepare for natural disasters.[81] In addition to its agritech innovations, Sagri is endeavoring to expand access of farmers in the developing world to microfinancing opportunities, including small loans that Sagri itself provides through an online platform. In 2019, Sagri also established a start-up in India to collaborate with partners to use satellite and other data to help predict harvest yields, information that the Indian farmers can use to show creditworthiness.[82]

Sagri has received recognition for its work, including winning the Sompo Japan Award at the 7th Agritech Grand Prix and the Global Student Entrepreneur Award for Innovation, each in 2020, as well as the SingularityU Japan Global Impact Challenge in 2019.[83] Most recently, Sagri was selected to participate in the United Nations Office for Project Services (UNOPS)'s "Global Innovator Acceleration Program."[84] While it has

---

79 *See* Visualize abandoned cultivated land ACTABA, at https://sagri.tokyo/service-actaba/.
80 *See* Asia Digital Transformation (ADX) Project (Support for Joint Pilot Projects (PoC) between ASEAN and Japan), at https://www5.jetro.go.jp/newsletter/jaa/2020/ADX_projects_lists_English.pdf.
81 *See* Introduction of ADX Project Sagri (Agriculture), November 25, 2020, at https://www.ameicc.org/disg/file/201125_Introduction-of-ADX-project-1-Sagri-(Agriculture).pdf.
82 *See* "Japan's Sagri Looks To Revitalise Agri Credit In India With Satellite Imagery, Soil Data," June 9, 2020, at https://inc42.com/features/sagri-looks-to-revitalise-agri-credit-with-satellite-imagery-soil-data/; SAgri is the first startup to enter India through Japan India Startup Hub, at https://www.wtcmumbai.org/images/covid/5May-Sagri.pdf.
83 *See* "Breaking News of the 7th Agritech Grand Prix," September 28, 2020, at https://lne.st/2020/09/28/atg2020/; Global Student Entrepreneur Awards Shunsuke Tsuboi, March 11, 2020, at https://gsea.org/finalists/shunsuke-tsuboi/; https://su.org/gic/japan-2019/.
84 *See* UNOPS Global Innovator Acceleration Program, November 9, 2020, at https://sagri.tokyo/2020/11/09/%e5%9b%bd%e9%80%a3%e3%83%97%e3%83%ad%e3%82%b8%e3%82%a7%e3%82%af%e3%83%88%e3%82%b5%e3%83%bc%e3%83%93%e3%82%b9%e6%a9%9f%e9%96%a2%ef%bc%88unops%ef%bc%89%e3%81%aeglobal-innovation-accelerator-program/.

raised initial funds through traditional start-up equity financing, Sagri remains a relatively small operation thus far, reporting capital reserves of around USD 430,000.[85]

* * *

The foregoing provides a window into the range of Japanese business innovation in the use and development of space. Large corporations such as MHI and IHI Aerospace are offering space services different from their traditional industry lines. Mid-sized firms such as ispace and ALE are seeking to create new commercial opportunities. Start-ups such as Infostellar and Sagri are designing novel ways to use existing space technologies. Such diversity should embolden others in Japan's private sector (as well as potential participants elsewhere) to consider the commercial possibilities of space. Small-scale entrepreneurs with trepidation about entering this field should recall that the origin of Japan's space program was a rocket just 23 centimeters long.

It is to be hoped, however, that any such concerns might be allayed by the consolidation and enhanced coordination of Japan's space program at the highest levels of Japan's government since the early 2000s, the strong national support for and promotion of space development (*e.g.*, in the Basic Space Plan), the recent governmental programs to facilitate private sector participation in space ventures (*e.g.*, JAXA's Space Exploration Innovation Hub Center and J-SPARC program), and the considerable contribution to establishing a clear legal regime as evidenced by enactment of the Basic Space Law in 2008 and passage of the Space Activities Act and Remote Sensing Act in 2016.

All these steps have created a commercial and regulatory environment in Japan that is conducive to encouraging space-related business innovation, including by start-ups and other small and mid-sized firms. That such a diverse cross section of the private sector has responded in the creative ways described above is testimony to the success of these various efforts. The combination of robust governmental support and resourceful private sector innovation bodes well for Japan's space program in the coming decade and beyond.

---

85 *See* Sagri News, April 30, 2020, at https://sagri.tokyo/2020/04/30/%e7%ac%ac%e4%ba%8c%e5%9b%9e%e7%ac%ac%e4%b8%89%e8%80%85%e5%89%b2%e5%bd%93%e5%a2%97%e8%b3%87%e3%81%ab%e3%82%88%e3%82%8b%e8%b3%87%e9%87%91%e8%aa%bf%e9%81%94%e3%82%92%e5%ae%9f%e6%96%bd%e3%81%97%e3%81%be-2/; Corporate Information, at https://sagri.tokyo/company/.

Select Sources

Books and Select Articles

Aoki, S., "Current Status and Recent Developments in Japan's National Space Law and its Relevance to Pacific Rim Space Law and Activities," 35 *Journal of Space Law*, Issue 2, 2009, at https://spacelaw.sfc.keio.ac.jp/sitedev/archive/current.pdf.

Aoki, S., "New Law Aims to Expand Japan's Space Business," March 3, 2017, at https://www.nippon.com/en/currents/d00294/new-law-aims-to-expand-japan%E2%80%99s-space-business.html.

Aoki, S., "Symposium on the New Space Race – Domestic Legal Conditions for Space Activities in Asia," *AJIL Unbound*, Vol. 113, April 1, 2019, at https://www.cambridge.org/core/services/aop-cambridge-core/content/view/093CF942F1D5A2F04AC3C2A9B7FC3D93/S239877231900014Xa.pdf/domestic_legal_conditions_for_space_activities_in_asia.pdf.

Benkö, M. and K.H. Böcksteigel, eds., *Space Law Basic Legal Documents* (Eleven International Publishing, 2019).

Center for Information on Security Trade Control, *Overview of Japan's Export Controls*, (4th ed., 2015), at https://www.cistec.or.jp/english/export/Overview4th.pdf.

Dempsey, P.S., "National Laws Governing Space Activities: Legislation, Regulation, & Enforcement," *Northwest. J. Int. Law Bus.*, v. 36, no. 1, 2016.

Dunphy, R., *Space Industry Business Opportunities in Japan: Analysis on the Market Potential for EU SMEs Involved in the Earth-Observation Products & Services*, EU-Japan Centre for Industrial Cooperation, October 2016, at https://www.eu-japan.eu/sites/default/files/publications/docs/2016-10-space-industry-business-opportunities-japan-ryuichi-dunphy-min.pdf.

Gerhard, M. and K. Gungaphul-Brocard, "The Impact of National Space Legislation on Space Industry Contracts," in *Contracting for Space – Contract Practice in the European Space Sector* (Ashgate Publishing, 2016).

Kozuka, S., and M. Sato, *Uchū business notameno uchū hō nyumon* [An Introduction to Space Law for Space Business] (Yuhikaku, 2018).

Mainura, T.I., "The United Kingdom's Outer Space Act 1986," *Int. J. of Social Sciences and Humanities Invention*, v. 5, no. 5, May 30, 2018, at https://valleyinternational.net/index.php/theijsshi/article/view/1278.

Pekannen, S. and P. Kallender-Umezu, *In Defense of Japan: From the Market to the Military in Space Policy* (Stanford University Press, 2010).

Robinson, G., "Impact of the U.S. International Traffic in Arms Regulations (ITAR) on International Collaboration Involving Space Research, Exploration, and Commercialization," *ZLW 58. Jg.* 3, 2009.

Smith, M., "United States," *The Space Law Review*, ed. 2, December 2020, at https://thelawreviews.co.uk/edition/the-space-law-review-edition-2/1235331/united-states.

Uga, K., *Chikujo kaisetsu uchū nihō* [Commentary on Two Space Laws] (Kobundo 2019).

von der Dunk, F.G., "Launching from 'Down Under': The New Australian Space Activities Act of 1998," 2000, Space, Cyber, and Telecommunications Law Program Faculty Publications, University of Nebraska, at https://digitalcommons.unl.edu/spacelaw/37/.

Wakimoto, T., *A Guide to Japan's Space Policy Formulation: Structures, Roles and Strategies of Ministries and Agencies for Space*, (Pacific Forum, April 2019), at https://pacforum.org/wp-content/uploads/2019/04/issuesinsights_Vol19WP3_0.pdf.

Wilson, R., "Japan's Gradual Shift Toward Space Security," May 2020, at https://aerospace.org/sites/default/files/2020-05/Wilson_JapansGradualShift_20200428_0.pdf.

Yotsumoto, H. and D. Ishikawa, "Japan," *The Space Law Review*, Edition 1, December 2019, at https://thelawreviews.co.uk/edition/the-space-law-review-edition-1/1211969/japan.

SELECT SOURCES

Treaties and Select Japanese Laws and Regulations

Dokuritsughōseihōjin uchūkōkū kenkyūkaihatsu kikōhō [Law on the Japan Aerospace Exploration Agency] (Law No. 161 of December 13, 2002), unofficial English translation at http://law.e-gov.go.jp/htmldata/H14/H14HO161.html.

Eisei rimōtosenshingu kiroku no tekiseina toriatsukai no kakuho ni kansuru hōritsu [Act on Securing Proper Handling of Satellite Remote Sensing Records] (Act No.77 of 2016), in Japanese and English at https://www8.cao.go.jp/space/english/rs/rs_act.pdf.

Eisei rimōtosenshingu kiroku no tekiseina toriatsukai no kakuho ni kansuru hōritsu sekoukisoku [Regulation for Enforcement of the Act on Ensuring Appropriate Handling of Satellite Remote Sensing Data] (Cabinet Office Order No. 41 of August 9, 2017), in English at https://www8.cao.go.jp/space/english/rs/rs_cabinetofficeorder.pdf.

Gaikoku kawase oyobi gaikoku bōeki hō [Foreign Exchange and Foreign Trade Act] (Act No. 228 of December 1, 1949), Japanese and unofficial English translation at http://www.japaneselawtranslation.go.jp/law/detail/?printID=&id=21&re=02&vm=03.

Jinkō eisei-tō no uchiage oyobi jinkō eisei no kanri ni kansuru hōritsu [Act on Launching Artificial Satellites and Managing Satellites] (Act No. 76 of 2016), in Japanese and English at https://www8.cao.go.jp/space/english/activity/documents/space_activity_act.pdf.

Jinkō eisei-tō no uchiage oyobi jinkō eisei no kanri ni kansuru hōritsu ni motozuku shinsa kijun & hyōjun shorikikan [Review Standards and Standard Period of Time for Process Relating to Procedures under the Act on Launching of Spacecraft, etc. and Control of Spacecraft], November 15, 2017, in English at https://www8.cao.go.jp/space/english/activity/documents/reviewstand.pdf.

Kaisha hō [Companies Act] (Act No. 86 of 2005), Japanese and unofficial English translation at http://www.japaneselawtranslation.go.jp/law/detail/?id=3206&vm=04&re=02 and http://www.japaneselawtranslation.go.jp/law/detail/?re=02&vm=02&al[]=C&ky=%E6%A0%AA%E5%BC%8F&page=18.

Treaty on Principles Governing the Activities of States in the Exploration and Use of Outer Space, including the Moon and Other Celestial Bodies, at https://www.unoosa.org/oosa/en/ourwork/spacelaw/treaties/outerspacetreaty.html.

Uchū kaihatsu jigyodan hō [Law Concerning the National Space Development Agency of Japan] (Law No. 50 of June 23, 1969, as amended), unofficial English translation at https://www.unoosa.org/oosa/en/ourwork/spacelaw/nationalspacelaw/japan/nasda_1969E.html.

Uchū kihon hō [Basic Space Law] (Law No. 43 of 2008), at https://www8.cao.go.jp/space/law/law.html (Japanese) (unofficial English translation by ZeLo in Annex 7).

Select Japanese Government and International Documents

Inter-Agency Space Debris Coordination Committee, IADC Space Debris Mitigation Guidelines (September 2007), at https://orbitaldebris.jsc.nasa.gov/library/iadc_mitigation_guidelines_rev_1_sep07.pdf.

Japan External Trade Organization, Asia Digital Transformation (ADX) Project (Support for Joint Pilot Projects (PoC) between ASEAN and Japan), at https://www5.jetro.go.jp/newsletter/jaa/2020/ADX_projects_lists_English.pdf.

Japan External Trade Organization, Steps of Setting Up Business, at https://www.jetro.go.jp/en/invest/setting_up/.

National Defense Program Guidelines for FY2019 and beyond, in English at https://www.mod.go.jp/e/publ/w_paper/wp2019/pdf/DOJ2019_Special_1.pdf.

Uchū kihon keikaku [Basic Plan on Space Policy] approved by the Cabinet office on June 30, 2020, in Japanese at https://www8.cao.go.jp/space/plan/kaitei_fy02/fy02.pdf (unofficial English translation by ZeLo in Annex 9).

Uchū sangyō bijon 2030 yōyaku [Summary of the Space Industry Vision 2030], May 29, 2017, in Japanese at https://www8.cao.go.jp/space/vision/summary.pdf (unofficial English translation by ZeLo in Annex 8).

United Nations Office for Outer Space Affairs, Space Debris Mitigation Guidelines of the Committee on the Peaceful Uses of Outer Space (2010), at https://www.unoosa.org/pdf/publications/st_space_49E.pdf.

Select Japanese Government Websites

Japan Aerospace Exploration Agency, at https://global.jaxa.jp/.

Japan Cabinet Office, Space Policy, at https://www8.cao.go.jp/space/english/index-e.html.

ANNEXES

Annexes

Annex 1   Act on Launching of Spacecraft, etc. and Control of Spacecraft (Act No. 76 of 2016) (Translation from the Japan Cabinet Office, at https://www8.cao.go.jp/space/english/activity/documents/space_activity_act.pdf; Table of Contents has been omitted)

Annex 2   Review Standards and Standard Period of Time for Process Relating to Procedures under the Act on Launching of Spacecraft, etc. and Control of Spacecraft (Translation from the Japan Cabinet Office, at https://www8.cao.go.jp/space/english/activity/documents/reviewstand.pdf)

Annex 3   Application Manual for Act on Launching of Spacecraft, etc. and Control of Spacecraft (Translation from the Japan Cabinet Office, at https://www8.cao.go.jp/space/english/activity/documents/apmnl.pdf; pagination in Table of Contents from original English version)

Annex 4   Act on Ensuring Appropriate Handling of Satellite Remote Sensing Data (Act No. 77 of November 16, 2016) (Translation from the Japan Cabinet Office, at https://www8.cao.go.jp/space/english/rs/rs_act.pdf; Table of Contents has been omitted)

Annex 5   Guidelines on Measures, etc. Under Act on Ensuring Appropriate Handling of Satellite Remote Sensing Data (Translation from the Japan Cabinet Office, at https://www8.cao.go.jp/space/english/rs/rs_guideline.pdf; pagination in Table of Contents from original English version)

Annex 6   Application Manuals for Act on Ensuring Appropriate Handling of Satellite Remote Sensing Data (Translation from the Japan Cabinet Office, at https://www8.cao.go.jp/space/english/rs/rs_manual.pdf; pagination in Table of Contents from original English version)

Annex 7   Basic Space Law (Law No. 43 of 2008) (Translation by ZeLo)

Annex 8   Summary of Space Industry Vision 2030 - Creating Space Use in the Fourth Industrial Revolution (Translation by ZeLo)

Annex 9 Changes to the Basic Space Plan, June 30, 2020 (Translation by ZeLo except for Preamble; translation of Preamble from the Japan Cabinet Office, at https://www8.cao.go.jp/space/english/basicplan/2020/abstract_0701.pdf; pagination in Japanese version Table of Contents has been omitted)

Annex 1 – Act on Launching of Spacecraft, etc. and Control of Spacecraft (Act No. 76 of 2016)

Chapter I General Provisions

(Purpose)
Article 1
The purpose of this Act is to ensure the accurate and smooth implementation of conventions concerning the development and use of outer space, ensure public safety and protect people affected by relevant damage by establishing a system for permission and license related to the launching of spacecraft, etc. and the control of spacecraft in Japan, as well as a system for compensation for damage caused by a fall, etc. of a spacecraft, etc. in Japan, in accordance with the basic principles of the Space Basic Act (Act No. 43 of 2008) (hereinafter simply referred to as the "basic principles"), thereby to contribute to the improvement of the lives of the citizens as well as the development of the economy and society.

(Definitions)
Article 2
In this Act, the meanings of the terms set forth in the following items are as provided respectively in those items:
i. "conventions on development and use of outer space" collectively means the Treaty on Principles Governing the Activities of States in the Exploration and Use of Outer Space, including the Moon and Other Celestial Bodies (referred to as the "Outer Space Treaty" in Article 22, item (ii)), the Agreement on the Rescue of Astronauts, the Return of Astronauts and the Return of Objects Launched into Outer Space, the Convention on International Liability for Damage Caused by Space Objects and the Convention on Registration of Objects Launched into Outer Space.
ii. "spacecraft" means an artificial object which is used by putting it into Earth orbit or beyond, or placed on a celestial body other than the Earth.
iii. "spacecraft, etc." means spacecraft and a vehicle for launching the spacecraft.
iv. "launch site" means a facility equipped with a function to lift off a launch vehicle.
v. "launching of spacecraft, etc." means loading spacecraft onto a launch vehicle, lifting off and accelerating the launch vehicle until it reaches a certain speed and altitude,

and separating the spacecraft at that point, using a launch site managed and operated by the person or another person.

vi. "spacecraft control facility" means radio equipment equipped with functions to detect signals indicating the position, attitude and condition of a spacecraft transmitted by spacecraft-borne radio equipment (meaning electrical equipment for transmitting or receiving codes using electromagnetic waves, and a computer connected to the equipment via telecommunication lines; the same applies in this item and Article 6, item (ii)) either directly or by receiving it via other radio equipment using electromagnetic waves, or to detect the position of the spacecraft by transmitting signals to the spacecraft either directly or via other radio equipment and then receiving the reflected signals from the spacecraft directly or via other radio equipment, or by other means, and to transmit signals to control the position, attitude and condition of the spacecraft to the spacecraft-borne radio equipment directly or via other radio equipment using electromagnetic waves.

vii. "control of spacecraft" means to detect the position, attitude and condition of a spacecraft and to control these using a spacecraft control facility.

viii. "launch vehicle fall damage" means damage caused to human life or body, or property on the ground surface or water surface, or an aircraft in flight or other flying objects caused by the fall, collision or explosion of a spacecraft, etc. in whole or part, that has not been successfully separated from the launch vehicle after the lift-off, or the launch vehicle after all of the spacecraft have been successfully separated; provided, however, that damages suffered by workers of the person implementing the launching of the spacecraft, etc. or other persons specified by Cabinet Office Order as those having a close business relationship with the person implementing the launching of the spacecraft, etc. in the course of their businesses are excluded.

ix. "launch vehicle fall damage liability insurance contract" means a contract wherein an insurer (limited to a non-life insurance company provided in Article 2, paragraph (4) of the Insurance Business Act (Act No. 105 of 1995) or a foreign non-life insurance company, etc. provided in paragraph (9) of that Article that underwrites a liability insurance policy; the same applies hereinafter) promises to compensate for the loss suffered by a person implementing the launching of a spacecraft, etc. as a result of accruing for a liability to compensate for launch vehicle fall damage (excluding launch vehicle fall damage caused by the fall, collision or explosion of a spacecraft, etc. primarily caused by an act of terrorism or any other events specified by Cabinet Office Order as those for which the calculation of the reasonable amount of insurance premiums would be difficult if the proprietary benefit under the insurance contract is to be provided contingent upon the occurrence of the event (referred to as "specific launch vehicle fall damage" in Article 9, paragraph (2) and

Article 40, paragraph (1))) and making the compensation therefor, and the policyholder promises to pay insurance premiums to the insurer.
x. "launch vehicle fall damage liability indemnification contract" means a contract wherein the government promises to indemnify the person implementing the launching of spacecraft, etc. against losses caused by making a compensation for the launch vehicle fall damage for which the person became liable that cannot be covered by the launch vehicle fall damage liability insurance contract or any other means for compensation for launch vehicle fall damage.
xi. "spacecraft fall damage" means damage to human life or body, or property on the ground surface or water surface, or an aircraft in flight or other flying object caused by the fall or explosion of a spacecraft successfully separated from the launch vehicle; provided, however, that damages suffered by workers of the person implementing the control of the spacecraft or other persons specified by Cabinet Office Order as those having a close business relationship with the person implementing the control of the spacecraft in the course of their business are excluded.

(Consideration for Enforcement of this Act)
Article 3
For the enforcement of this Act, the national government is to pay due consideration for the strengthening of technical competence and international competitiveness of Japanese industries related to the launching of spacecraft, etc., and the control of spacecraft, as a part of policy measures for the promotion of the development and use of outer space by private businesses as provided in Article 16 of the Space Basic Act.

Chapter II Permission, etc. Related to the Launching of Spacecraft, etc.

Section 1 Permission Related to the Launching of Spacecraft, etc.

(Permission)
Article 4
(1) A person who intends to implement the launching of spacecraft, etc. using a launch site located in Japan or onboard a ship or aircraft with Japanese nationality must obtain permission from the Prime Minister for each instance of launching.
(2) A person who intends to obtain permission under the preceding paragraph must submit a written application to the Prime Minister, pursuant to the provisions of Cabinet Office Order, specifying the following information, attaching the documents specified by Cabinet Office Order:
  (i) the person's name and address;
  (ii) the design of the launch vehicle (in the case of a launch vehicle which obtained a type certification under Article 13, paragraph (1), its type certification number; or

in the case of a launch vehicle certified by the government of the foreign state specified by Cabinet Office Order as having a system for the certification for designs of a launch vehicle which is considered to be of an equivalent level to that of Japan for ensuring the safety of the vicinity of the trajectory and launch site of a launch vehicle (hereinafter referred to as a "foreign certification"), to the effect that the foreign certification has been obtained);

(iii) the location of the launch site (in the case of a launch site onboard a ship or aircraft, the name or registration code of the ship or aircraft), as well as its design and equipment (in the case of a launch site which has obtained a compliance certification under Article 16, paragraph (1), the compliance certification number);

(iv) a plan setting forth the methods for the launching of the spacecraft, etc., including the schedule for launching the spacecraft, etc., the trajectory of the launch vehicle, and the method of ensuring the safety of the vicinity of the trajectory and launch site (hereinafter referred to as a "launch plan");

(v) the number of spacecraft to be placed on the launch vehicle, as well as the purposes and methods of use of the respective spacecraft; and

(vi) other matters specified by Cabinet Office Order.

(Grounds for Disqualification)
Article 5
A person who falls under any of the following items may not obtain permission under paragraph (1) of the preceding Article:

i. a person who has violated the provisions of this Act or an order based on this Act or the laws and regulations of a foreign country (meaning countries or regions outside Japan; the same applies hereinafter) equivalent thereto, and has been sentenced to a fine or severer punishment (including a punishment under the laws and regulations of a foreign country equivalent thereto), and for whom three years have not elapsed since the date on which execution of the sentence was completed or since the date on which that person ceased to be subject to the execution of the sentence;

ii. a person whose permission has been rescinded pursuant to Article 12, and for whom three years have not elapsed since the date of that rescission;

iii. an adult ward or a person who is treated in the same manner under the laws and regulations of a foreign country;

iv. a corporation whose officers engaged in the business thereof or employees specified by Cabinet Office Order fall under any of the preceding three items; and

v. an individual whose employees specified by Cabinet Office Order fall under any of items (i) through (iii).

(Requirements for Permission)

Article 6

The Prime Minister must not grant the permission under Article 4, paragraph (1), unless the Prime Minister determines that the application for the permission under that paragraph meets all of the following requirem

(i) the design of the launch vehicle complies with the standard specified by Cabinet Office Order as the safety standard concerning a launch vehicle for ensuring the safety of the vicinity of the trajectory and launch site of the launch vehicle (hereinafter referred to as a "launch vehicle safety standard"), or the design has obtained a type certification under Article 13, paragraph (1) or a foreign certification;

(ii) the launch site is equipped with radio equipment set forth in the following items (a) and (b) or otherwise complies with the standard specified by Cabinet Office Order according to the type of launch vehicle as the safety standard concerning a launch site for ensuring the safety of the vicinity of the trajectory and launch site of the launch vehicle (hereinafter referred to as a "type-specific site safety standard"), or the launch site has obtained a compliance certification under Article 16, paragraph (1):

  (a) radio equipment equipped with a function to detect signals indicating the position, attitude and condition of a launch vehicle transmitted by radio equipment onboard the launch vehicle either directly or by receiving it via other radio equipment using electromagnetic waves, or to detect the position of the launch vehicle by transmitting signals to the launch vehicle either directly or via other radio equipment and then receiving the reflected signals either directly or via other radio equipment;

  (b) radio equipment equipped with a function of transmitting signals necessary for the destruction of a launch vehicle or any other measures to terminate the flight in the case of the deviation of the launch vehicle from the scheduled trajectory or any other extraordinary circumstances (referred to as "flight termination measures" in the following item and Article 16, paragraph (2), item (iv)) to the radio equipment onboard the launch vehicle either directly or via other radio equipment using electromagnetic waves;

(iii) the launch plan sets forth the flight termination measures or other means of ensuring the safety of the vicinity of the trajectory and launch site of the launch vehicle, that the details of the plan are appropriate in light of ensuring public safety, and that the applicant has a sufficient ability to execute the launch plan; and

(iv) the purposes and methods of use of the spacecraft loaded on the launch vehicle are in compliance with the basic principles, and are not likely to cause any adverse effect on the accurate and smooth implementation of the conventions on development and use of outer space and ensuring public safety.

(Permission, etc. Related to Change)
Article 7
(1) When a person who obtained the permission under Article 4, paragraph (1) (hereinafter referred to as a "launch operator") intends to change any matter set forth in items (ii) through (v) of paragraph (2) of that Article (including when a change has been made to the launch vehicle safety standard and the design of the launch vehicle for which the permission was granted no longer satisfies the launch vehicle safety standard, and when a change has been made to the type-specific site safety standard and the launch site for which the permission was granted no longer satisfies the type-specific site safety standard), the person must obtain permission from the Prime Minister pursuant to the provisions of Cabinet Office Order; provided, however, that this does not apply to minor changes specified by Cabinet Office Order.
(2) When there has been a change to any of the items set forth in Article 4, paragraph (2), item (i) or (vi), or any minor change specified by Cabinet Office Order as referred to in the proviso to the preceding paragraph, a launch operator must make a notification to the Prime Minister to that effect without delay.
(3) The provisions of the preceding Article apply mutatis mutandis to the permission under paragraph (1).

(Obligation of Conformity of Designs, etc.)
Article 8
(1) When a launch operator implements the launching of spacecraft, etc., the launch operator must ensure that the relevant launch vehicle pertaining to the launching of the spacecraft, etc. complies with the design for which the permission under Article 4, paragraph (1) was granted.
(2) When a launch operator implements the launching of spacecraft, etc., the launch operator must comply with the launch plan for which the permission under Article 4, paragraph (1) was granted, except in the case of a disaster or any unavoidable situation.

(Obligation to Provide Security Measures for Compensation for Damages)
Article 9
(1) A launch operator must not implement the launching of spacecraft, etc. for which the permission under Article 4, paragraph (1) was granted, unless it has taken security measures for compensation for damages.
(2) The "security measures for compensation for damages" provided in the preceding paragraph means the execution of a launch vehicle fall damage liability insurance contract and a launch vehicle fall damage liability indemnification contract (limited

to a contract pertaining to specific launch vehicle fall damages) or deposit with official depository as approved by the Prime Minister, which enables an amount specified by Cabinet Office Order as the appropriate amount in light of the protection of victims of the launch vehicle fall damage, considering the design of the launch vehicle, the location of the launch site or other situations (referred to as an "amount covered by compensation measures" in Article 40, paragraphs (1) and (2)) to be appropriated to the compensation for launch vehicle fall damage, or a measure equivalent to these as approved by the Prime Minister (referred to as an "equivalent measure" in Article 40, paragraph (2)).

(Succession)
Article 10
(1) When a launch operator transfers the business with respect to the launching of spacecraft, etc. permitted under Article 4, paragraph (1), if the transferrer and the transferee have obtained authorization on that transfer from the Prime Minister in advance pursuant to the provisions of Cabinet Office Order, the transferee succeeds to the status of the launch operator under the provisions of this Act.
(2) When a corporation that is a launch operator is to be extinguished by a merger and if the merger has been authorized in advance by the Prime Minister pursuant to the provisions of Cabinet Office Order, a corporation surviving the merger or a corporation established as a result of the merger succeeds to the status of the launch operator under the provisions of this Act.
(3) When a corporation that is a launch operator has the business with respect to the launching of spacecraft, etc. permitted under Article 4, paragraph (1) succeeded to by corporate split, and that corporate split has been authorized in advance by the Prime Minister pursuant to the provisions of Cabinet Office Order, a corporation which succeeded to the business as a result of the corporate split succeeds to the status of the launch operator under the provisions of this Act.
(4) Articles 5 and 6 (limited to the part concerning item (iii) (limited to the part concerning the ability to execute the launch plan)) apply mutatis mutandis to the authorization under the preceding three paragraphs.
(5) When a launch operator transfers its business with respect to the launching of spacecraft, etc. permitted under Article 4, paragraph (1), or when a corporation that is a launch operator is extinguished by a merger or has the business succeeded to through a corporate split, if a disposition to refuse the authorization under paragraphs (1) through (3) is rendered (if an application for the authorization is not submitted, when the transfer of business, merger or corporate split takes place), the permission under Article 4, paragraph (1) ceases to be effective.

(Expiration of Permission Due to Death or Other Reasons)
Article 11

In addition to the case provided in paragraph (5) of the preceding Article, if a launch operator falls under any of the following items, the permission under Article 4, paragraph (1) ceases to be effective; in this case, the persons respectively specified in these items must notify the Prime Minister to that effect within thirty days from the date when they fall under the relevant item:
i. when the person has deceased: the heir;
ii. when the corporation has dissolved due to an order commencing bankruptcy proceedings: its bankruptcy trustee;
iii. when the corporation has dissolved for reasons other than a merger or an order commencing bankruptcy proceedings: its liquidator; and
iv. when the launching of spacecraft, etc. has been completed: an individual who was formerly the launch operator, or an officer representing the corporation which was formerly the launch operator.

(Rescission of Permission)
Article 12

If a launch operator falls under any of the following items, the Prime Minister may rescind the permission under Article 4, paragraph (1):
i. the launch operator has obtained the permission under Article 4, paragraph (1) or Article 7, paragraph (1), or the authorization under Article 10, paragraphs (1) through (3), by deception or other wrongful means;
ii. the launch operator has come to fall under any of the items of Article 5, item (i) or (iii) through (v);
iii. the design of the launch vehicle to be used for the launch operator's launching of spacecraft, etc. no longer complies with the launch vehicle safety standard;
iv. the launch site to be used for the launch operator's launching of spacecraft, etc. no longer complies with the type-specific site safety standard;
v. the launch operator has changed a matter for which permission must be obtained pursuant to Article 7, paragraph (1) without obtaining the permission under that paragraph;
vi. the Prime Minister determines the launch operator to have violated the provisions of Article 8; or
vii. the launch operator has violated the conditions attached to the permission under Article 4, paragraph (1) or Article 7, paragraph (1), or the authorization under Article 10, paragraphs (1) through (3), pursuant to Article 34, paragraph (1).

Section 2 Type Certification for Launch Vehicle

(Type Certification)
Article 13
(1) The Prime Minister grants a type certification for the design of a launch vehicle upon application.
(2) A person who intends to obtain a type certification referred to in the preceding paragraph must submit a written application to the Prime Minister, pursuant to the provisions of Cabinet Office Order, specifying the following information, attaching a document certifying that the design of the launch vehicle complies with the launch vehicle safety standard and other documents specified by Cabinet Office Order.
  (i) the person's name and address;
  (ii) the design of the launch vehicle; and
  (iii) other matters specified by Cabinet Office Order.
(3) When the application under paragraph (1) is submitted, the Prime Minister must grant the type certification under that paragraph if the Prime Minister determines the design of the launch vehicle to which the application pertains to be in compliance with the launch vehicle safety standard.
(4) The type certification under paragraph (1) is to be granted by delivering a type certificate specifying the type certification number to the applicant.

(Change of Design, etc.)
Article 14
(1) When a person who obtained a type certification under paragraph (1) of the preceding Article intends to change any matter set forth in item (ii) of paragraph (2) of that Article (including when a change has been made to the launch vehicle safety standard and the design of the launch vehicle for which the type certification was granted no longer satisfies the launch vehicle safety standard), the person must obtain authorization from the Prime Minister pursuant to the provisions of Cabinet Office Order; provided, however, that this does not apply to minor changes specified by Cabinet Office Order.
(2) When there has been a change to any of the items set forth in Article 13, paragraph (2), item (i) or (iii), or any minor change specified by Cabinet Office Order as referred to in the proviso to the preceding paragraph, the person who obtained a type certification under Article 13, paragraph (1) must make a notification to that effect to the Prime Minister without delay.
(3) The provisions of paragraph (3) of the preceding Article apply mutatis mutandis to the approval under paragraph (1).

(Rescission of Type Certification)
Article 15
(1) When a person who obtained the type certification under Article 13, paragraph (1) falls under any of the following items, the Prime Minister may rescind the type certification:
　(i) the design of the launch vehicle no longer complies with the launch vehicle safety standard; or
　(ii) the person has violated an order under Article 33, paragraph (1).
(2) When the type certification is rescinded pursuant to the preceding paragraph, the person who obtained the type certification under Article 13, paragraph (1) must return the type certificate to the Prime Minister without delay.

Section 3 Compliance Certification for Launch Site

(Compliance Certification)
Article 16
(1) The Prime Minister grants a compliance certification with respect to a launch site located in Japan or onboard a ship or aircraft with Japanese nationality, according to the type of launch vehicle related to the launching of spacecraft, etc. to be implemented using the relevant launch site (limited to a type of launch vehicle whose design has obtained a type certification under Article 13, paragraph (1) or foreign certification), upon application.
(2) A person who intends to obtain a compliance certification under the preceding paragraph must submit a written application to the Prime Minister, pursuant to the provisions of Cabinet Office Order, specifying the following information, attaching a document certifying that the launch site complies with the type-specific site safety standard and other documents specified by Cabinet Office Order.
　(i) the person's name and address;
　(ii) the location of launch site (in the case of a launch site onboard the ship or aircraft, the name or registration code of the ship or aircraft), as well as its design and equipment;
　(iii) the type certification number pertaining to the type certification under Article 13, paragraph (1) or the fact that a foreign certification has been obtained for the type of launch vehicle;
　(iv) the flight termination measures or other means of ensuring the safety of the vicinity of the trajectory and launch site of the launch vehicle; and
　(v) other matters specified by Cabinet Office Order.
(3) When the application under paragraph (1) is submitted, the Prime Minister must grant the compliance certification under that paragraph if the Prime Minister

determines the launch site to which the application pertains to be in compliance with the type-specific site safety standard.
(4) The compliance certification under paragraph (1) is to be granted by delivering a launch site certificate specifying the compliance certification number to the applicant.

(Change of Location, etc. of Launch Site)
Article 17
(1) When a person who obtained a compliance certification under paragraph (1) of the preceding Article intends to change any matter set forth in item (ii) or (iv) of paragraph (2) of that Article (including the case when a change has been made to the type-specific site safety standard and the launch site for which the compliance certification was granted no longer satisfies the type-specific site safety standard), the person must obtain authorization from the Prime Minister pursuant to the provisions of Cabinet Office Order; provided, however, that this does not apply to minor changes specified by Cabinet Office Order.
(2) When there has been a change to any of the matters set forth in Article 16, paragraph (2), item (i) or (v), or any minor changes specified by Cabinet Office Order as referred to in the proviso to the preceding paragraph, the person who obtained a compliance certification under Article 16, paragraph (1) must make a notification to that effect to the Prime Minister without delay.
(3) The provisions of paragraph (3) of the preceding Article apply mutatis mutandis to the approval under paragraph (1).

(Rescission of Compliance Certification)
Article 18
(1) When a person who obtained the compliance certification under Article 16, paragraph (1) falls under any of the following items, the Prime Minister may rescind the compliance certification.
　(i) the launch site no longer complies with the type-specific site safety standard; or
　(ii) the person has violated an order under Article 33, paragraph (2).
(2) When the compliance certification is rescinded pursuant to the preceding paragraph, the person who obtained the compliance certification under Article 16, paragraph (1) must return the launch site certificate to the Prime Minister without delay.

Section 4 Special Provisions on Application Procedures for Japan Aerospace Exploration Agency, National Research and Development Agency

Article 19

(1) If the Japan Aerospace Exploration Agency, National Research and Development Agency (hereinafter referred to as "JAXA") submits an application for the type certification under Article 13, paragraph (1) for its design of a launch vehicle, notwithstanding the provisions of paragraph (2) of that Article, procedures without requiring a part of information or attachments relating to the application or other simplified procedures specified by Cabinet Office Order may be applied.

(2) If JAXA submits an application for the compliance certification under Article 16, paragraph (1) for the launch site managed and operated by it, notwithstanding the provisions of paragraph (2) of that Article, procedures without requiring a part of information or attachments relating to the application or other simplified procedures specified by Cabinet Office Order may be applied.

Chapter III License, etc. Relating to the Control of Spacecraft

(License)

Article 20

(1) A person who intends to implement the control of a spacecraft using a spacecraft control facility located in Japan must obtain a license from the Prime Minister for each of the spacecraft.

(2) A person who intends to obtain the license under the preceding paragraph must submit a written application to the Prime Minister, pursuant to the provisions of Cabinet Office Order, specifying the following information, attaching the documents specified by Cabinet Office Order:
   (i) the person's name and address;
   (ii) the location of the spacecraft control facility;
   (iii) if the spacecraft is to be used by putting it into Earth orbit, the orbit of the spacecraft;
   (iv) the purposes and methods of use of the spacecraft;
   (v) the configuration of the spacecraft;
   (vi) the details of the measures to be taken upon the termination of the control of the spacecraft (hereinafter referred to as "termination measures");
   (vii) beyond what is set forth in the preceding items, a plan setting forth the methods of the control of the spacecraft (hereinafter referred to as a "control plan");
   (viii) if the applicant is an individual, the name and address of the person who will implement the control of the spacecraft upon the death of the applicant on

behalf of the applicant (hereinafter referred to as the "representative in case of death"); and
(ix) other matters specified by Cabinet Office Order.

(Grounds for Disqualification)
Article 21
A person who falls under any of the following items may not obtain a license under paragraph (1) of the preceding Article:
i. a person who has violated the provisions of this Act or orders based on this Act or the laws and regulations of a foreign country equivalent thereto, and has been sentenced to a fine or severer punishment (including a punishment under the laws and regulations of a foreign country equivalent thereto), and for whom three years have not elapsed since the date on which execution of the sentence was completed or since the date on which that person ceased to be subject to the execution of the sentence;
ii. a person whose license has been rescinded pursuant to Article 30, paragraph (1), and for whom three years have not elapsed since the date of that rescission;
iii. an adult ward or a person who is treated in the same manner under the laws and regulations of a foreign country;
iv. a corporation whose officers engaged in the business thereof or employees specified by Cabinet Office Order fall under any of the preceding three items;
v. an individual whose employees specified by Cabinet Office Order fall under any of items (i) through (iii); and
vi. an individual whose representative in case of death falls under any of the preceding items.

(Requirements for License)
Article 22
The Prime Minister must not grant the license under Article 20, paragraph (1), unless the Prime Minister determines that the application for the license under that paragraph meets all of the following requirements:
(i) the purposes and methods of use of the spacecraft are in compliance with the basic principles, and are not likely to cause any adverse effect on the accurate and smooth implementation of the conventions on development and use of outer space and the ensuring of public safety;
(ii) the configuration of the spacecraft, a mechanism for the prevention of dispersion of its components and parts has been implemented, or that the configuration of the spacecraft otherwise complies with the standard specified by Cabinet Office Order as being those that are not likely to cause an adverse effect on the prevention of the harmful contamination of outer space including the Moon and other celestial bodies and the prevention of potentially harmful interference with activities of other

countries in the peaceful exploration and use of outer space provided in Article 9 of the Outer Space Treaty (referred to as "harmful contamination of outer space, etc." in the following item and item (iv)(d)) and the ensuring of public safety;
(iii) the control plan requires the implementation of measures to avoid collision with other spacecraft or other measures specified by Cabinet Office Order which are necessary for the prevention of harmful contamination of outer space, etc. as well as termination measures, and the applicant (in the case of an individual, including the representative in case of death) has sufficient ability to execute the control plan; and
(iv) the details of the termination measures satisfy any of the conditions specified in the following items (a) through (d):
   (a) to control the position, attitude and condition of the spacecraft to descend its altitude and burn in the atmosphere (including the retrieval of a part of the components by guiding them to fall on the ground surface or water surface without burning), while ensuring the safety of the vicinity of the trajectory of the spacecraft and an expected point of landing or water landing of the part of the components;
   (b) to control the position, attitude and condition of the spacecraft to increase its altitude and put it into an Earth orbit from which its altitude will not decrease as time passes, without any risk of causing any adverse effect on the control of other spacecraft;
   (c) to control the position, attitude and condition of the spacecraft to put it into the orbit around a celestial body other than the Earth or guide it to fall to the celestial body, without any risk of significantly deteriorating the environment of the celestial body; or
   (d) if it is impossible to implement the measures set forth in items (a) through (c), to suspend the control of the spacecraft after taking measures to prevent the unexpected activation and explosion or other measures that are necessary for the prevention of harmful contamination of outer space, etc. as specified by Cabinet Office Order, and notifying the Prime Minister of the position, attitude and condition of the spacecraft.

(Permission, etc. Related to Change)
Article 23
(1) When a person who obtained the license under Article 20, paragraph (1) (hereinafter referred to as a "spacecraft control operator") intends to change any matter set forth in items (iv) through (viii) of paragraph (2) of that Article, the person must obtain a license from the Prime Minister pursuant to the provisions of Cabinet Office Order; provided, however, that this does not apply to minor changes specified by Cabinet Office Order.

(2) When there has been a change to any of the matters set forth in Article 20, paragraph (2), items (i) through (iii) or item (ix), or any minor changes specified by Cabinet Office Order as referred to in the proviso to the preceding paragraph, a spacecraft control operator must make a notification to the Prime Minister to that effect without delay.

(3) The provisions of the preceding Article apply mutatis mutandis to the license under paragraph (1).

(Compliance with Control Plan)
Article 24
When a spacecraft control operator implements the control of a spacecraft, it must comply with the control plan for which the license under Article 20, paragraph (1) was granted, except in the case of disasters or other unavoidable situations.

(Measures in Case of Accident)
Article 25
If a spacecraft control operator becomes unable to implement the control of the spacecraft without taking any termination measures pertaining to the license under Article 20, paragraph (1) due to the collision of a spacecraft pertaining to the license under that paragraph with another object or any other accidents, and if there is no prospect of recovery, the spacecraft control operator must promptly make a notification to the Prime Minister to that effect, the circumstances of the accident and the matters specified by Cabinet Office Order which assist with the identification of the position of the spacecraft after the occurrence of the accident, pursuant to the provisions of Cabinet Office Order. In this case, the license under that paragraph ceases to be effective.

(Succession)
Article 26
(1) When a spacecraft control operator intends to transfer its business with respect to the control of spacecraft licensed under Article 20, paragraph (1) to a person who intends to implement the control of spacecraft using a spacecraft control facility located in Japan, if the transferrer and transferee have obtained authorization for that transfer and acquisition from the Prime Minister in advance pursuant to the provisions of Cabinet Office Order, the transferee succeeds to the status of the spacecraft control operator under the provisions of this Act.
(2) When a spacecraft control operator intends to transfer the business with respect to the control of spacecraft licensed under Article 20, paragraph (1) to a person who intends to implement the control of spacecraft without using a spacecraft control facility located in Japan, the spacecraft control operator must make a notification to the

Prime Minister in advance to that effect pursuant to the provisions of Cabinet Office Order.

(3) When a corporation that is a spacecraft control operator is to be extinguished by merger and that merger has been authorized in advance by the Prime Minister pursuant to the provisions of Cabinet Office Order, a corporation surviving the merger or a corporation established as a result of the merger succeeds to the status of the spacecraft control operator under this Act.

(4) When a corporation that is a spacecraft control operator has the business with respect to the control of spacecraft licensed under Article 20, paragraph (1) succeeded to by corporate split, and that corporate split has been authorized in advance by the Prime Minister pursuant to the provisions of Cabinet Office Order, a corporation which succeeded to the business as a result of the corporate split succeeds to the status of the spacecraft control operator under the provisions of this Act.

(5) Articles 21 and 22 (limited to the part concerning item (iii) (limited to the part concerning ability to implement the control plan)) apply mutatis mutandis to the authorization under paragraph (1) and the preceding two paragraphs.

(6) When a spacecraft control operator transfers its business with respect to the control of spacecraft licensed under Article 20, paragraph (1), or when a corporation that is a spacecraft control operator is extinguished by a merger or has the business succeeded to through a corporate split, if a disposition to refuse the authorization under paragraphs (1), (3) or (4) is rendered (if an application for the authorization is not submitted, when the transfer of business, merger or corporate split takes place), the license under Article 20, paragraph (1) ceases to be effective, and its transferee (except for the transferee with respect to business transfer provided in paragraph (2)), a corporation surviving the merger or a corporation established as a result of the merger, or a corporation which succeeded to that business by the corporate split, must take termination measures for which the license under paragraph (1) of that Article was granted, within 120 days from the day of the disposition (if an application for the authorization is not submitted, the day of the transfer of business, merger or corporate split). In this case, until the termination measures are completed (in the case provided in the preceding Article, until the notification under that Article is submitted),the provisions of Article 24, the first sentence of the preceding Article, Article 31, Article 32 and Article 33, paragraph (3) (including penal provisions relating to these provisions) apply by deeming these persons as spacecraft control operators.

(Notification and Other Procedures Related to Death)
Article 27

(1) When a spacecraft control operator has deceased, the heir must make a notification to the Prime Minister to that effect without delay.

(2) When a spacecraft control operator has deceased, the license under Article 20, paragraph (1) ceases to be effective, and the representative in case of death must take termination measures pertaining to the license under Article 20, paragraph (1), within 120 days from the date of death, unless the transfer of business with respect to the control of the spacecraft has been authorized under paragraph (1) of the preceding Article. In this case, until the implementation of the business transfer or the completion of the termination measures (in the case provided in Article 25, until the notification provided therein is submitted), provisions of Article 24, the first sentence of Article 25, Article 26, paragraphs (1) and (5), Article 31, Article 32 and Article 33, paragraph (3) (including penal provisions relating to these provisions) apply by deeming the representative in case of death as a spacecraft control operator.

(Termination Measures)
Article 28
(1) When a spacecraft control operator intends to terminate the control of a spacecraft pursuant to the control plan pertaining to the license under Article 20, paragraph (1), the spacecraft control operator must make a notification to the Prime Minister to that effect in advance and take the termination measures pertaining to the license under that paragraph, pursuant to the provisions of Cabinet Office Order.
(2) When the termination measures are taken pursuant to the preceding paragraph, the license under Article 20, paragraph (1) ceases to be effective.

(Notification and Other Procedures Related to Dissolution)
Article 29
(1) When a corporation that is a spacecraft control operator dissolves for reasons other than a merger, a liquidator or a bankruptcy trustee must make a notification to the Prime Minister to that effect without delay.
(2) When a corporation that is a spacecraft control operator dissolves for reasons other than a merger, the license under Article 20, paragraph (1) ceases to be effective, and the corporation under liquidation (meaning a corporation under liquidation or special liquidation, or a corporation against which bankruptcy proceedings have been commenced; hereinafter the same applies in this paragraph) must take termination measures pertaining to the license under Article 20, paragraph (1) within 120 days from the date of dissolution, unless the transfer of business with respect to the control of spacecraft has been authorized under Article 26, paragraph (1). In this case, until the implementation of the business transfer or the completion of the termination measures (in the case provided in Article 25, until the notification provided therein is submitted), provisions of Article 24, the first sentence of Article 25, Article 26, paragraphs (1) and (5), Article 31, Article 32 and Article 33,

paragraph (3) (including penal provision relating to these provisions) apply by deeming the corporation under liquidation as a spacecraft control operator.

(Rescission and Other Procedures Related to License)
Article 30
(1) If a spacecraft control operator falls under any of the following items, the Prime Minister may rescind the license under Article 20, paragraph (1):
   (i) the spacecraft control operator has obtained the license under Article 20, paragraph (1) or Article 23, paragraph (1), or the authorization under Article 26, paragraph (1), (3) or (4) by deception or other wrongful means;
   (ii) the spacecraft control operator has come to fall under any of the items of Article 21, item (i) or items (iii) through (iv);
   (iii) the spacecraft control operator has changed a matter for which permission must be obtained pursuant to Article 23, paragraph (1) without obtaining the permission under that paragraph;
   (iv) the spacecraft control operator has violated an order under Article 33, paragraph (3); or
   (v) the spacecraft control operator has violated the conditions attached to the license under Article 20, paragraph (1) or Article 23, paragraph (1), or the authorization under Article 26, paragraph (1), (3) or (4), pursuant to Article 34, paragraph (1).
(2) If the license of a spacecraft control operator under Article 20, paragraph (1) is rescinded pursuant to the preceding paragraph, the spacecraft control operator must take termination measures pertaining to the license under Article 20, paragraph (1) within 120 days from the date of the rescission unless transfer of business with respect to the control of spacecraft has been authorized under Article 26, paragraph (1). In this case, until the implementation of the business transfer or the completion of the termination measures (in the case provided in Article 25, until the notification provided therein is submitted), provisions of Article 24, the first sentence of Article 25, Article 26, paragraphs (1) and (5), Article 31, Article 32 and Article 33, paragraph (3) (including penal regulations relating to these provisions) apply by deeming the person as a spacecraft control operator.

Chapter IV Supervision by the Prime Minister

(On-site Inspection, etc.)
Article 31
(1) The Prime Minister may, to the extent necessary for the enforcement of this Act, request a launch operator, a person who obtained a type certification under Article 13, paragraph (1), a person who obtained a compliance certification under

Article 16, paragraph (1) or a spacecraft control operator to provide necessary reports or have Cabinet Office officials enter its office or any other places of business to inspect books, documents or other items or to question relevant persons.

(2) An official who conducts an on-site inspection under the preceding paragraph must carry an identification card and present this at the request of relevant persons.

(3) The authority to conduct an on-site inspection under paragraph (1) must not be construed as being granted for a criminal investigation.

(Guidance, etc.)
Article 32
The Prime Minister may give necessary guidance, advice and recommendations to launch operators, persons who obtained a type certification under Article 13, paragraph (1), persons who obtained a compliance certification under Article 16, paragraph (1) or spacecraft control operators in order to ensure the accurate and smooth implementation of the conventions on development and use of outer space and public safety, in accordance with the basic principles.

(Correction Order)
Article 33
(1) When the Prime Minister determines that design of a launch vehicle which obtained a type certification under Article 13, paragraph (1) fails or is likely to fail to comply with the launch vehicle safety standard, the Prime Minister may order the person who obtained the type certification to make necessary changes to the design to conform to the launch vehicle safety standard or preclude the possibility of failing to comply with the launch vehicle safety standard.

(2) When the Prime Minister determines that a launch site which obtained the compliance certification under Article 16, paragraph (1) fails or is likely to fail to comply with the type-specific site safety standard, the Prime Minister may order the person who obtained the compliance certification to take measures to the extent necessary for ensuring compliance with the type-specific site safety standard or precluding the possibility of failing to comply with the type-specific site safety standard.

(3) When the Prime Minister determines that a spacecraft control operator is in violation of the provisions of Article 24, the Prime Minister may order the spacecraft control operator to take necessary measures for correcting that violation.

(Conditions of the Permission, etc.)
Article 34
(1) Conditions may be imposed on the permission under Article 4, paragraph (1), Article 7, paragraph (1), the license under Article 20, paragraph (1), the permission

under Article 23, paragraph (1), or the authorization under Article 10, paragraphs (1) through (3) or Article 26, paragraph (1), (3) or (4), and changes may be made to the conditions.
(2) The conditions referred to in the preceding paragraph must be limited to the minimum extent required for ensuring the proper implementation of the matters pertaining to the permissions, licenses or authorizations and must not result in imposing unreasonable duties on the person obtaining the permission, license or authorization.

Chapter V Compensation for Launch Vehicle Fall Damage

Section 1 Liability for Compensation for Launch Vehicle Fall Damage

(Strict Liability)
Article 35
A person who implements the launching of spacecraft, etc. using a launch site located in Japan or onboard a ship or aircraft with Japanese nationality is liable to compensate for any launch vehicle fall damage caused by the person in relation to the launching of the spacecraft, etc.

(Channeling of Liability)
Article 36
(1) In the case referred to in the preceding Article, no person other than the person implementing the launching of the spacecraft, etc., who is to be held liable to compensate for the damage pursuant to that Article, is liable to compensate for the damage.
(2) The provisions of the Product Liability Act (Act No. 85 of 1994) do not apply to launch vehicle fall damage.
(3) The provisions of paragraph (1) must not be construed to preclude the application of the Act on Compensation for Nuclear Damage (Act No. 147 of 1961).

(Consideration for Determination of Compensation for Damages)
Article 37
Notwithstanding the provisions of the preceding two Articles, if a natural disaster or other force majeure also contributed to the occurrence of the launch vehicle fall damage, the court may also take those into consideration to determine the liabilities and amount of compensation for damages.

(Right to Reimbursement)
Article 38

(1) In the case referred to in Article 35, if any other person is to be held liable for the cause of the damage, the person who made a compensation for the damage pursuant to that Article has a right to reimbursement from that person; provided, however, that if the person who is to be held liable was the supplier of materials or other goods or services used for the launching of the spacecraft, etc. (excluding the manager and operator of the launch site used for the launching of the spacecraft, etc.), the person who compensated for the damage has a right to reimbursement from the supplier only if the damage was caused by the intentional conduct of the supplier or its workers.

(2) The provisions of the preceding paragraph do not preclude the execution of a written special agreement on the right to reimbursement.

Section 2 Launch Vehicle Fall Damage Liability Insurance Contract

Article 39

(1) Victims of launch vehicle fall damage are entitled to receive the payment from insurance proceeds under the launch vehicle fall damage liability insurance contract, in relation to their respective claims for damages, prior to other creditors.

(2) In relation to the amount of damages payable to the victims of launch vehicle fall damage, the insured may make an insurance claim against the insurer only to the extent of the amount paid by the insured itself or the amount approved by the victim.

(3) An insurance claim under a launch vehicle fall damage liability insurance contract may not be assigned, provided for security or attached; provided, however, that this does not apply if a victim of the launch vehicle fall damage attaches the insurance claim in relation to the victim's own claim for damages.

Section 3 Launch Vehicle Fall Damage Liability Indemnification Contract

(Launch Vehicle Fall Damage Liability Indemnification Contract)
Article 40

(1) The government may execute a launch vehicle fall damage liability indemnification contract with a launch operator whereby the government promises to indemnify the launch operator for the losses arising from making the compensation for damages in the case of accruing a liability to compensate for the specific launch vehicle fall damages, to the extent not exceeding an amount equivalent to the amount covered by compensation measures in relation to the security measures for compensation for damages provided in Article 9, paragraph (2) to be allocated to the compensation for

the specific launch vehicle fall damages (hereinafter simply referred to as "security measures for compensation for damages").

(2) Beyond what is provided in the preceding paragraph, the government may execute a launch vehicle fall damage liability indemnification contract with a launch operator whereby the government promises to indemnify the launch operator for the loss caused by making compensation for launch vehicle fall damage in the case of accruing a liability to compensate for the launch vehicle fall damage, that is not covered by a launch vehicle fall damage liability insurance contract, a launch vehicle fall damage liability indemnification contract under that paragraph or any other measures for compensation for launch vehicle fall damage, in an amount not exceeding the amount specified by Cabinet Office Order which is appropriate from the standpoint of the reinforcement of international competitiveness of Japanese industries related to the launching of spacecraft, etc., after deduction of an amount equivalent to the amount covered by compensation measures relating to the security measures for compensation for damages to be allocated to the launch operator's compensation for launch vehicle fall damage (or, when equivalent measures have been taken in relation to the launch vehicle fall damage, an amount equivalent to the amount covered by compensation measures or the amount covered by the equivalent measures that can be allocated to the compensation for the launch vehicle fall damage, whichever is higher).

(3) The provisions of the preceding Article apply mutatis mutandis to the indemnification payment under a launch vehicle fall damage liability indemnification contract.

(Term of Launch Vehicle Fall Damage Liability Indemnification Contract)
Article 41
The term of a launch vehicle fall damage liability indemnification contract commences from the time of execution thereof and terminates upon the completion of the launching of the spacecraft, etc. pertaining to the relevant launch vehicle fall damage liability indemnification contract.

(Indemnification Payment)
Article 42
The amount to be indemnified by the government in relation to a launch vehicle fall damage liability indemnification contract must not exceed the contract amount of the launch vehicle fall damage liability indemnification contract relating to the losses arising from the compensation by the launch operator for launch vehicle fall damage caused by the launching of spacecraft, etc. during the term of the launch vehicle fall damage liability indemnification contract.

(Limitation on Execution of Launch Vehicle Fall Damage Liability Indemnification Contract)
Article 43
The government is to execute a launch vehicle fall damage liability indemnification contract within the limit so that the total contract amount of launch vehicle fall damage liability indemnification contracts executed in a single fiscal year will not exceed the amount approved by the Diet for that fiscal year.

(Prescription)
Article 44
The right to receive the indemnification payment is extinguished by prescription if three years elapse from the time when it became possible to exercise the right.

(Subrogation)
Article 45
When the government made an indemnification under a launch vehicle fall damage liability indemnification contract, and if the launch operator that is the counterparty to the launch vehicle fall damage liability indemnification contract has a right to reimbursement from a third party, the government acquires the right to reimbursement up to the amount specified in the following, whichever is lower:
i. the amount indemnified by the government; or
ii. the amount of the right to reimbursement.

(Redemption of Indemnification Payment)
Article 46
When the government made an indemnification payment under a launch vehicle fall damage liability indemnification contract, and if the launch operator that is the counterparty to the launch vehicle fall damage liability indemnification contract falls under any of the following items, the government is to order the launch operator to redeem the amount of indemnification pursuant to the provisions of Cabinet Order:
(i) the launch operator implemented the launching of spacecraft, etc. in violation of the provisions of Article 8; or
(ii) the launch operator fell under Article 12, item (i) or (v) when it implemented the launching of spacecraft, etc.

(Administration of Businesses)
Article 47
(1) The business of the government provided in this Section is to be administered by the Prime Minister.

(2) When the Prime Minister intends to execute a launch vehicle fall damage liability indemnification contract, the Prime Minister must consult with the Minister of Finance in advance.

(Entrustment of Business)
Article 48
(1) The government may entrust a part of its business relating to a launch vehicle fall damage liability indemnification contract to insurers, pursuant to the provisions of Cabinet Order.
(2) When the Prime Minister made an entrustment of businesses under the preceding paragraph, the Prime Minister must make a public notice of the name of the entrusted person and other matters specified by Cabinet Office Order.

Section 4 Deposit with Official Depository

(Deposit as Security Measure for Compensation for Damages)
Article 49
A deposit with official depository as a security measure for compensation for damages is to be made by depositing cash or securities specified by Cabinet Office Order (including book-entry transfer bonds provided in Article 278, paragraph (1) of the Act on Book-Entry Transfer of Corporate Bonds and Shares (Act No. 75 of 2001); the same applies in the following Article and Article 51) with the Legal Affairs Bureau or District Legal Affairs Bureau nearest to the principal office of the launch operator (or, if the launch operator has no offices in Japan, the location of the launch site pertaining to the permission under Article 4, paragraph (1) (for a launch site onboard a ship, the location of the port of registry of the ship, or for a launch site onboard an aircraft, the location of the home base of the aircraft)).

(Payment from Deposited Properties)
Article 50
Victims of launch vehicle fall damage are entitled to receive the payment from the cash or securities deposited with official depository by the launch operator pursuant to the preceding Article, in relation to their respective claims for damages, in priority to other creditors.

(Recovery of Deposited Properties)
Article 51
In the following cases, a launch operator may recover cash or securities deposited with official depository pursuant to Article 49, with approval from the Prime Minister:

Annex 1 – Act on Launching of Spacecraft, etc. and Control of Spacecraft (Act No. 76 of 2016)

(i) the launch operator has completed the launching of a spacecraft, etc., and it is clear that no launch vehicle fall damage will arise;
(ii) launch vehicle fall damage occurred and the launch operator completed the compensation for damages arising therefrom; or
(iii) the launch operator has taken other security measures for compensation for damages in place of making a deposit with official depository.

(Delegation to Cabinet Office Order and Ministry of Justice Order)
Article 52
Beyond what is provided for in this Section, the matters relating to the deposit with official depository are specified by Cabinet Office Order and Ministry of Justice Order.

Chapter VI Compensation for Spacecraft Fall Damage

(Strict Liability)
Article 53
A person who implements the control of a spacecraft using a spacecraft control facility located in Japan is liable to compensate for spacecraft fall damage caused by the person in relation to the control of the spacecraft.

(Consideration for Determination of Compensation for Damages)
Article 54
Notwithstanding the provisions of the preceding Article, if a natural disaster or other force majeure also contributed to the occurrence of the spacecraft fall damage, the court may also take those into consideration to determine the liabilities and amount of compensation for damages.

Chapter VII Miscellaneous Provisions

(Hearing of Opinions of Committee on National Space Policy)
Article 55
When the Prime Minister intends to establish, revise or abolish the Cabinet Office Order referred to in Article 4, paragraph (2), item (ii), Article 6, item (i) or (ii) or Article 22, item (ii) or (iii), the Prime Minister must seek opinions from the Committee on National Space Policy in advance.

(Consultation with Minister of Finance)
Article 56
When the Prime Minister intends to establish, revise or abolish the Cabinet Office Order referred to in Article 9, paragraph (2) or Article 40, paragraph (2), the Prime Minister must consult with the Minister of Finance in advance.

(Exemption for National Government)
Article 57
(1) The provisions of Article 4, paragraph (1) do not apply to the launching of spacecraft, etc. to be implemented by the national government.
(2) The provisions of Article 20, paragraph (1) do not apply to the control of spacecraft to be implemented by the national government.

(Transitional Measures)
Article 58
Where an order is established, revised or abolished pursuant to the provisions of this Act, the necessary transitional measures (including transitional measures relating to penal provisions) may be stipulated in the order to the extent considered reasonably necessary for the establishment, revision or abolition.

(Delegation to Cabinet Office Order)
Article 59
Beyond what is provided for in this Act, the procedures for the enforcement of this Act and any other matters necessary for the enforcement of this Act are specified by Cabinet Office Order.

Chapter VIII Penal Provisions

Article 60
A person falling under any of the following items is to be punished by imprisonment of not more than three years or a fine of not more than 3,000,000 yen, or both:
(i) a person who has implemented the launching of spacecraft, etc. in violation of the provisions of Article 4, paragraph (1);
(ii) a person who has obtained the permission under Article 4, paragraph (1), Article 7, paragraph (1), Article 20, paragraph (1) or Article 23, paragraph (1), the authorization under Article 10, paragraphs (1) through (3) or Article 26, paragraph (1), (3) or (4), the type certification under Article 13, paragraph (1), the authorization under Article 14, paragraph (1) or Article 17, paragraph (1) or the compliance certification under Article 16, paragraph (1) by deception or other wrongful means;

(iii) a person who has changed the matters set forth in Article 4, paragraph (2), items (ii) through (v) in violation of the provisions of Article 7, paragraph (1);
(iv) a person who has implemented the control of a spacecraft in violation of the provisions of Article 20, paragraph (1);
(v) a person who has changed the matters set forth in Article 20, paragraph (2), items (iv) through (viii) in violation of the provisions of Article 23, paragraph (1); or
(vi) a person who has violated an order under Article 33, paragraph (3).

Article 61

A person falling under any of the following items is to be punished by imprisonment of not more than one year or a fine of not more than 1,000,000 yen, or both:
(i) a person who has implemented the launching of spacecraft, etc. in violation of the provisions of Article 8 or Article 9, paragraph (1); or
(ii) a person who has failed to take the termination measures pertaining to the license under Article 20, paragraph (1), in violation of Article 26, paragraph (6), Article 27, paragraph (2), Article 28, paragraph (1), Article 29, paragraph (2) or Article 30, paragraph (2).

Article 62

A person falling under any of the following items is to be punished by a fine of not more than 1,000,000 yen:
(i) a person who has changed the matters set forth in Article 13, paragraph (2), item (ii) in violation of the provisions of Article 14, paragraph (1);
(ii) a person who has changed the matters set forth in Article 16, paragraph (2), item (ii) or (iv) in violation of the provisions of Article 17, paragraph (1);
(iii) a person who has failed to make a report under Article 31, paragraph (1) or makes a false report, or refuses, obstructs or avoids the inspection under that paragraph, or fails to answer or gives a false answer to the questions under that paragraph; or
(iv) a person who has violated an order under Article 33, paragraph (1) or (2).

Article 63

A person falling under any of the following items is to be punished by a fine of not more than 500,000 yen:
(i) a person who has failed to make the notification under Article 7, paragraph (2), Article 14, paragraph (2), Article 17, paragraph (2), Article 23, paragraph (2), Article 25, Article 26, paragraph (2) or Article 28, paragraph (1) or who makes a false notification;
(ii) a person who has failed to return a type certificate in violation of the provisions of Article 15, paragraph (2); or

(iii) a person who has failed to return a launch site certificate in violation of the provisions of Article 18, paragraph (2).

Article 64

If the representative of a corporation or an agent, employee or other worker of a corporation or an individual commits any of the violations prescribed in Article 60 through the preceding Article with regard to the business of the corporation or individual, not only the offender but also the corporation or individual is subject to the fine prescribed in the respective Articles.

Article 65

A person who fails to make the notification under Article 11, Article 27, paragraph (1) or Article 29, paragraph (1) or makes a false notification is to be punished by a civil fine of not more than 100,000 yen.

Supplementary Provisions

(Effective Date)
Article 1

This Act comes into effect as of the date specified by Cabinet Order within a period not exceeding two years from the date of promulgation; provided, however, that the provisions set forth in the following items come into effect as of the dates specified respectively in the following items:
(i) the provisions of Articles 3 and 10 of the Supplementary Provisions: the date of promulgation; and
(ii) the provisions of the following Article: the date specified by Cabinet Order within a period not exceeding one year from the date of promulgation.

(Preparatory Actions)
Article 2

(1) A person who intends to obtain the permission under Article 4, paragraph (1) or Article 20, paragraph (1) may submit an application even prior to the enforcement of this Act in accordance with the provisions of Article 4, paragraph (2) or Article 20, paragraph (2).
(2) A person who intends to obtain the type certification under Article 13, paragraph (1) or the compliance certification Article 16, paragraph (1) (excluding JAXA) may submit an application even prior to the enforcement of this Act in accordance with the provisions of Article 13, paragraph (2) or Article 16, paragraph (2).

(3) JAXA may submit an application for the type certification under Article 13, paragraph (1) in relation to its design of a launch vehicle even prior to the enforcement of this Act, in accordance with the provisions of Article 19, paragraph (1).
(4) JAXA may submit an application for the compliance certification under Article 16, paragraph (1) in relation to the launch site managed and operated by it even prior to the enforcement of this Act, in accordance with the provisions of Article 19, paragraph (2).

Article 3
(1) When the Prime Minister intends to establish the Cabinet Office Order referred to in Article 4, paragraph (2), item (ii), Article 6, item (i) or (ii), or Article 22, item (ii) or (iii), the Prime Minister may seek opinions from the Committee on National Space Policy even prior to the enforcement of this Act.
(2) When the Prime Minister intends to establish the Cabinet Office Order referred to in Article 9, paragraph (2) or Article 40, paragraph (2), the Prime Minister may consult with the Minister of Finance even prior to the enforcement of this Act.

(Transitional Measures)
Article 4
The provisions of Article 20, paragraph (1) do not apply to the control of spacecraft being implemented at the time of the enforcement of this Act.

(Review)
Article 5
When five years have passed after this Act comes into effect, the government is to review the state of enforcement of this Act, and, when it determines it necessary, take necessary measures based on the findings of the review.

(Transitional Measures Relating to Penal Provisions Associated with Partial Revision to Act on the Japan Aerospace Exploration Agency, National Research and Development Agency)
Article 7
For the application of the penal provisions to acts in violation of the provisions of the Act on the Japan Aerospace Exploration Agency, National Research and Development Agency prior to the revision under the preceding Article, which were committed before the enforcement of this Act, the provisions then in force remain applicable.

(Delegation to Cabinet Order)
Article 10
Beyond what is provided for in these Supplementary Provisions, any necessary transitional measures (including transitional measures concerning penal provisions) for the enforcement of this Act are specified by Cabinet Order.

ANNEX 2

Review Standards and Standard Period of Time for Process Relating to Procedures under the Act on Launching of Spacecraft, etc. and Control of Spacecraft

November 15, 2017

Cabinet Office
National Space Policy Secretariat

The review standards under Article 5, paragraph (1) of the Administrative Procedure Act (Act No. 88 of 1993) in relation to permissions, etc. related to the launching of spacecraft, etc., and the control of spacecraft, pursuant to the Act on Launching of Spacecraft, etc. and Control of Spacecraft (Act No. 76 of 2016), and the standard period of time for process under Article 6 of the Administrative Procedure Act, are as set out in the appended table.

The terms used in these review standards are governed by the definitions under the Act on Launching of Spacecraft, etc. and Control of Spacecraft (hereinafter referred to as the "Act") and the Regulation for Enforcement of the Act on Launching of Spacecraft, etc. and Control of Spacecraft (Cabinet Office Order No. 50 of 2017; hereinafter referred to as the "Cabinet Office Order").

Supplementary Provisions
These regulations come into effect from the date on which the Act comes into effect.

ANNEX 2

(Appended Table)

[Permission Related to Launching of Spacecraft, etc.]

Category	Provisions	Review standards		Standard period of time for processing
Permission related to launching of spacecraft, etc.	Article 4, paragraph (1) of the Act	Article 6, item (i) of the Act (Design of launch vehicle)	Same as the review standards for "Design of launch vehicle" as referred to in [Type certification for launch vehicle]	When a type certification is obtained: 1-3 months
		Article 6, item (ii) of the Act (Launch site)	Same as the review standards for "Launch site" as referred to in [Compliance certification for launch site]	Other cases: 4-6 months
		Article 6, item (iii) of the Act (Launch plan and sufficient ability to execute the launch plan)	1. Safety and security measures - In connection with the launching of a spacecraft, etc. (hereinafter simply referred to as a "launch"), the applicant takes appropriate safety and security measures during the period from the prelaunch processing phase to the completion of launch. 2. Development of disaster prevention plan, etc. - The applicant prepares a disaster prevention plan for preventing the occurrence of a disaster at the launch site, and complies with the related laws and regulations for ensuring the safety of facilities and handling thereof necessary for the disaster prevention. - The applicant ensures that information including the detection of fire and gas, security alarm, etc. is centralized and is kept updated on the status, and performs a sufficient facility inspection for fire prevention, firefighting and protection facilities before doing any hazardous activities. 3. Safety measures related to handling of propellant, etc. - The applicant establishes measures based on the applicable laws and regulations, so as to ensure the safety of handling of propellant, etc. (meaning explosives, high-pressure gases and hazardous materials, etc.) at the launch site. 4. Design of trajectory considering the planned impact area, etc. - The planned impact area of objects to be separated and jettisoned from a spacecraft launch vehicle (hereinafter	

simply referred to as a "launch vehicle") in the course of its normal flight, including its combustion residue, is separated from the land or its surrounding sea waters to the possible extent.
- The planned impact area would not interfere with any territory or territorial waters of a foreign state. If any interference is expected, the applicant obtains consent from the relevant state.
- For a dispersion area of a trajectory of the estimated impact point in the case of a sudden thrust termination of a launch vehicle in thrust flight (a trajectory of estimated impact point), a trajectory has been determined so that the launch vehicle will pass an area as far as possible from densely populated areas, and the necessary measures have been taken so that the level of risk to the vicinity of the trajectory and launch site will not exceed the international standard or standard provided by the space agency of each state, even in emergency situations.

5. Design of the appropriate impact limit lines
- The applicant establishes a set of lines indicating the boundary limit beyond which the launch vehicle must not cause any harmful effect in case of suspension of flight of the launch vehicle to ensure safety (i.e. impact limit lines).

6. Creation of restricted area and establishment of system to prevent entry of third parties
- The applicant designates a restricted area according to each stage during the preparation period for launch considering the situations surrounding the launch site, and restricts entry of non-related parties.
  (1) Restricted area for preparation period
      For each stage of launch vehicle assembling work, the applicant designates a warning area to minimize the effect of an accident, etc.
  (2) Restricted area for launch
      Restricted areas for launch cover all of the areas which are, at least, included in the following restricted areas for ground safety and flight safety.
      [Restricted area for ground safety]
      At least blast, scattering objects, gas,

| | | | | radiation heat generated by a fireball, etc. are taken into consideration. [Restricted area for flight safety] The following matters are taken into consideration:
 (A) The area is capable of preventing the occurrence of the following hazards in the vicinity of the launch site.
  (i) Collision of falling objects
  (ii) Blast in case of explosion in flight
  (iii) Blast and secondary scattering of fragments caused by a secondary explosion when there is a risk of a solid propellant falling to and impacting the ground surface, etc. and exploding (secondary explosion)
  (iv) Leak and dispersion of propellant onboard
 (B) In addition, in relation to the sea waters in the vicinity of the launch site, the distribution of impact of fragments generated due to the termination of flight immediately after the lift-off is evaluated, and the hazard to ships, etc. that may be caused by the impact of fragments is prevented to the possible extent.

 7. Measures to be taken at the time of natural disaster warning, etc.

 - The applicant provides measures in response to a warning of stormy weather, lightning strike, earthquake, etc.

 8. Prior notice to aircraft, ships, etc.
 - The applicant establishes communication means, etc. with relevant agencies so as to ensure the safety of aircraft, ships, etc. during the launch operation period.

 9. Determination of the appropriate date and time for launch
 - The applicant sets the date of launch such that a collision with international space stations or manned spacecraft on the orbit will be avoided, so as to ensure the safety of lives of persons operating on the orbit.

 10. Flight capability considering the spacecraft to be loaded
 - The flight capability of the launch vehicle can provide the ability to put the spacecraft into the planned orbit.
 - Safety-critical systems, etc. of the | |

				launch vehicle which constitute the function to ensure the safety of the vicinity of the trajectory and launch site will not receive any severe adverse effect by the spacecraft onboard the launch vehicle.
				
				11. Confirmation of feasibility for flight considering weather conditions
				- The applicant, immediately before the launch, confirms that the launch vehicle will not deviate from the planned trajectory and planned impact area due to the effect of weather conditions.
				- The applicant avoids any circumstances that may result in third-party damage, including a failure of equipment due to lightning, by predicting the possibility of lightning on the course of the trajectory.
				
				12. Prevention of third-party damage prior to the termination of designation of restricted areas
				- During the launch operation period, the applicant takes safety measures as necessary, including the suspension of the operation.
				
				13. Implementation of flight safety operation
				- The applicant takes measures for the monitoring of the conditions of the launch vehicle in flight and ensures that the flight can be terminated in a safe manner if necessary, so as to ensure the safety against falling objects in the case of a failure of the launch vehicle.
				
				14. Implementation of flight termination
				- The applicant terminates the flight of the launch vehicle in any of the following cases:
				  (i) When the estimated impact area of the launch vehicle and its fragments may cross the impact limit line; provided, however, that this does not apply to the case where the estimated impact area of the launch vehicle flying over the scheduled flight range passes the impact limit line, and where the flight is normal under the sufficient monitoring of the flight conditions immediately before the passage.
				  (ii) When it is impossible to monitor the estimated impact area of the launch vehicle, and there is a risk that the estimated impact area of the launch vehicle and its fragments may cross the impact limit line.	

ANNEX 2

| | | | | (iii) When there is a possibility of loss of the flight termination function of the launch vehicle, and there is a risk that the estimated impact area of the launch vehicle and its fragments may cross the impact limit line.
(iv) In other cases where any risk is considered to exist which may cause any adverse effect on the ensuring of safety if the thrust flight of the launch vehicle is continued.

15. Retrieval of objects remaining buoyant on the sea
· The applicant makes efforts to retrieve objects remaining buoyant on the sea generating from jettisoned objects from a launch vehicle which pose the risk of any severe adverse effects on the navigation of ships.

16. Mitigation of the generation of orbital debris
· The applicant takes the following measures in relation to the generation of objects which may become orbital debris.
  (i) Take measures to prevent the unexpected activation of pyrotechnic devices for command destruction of the orbital stage of launch vehicle.
  (ii) In the case of a launch vehicle for which the propellant is liquid fuel, vent the remaining propellant, gas, etc. to the possible extent, and take measures to install safety valves to avoid the increase of internal pressure or ensure the safety by way of design so as to avoid break-up even if the venting is not completed.

17. Removal of an orbital stage of a launch vehicle from protected regions
· Where possible, the orbital stage of a launch vehicle that completed the launching into an orbit passing through a low earth orbit region (meaning a spherical region up to the altitude of 2,000km from the Earth's surface) or an orbit that may interfere with a low earth orbit region must be transferred into an orbit for which the orbital life is shorter or must be disposed of by a reentry in a way to prevent damage to the ground, by controlling its position, attitude and conditions.
· Where possible, eternal or periodic | |

			crossing of the orbital stage of a launch vehicle and the geosynchronous orbit region (geostationary earth orbit altitude of 35,786km ± 200km and latitude within ±15 degrees) is to be avoided. 18. Establishment of organizational structures for the implementation of launch plan · In order to ensure the proper implementation of the measures set forth in items 1 through 17 above, appropriate organizational structures must be established as follows. 　· Organization for safety and its duties 　　The applicant establishes an organization dedicated to the ensuring of safety, takes measures to ensure that the organization will function in an organic way through closely connected means of communication, and establishes a reporting structure so that all safety issues will be communicated to the person responsible for launch. 　· Implementation of safety education training 　　The applicant implements safety education and training for persons engaged in launch, and ensures that these persons are fully aware of the matters relating to the ensuring of safety. 　· Response to emergency 　　The applicant establishes an organizational structure to accurately and immediately respond to emergencies, such as an accident occurred during the launch operation period.	
		Article 6, item (iv) of the Act (Purpose and method of spacecraft to be loaded on a launch vehicle)	Same as the review standards for "Purpose and method of use of spacecraft" as referred to in [License related to control of spacecraft]	
Permission to make changes	Article 7, paragraph (1) of the Act		As per the standards provided in Article 6, items (i) through (iv) of the Act.	No standard is set, as each application is to be reviewed on a case-by-case basis

ANNEX 2

				depending on the scope of change.
Authorization related to succession	Article 10, paragraph (1) of the Act (Transfer and acquisition)	Article 6, item (iii) of the Act (Limited to the part relating to the ability to execute the launch plan)	Same as the review standards of "18. Establishment of organizational structures for the implementation of launch plan" as referred to in "Launch plan and sufficient ability to execute the launch plan"	1 month
	Article 10, paragraph (2) of the Act (Merger)			
	Article 10, paragraph (3) of the Act (Corporate split)			

[Type certification for launch vehicle]

Category	Provisions		Review standard	Standard period of time for process
Type certification for launch vehicles	Article 13, paragraph (1) of the Act	Article 6, item (i) of the Act (Design of launch vehicle)	1. Flight capability (Article 7, item (i) of the Cabinet Office Order) - The launch vehicle is designed so that it has a flight capability sufficient for the launch, and the design has undergone verification. 2. Safety requirement for ignition device, etc. (Article 7, item (ii) of the Cabinet Office Order) - Measures are taken to ensure the safety of the vicinity of the trajectory and launch site of the launch vehicle for the case of any combinations of two types of failures, etc. This measure may also include the measures to be taken at the launch site. - Among the measures taken, two or more measures are always capable of being monitored. - Necessary measures have been taken to prevent any accidental ignition of pyrotechnic devices caused by stray lightning, etc. - Measures have been taken to prevent the easy occurrence of failures, etc. caused by the effect of the ambient electromagnetic waves, etc. 3. Function for flight safety operation (Article 7, item (iii) of the Cabinet Office Order) - The launch vehicle has a function of transmitting signals indicating its position, attitude and condition. 4. Flight termination function (Article 7, item (iv) of the Cabinet Office Order) - The launch vehicle has a function of receiving signals necessary for implementing the flight termination measures and a function of implementing flight termination, etc. In addition, the assessment on ensuring the safety is conducted according to the specific launch plan to be contemplated in the future, the risk to the vicinity of the trajectory and launch site does not exceed the level stipulated in the international standards or standards provided by the space agency of each state, and it is possible to prevent the risk of falling outside the impact limit line determined in advance. - Even in the case of another methods (including the methods, etc. to suspend sequences when signals are not received), the risk to the vicinity of the trajectory and	4-6 months

			launch site does not exceed the level stipulated in the international standards or standards provided by the space agency of each state, and it is possible to prevent the risk of violating the impact limit line established in advance. 5. Reliability and redundancy of safety-critical systems, etc. (Article 7, item (v) of the Cabinet Office Order) - For safety-critical systems, etc. which constitute the function to ensure the safety of the vicinity of the trajectory and launch site by flight termination measures of a launch vehicle, the reliability is 0.999 or more at a 95% confidence or an equivalent level, and the systems, etc. are fully redundant so that the they will function even in cases of failure, etc. 6. Mitigation of the generation of orbital debris relating to the separation of spacecraft, etc. (Article 7, item (vi) of the Cabinet Office Order) - The launch vehicle is designed so as to prevent the dispersion of fragments, etc. to the possible extent upon the operation of the stage separation system, spacecraft separation system, etc. of the launch vehicle; provided, however, that this does not apply to spacecraft support structures that are unavoidably released upon the launch of two or more spacecraft. 7. Mitigation of the generation of orbital debris relating to the orbital stage of launch vehicle (Article 7, item (vii) of the Cabinet Office Order) - Measures are taken to prevent the unexpected activation of pyrotechnic devices for command destruction of the orbital stage of the launch vehicle. - In the case of a launch vehicle for which the propellant is liquid fuel, the launch vehicle has a function to vent the remaining propellant, gas, etc. to the possible extent, and measures are taken to install safety valves to avoid the increase of internal pressure so as to avoid break-up even if the venting is not completed.	
Authorization related to change	Article 14, paragraph (1) of the Act	As per the standards provided in Article 6, items (i) of the Act.		No standard is set, as each application is to be reviewed on a case-by-case basis

				depending on the scope of change.

[Compliance certification for launch site]

Category	Provisions		Review standard	Standard period of time for process
Compliance certification for launch site	Article 16, paragraph (1) of the Act	Article 6, item (ii) of the Act (Launch site)	1. Securing of restricted areas and measures to prevent entry of third parties (Article 8, item (i) of the Cabinet Office Order) · The area is located in a place capable of securing an appropriate restricted area depending on the phases of the launch operation period. · The applicant makes efforts to establish the security measures for facilities, devices, information, etc. which would be important for the security of launch. 2. Installation of launcher (Article 8, item (ii) of Cabinet Office Order) · The launch site is capable of being equipped with a permanent or portable launcher appropriate for launch vehicles. · The launcher is capable of implementing the appropriate lift-off of a launch vehicle in a way to ensure the safety of the vicinity of the trajectory and launch site. 3. Safety requirement for ignition device, etc. (Article 8, item (iii) of the Cabinet Office Order) · Measures are taken to ensure the safety of the vicinity of the trajectory and launch site of the launch vehicle in case of any combinations of two types of failure, etc. This measure may also include the measures to be taken in relation to the launch vehicle. · Among the measures taken, two or more measures are always capable of being monitored. · Necessary measures have been taken to prevent any accidental ignition of pyrotechnics caused by stray lightning, etc. · Measures have been taken to prevent the easy occurrence of failures, etc. caused by the effect of ambient electromagnetic waves. 4. Radio equipment for flight safety operation (Article 8, item (iv) of the Cabinet Office Order) · The launch site is capable of being equipped with permanent or portable radio equipment having a function to detect a signal indicating the position, attitude and	1-3 months

ANNEX 2

			condition of the launch vehicle by the use of electromagnetic waves or other means; provided, however, that this does not apply if another place is used which is equipped with radio equipment having the same functions. · When the flight termination measures of a launch vehicle are to be implemented by way of receiving signals, the launch site is capable of being equipped with permanent or portable radio equipment having a function to transmit a signal necessary for the implementation of the flight termination measures to the radio equipment of the launch vehicle directly or via other radio equipment; provided, however, that this does not apply if another place is used which is equipped with radio equipment having the same functions. 5. Reliability and redundancy of safety-critical systems, etc. (Article 8, item (v) of the Cabinet Office Order) · For safety-critical systems, etc. which constitute the functions to ensure the safety of the vicinity of the trajectory and launch site by flight termination measures of a launch vehicle, the reliability is 0.999 or more at a 95% confidence or an equivalent level, and the systems, etc. are fully redundant so that the they will function even in cases of failure, etc.	
Authorization related to change	Article 17, paragraph (1) of the Act	As per the standards provided in Article 6, items (ii) of the Act.		No standard is set, as each application is to be reviewed on a case-by-case basis depending on the scope of change.

[License related to control of spacecraft]

Category	Provisions	Review standard		Standard period of time for process
License related to control of spacecraft	Article 20, paragraph (1) of the Act	Article 22, item (i) of the Act (Purpose and method of use of spacecraft)	· The purposes and methods of use of the spacecraft are in compliance with the basic principles (Articles 2 through 7 of the Space Basic Act). · The purposes and methods of use of the spacecraft are not likely to cause any adverse effect on the accurate	15 days - 3 months

				Article 22, item (ii) of the Act (Configuration of spacecraft)	and smooth implementation of the conventions on development and use of outer space and ensuring public safety.	
					1. Prevention of unintended release of objects (Article 22, item (i) of the Cabinet Office Order) - The spacecraft has a configuration to prevent the components, etc. from coming off or scattering easily. - The spacecraft has a configuration to prevent components, etc. from scattering easily upon the operation of the separation or deployment system, etc. of the spacecraft. - Due consideration is paid to the configuration so as to minimize the release of combustion products generated from pyrotechnics, etc. 2. Prevention of interference with the control of other spacecraft upon separation or docking (Article 22, item (ii) of the Cabinet Office Order) - The spacecraft has a configuration enabling it to be put into an appropriate orbit in a manner not having a severe adverse effect on the control of other spacecraft, including a manned spacecraft, upon the separation of components or parts of the spacecraft. - The spacecraft has a configuration enabling it to be docked with another spacecraft, etc. without the separation or release of components, etc. so as not to have any severe adverse effect on the control of other spacecraft. 3. Prevention of break-up in case of anomalies (Article 22, item (iii) of the Cabinet Office Order) - The spacecraft has a configuration to enable it to transmit signals indicating its position, attitude and condition to the spacecraft control facility directly or via other radio equipment. - The spacecraft has a configuration enabling the prevention of break-up, such as the venting of residual energy including residual propellant and electricity which may cause the break-up of the spacecraft. 4. Prevention of damage to the public upon reentry into Earth (Article 22, item (iv) of the Cabinet Office	

ANNEX 2

			Order) · The spacecraft or its component, etc. for reentry into Earth has a configuration to be completely ablated, or as a result of sufficient ablation, etc., the risk to the expected point of landing or water landing will not exceed the level stipulated in the international standards or standards provided by the space agency of each state. 5. Prevention of deterioration of the Earth's environment due to substances derived from other celestial bodies (Article 22, item (v) of the Cabinet Office Order) · In the case of retrieving a spacecraft, its components or parts which were put into orbit around a celestial body other than the Earth or which fell to the celestial body, by guiding them to fall to Earth, the spacecraft has a configuration for the prevention of the deterioration of the environment of the Earth that may be caused by the introduction of extraterrestrial substances. 6. Prevention of contamination of environment of other celestial bodies (Article 22, item (vi) of the Cabinet Office Order) · In the case of a spacecraft or its components, etc. which are put into the orbit around a celestial body other than the Earth or which are to be guided to fall to the celestial body, the spacecraft has a configuration for the prevention of the harmful contamination of the celestial body.	
		Article 22, item (iii) of the Act (Control plan and sufficient ability to execute the control plan)	1. Prevention of interference with the control of other spacecraft upon separation or docking (Article 23, item (i) of the Cabinet Office Order) · The applicant provides measures relating to the appropriate orbital insertion to ensure that there will be no severe adverse effect on the control of other spacecraft, including manned spacecraft, when separating components or parts of the spacecraft. · The applicant provides measures relating to the prevention of coming off or scattering of components, etc. to ensure that there will be no severe adverse effect on the control of other spacecraft when docking	

		the spacecraft with another spacecraft, etc.
		2. Prevention of break-up in case of anomalies (Article 23, item (ii) of the Cabinet Office Order) - The applicant provides methods and guidelines for the implementation, etc. of measures to prevent break-up in cases of detecting anomalies in the condition, etc. of the spacecraft.
		3. Prevention of collision with another spacecraft, etc. (Article 23, item (iii) of the Cabinet Office Order) - The applicant provides the methods of obtaining information on the possibility of collision with another spacecraft, etc. and the measures to be taken in case of obtaining the relevant information.
		4. Establishment of organizational structures for the implementation of the control of spacecraft - In order to ensure the control plan set forth in 1 through 3 above, the following appropriate organizational structures are established. · Organization for management and its duties · Response to anomalies · Establishment of security measures
	Article 22, item (iv) of the Act (Details of termination measures)	(In the case of Article 22, item (iv)(a) of the Act) - The applicant provides measures for the controlled reentry (e.g. trajectory, landing point) while ensuring the safety of an expected point of landing or water landing. (In the case of Article 22, item (iv)(b) of the Act) - The applicant provides measures to elevate the spacecraft to the altitude that would not have any adverse effect on the control of other spacecraft. (In the case of Article 22, item (iv)(c) of the Act) - The applicant provides measures to put the spacecraft into the orbit around a celestial body other than the Earth or guide it to fall to the celestial body, without any risk of significantly deteriorating the environment of the celestial body.

ANNEX 2

			(In the case of Article 22, item (iv)(d) of the Act (Article 24 of the Cabinet Office Order))
			- The applicant provides measures to vent residual energy, including residual propellant and electricity, which may cause break-up of the spacecraft or to prevent the break-up.
			- Upon the termination of the control of the spacecraft, the following measures are to be taken for the protected regions:
			  - Efforts must be made so that the spacecraft will be removed from the low earth orbit region within 25 years from the termination of the control.
			  - The spacecraft is to be removed from the geosynchronous orbit immediately.	

Permission to make changes	Article 23, paragraph (1) of the Act		As per the standards provided in Article 22, items (i) through (iv) of the Act.	No standard is set, as each application is to be reviewed on a case-by-case basis depending on the scope of change.
Authorization on succession	Article 26, paragraph (1) of the Act (Transfer and acquisition) Article 26, paragraph (3) of the Act (Merger) Article 26, paragraph (4) of the Act (Corporate split)	Article 22, item (iii) of the Act (Limited to the part relating to the ability to execute the control plan)	Same as the review standards of "4. Establishment of organizational structures for the implementation of the control of spacecraft" as referred to in "Control plan and sufficient ability to execute the control plan."	1 month

Annex 3

Application Manual for Act on Launching of Spacecraft, etc. and

Control of Spacecraft

Revised first edition dated March 30, 2018

Cabinet Office
National Space Policy Secretariat

History of revisions

Edition	Date of establishment	Detail of revisions
First Edition	November 15, 2017	New establishment
Revised first edition	March 30, 2018	Full revision

ANNEX 3

Table of Content

1. Introduction 3
2. Applications 3
   2.1. Types of applications 3
   2.2. Types of application processes 3
      2.2.1. Application processes for the launching of spacecraft, etc. 3
      2.2.2. Application processes related to the control of spacecraft 4
   2.3. Language to be used for written applications, etc. 11
3. Permission relating to launching of spacecraft, etc. 12
   3.1. Written application for permission 12
      3.1.1. Information items to be entered in the application form and guides for preparation 12
      3.1.2. Attachments 16
   3.2. Change related to permission 18
      3.2.1. Application for permission related to change 20
      3.2.2. Notification of change 21
   3.3. Succession 22
      3.3.1. Business transfer 23
      3.3.2. Merger 23
      3.3.3. Corporate split 24
   3.4. Notification of expiration of permission 25
4. Type certification for launch vehicle 27
   4. Application for type certification 27
      4.1.1. Information items to be entered in application form and guide for preparation 27
      4.1.2. Attachments 29
   4.2. Change of certification 34
      4.2.1. Application for authorization of change 36
      4.2.2. Notification of change 37
5. Compliance certification for launch site 38
   5.1. Application for compliance certification 38
      5.1.1. Information items to be entered in application form and guide for preparation 38
      5.1.2. Attachments 41
   5.2. Change of certification 44
      5.2.1. Application for authorization of change 46

5.2.2. Notification of change 46
6. License related to control of spacecraft 48
　6.1. Written application for license 49
　　6.1.1. Information items to be entered in the application form and guides for preparation 49
　　6.1.2. Attachments 55
　6.2. Change related to permission 58
　　6.2.1. Application for permission related to change 59
　　6.2.2. Notification of change 60
　6.3. Notification in case of accidents 61
　6.4. Succession 62
　　6.4.1. Business transfer 64
　　6.4.2. Notification of transfer of business to person who intends to implement the control of spacecraft without using a spacecraft control facility located in Japan 64
　　6.4.3. Merger 65
　　6.4.4. Corporate split 65
　6.5. Notification of death 66
　6.6. Termination measures 67
　6.7. Notification of dissolution 67
7. Scope of applicants 69
　7.1. Application for permission related to launching of spacecraft, etc. 69
　7.2. Application for license related to control of spacecraft 69
8. Example of preparation of application forms 70
9. List of main agencies and sections in charge 71
　9.1. List of main agencies and sections relating to launching of a spacecraft, etc. 71
　9.2. List of main agencies and sections relating to control of spacecraft 72
10. Checklist of documents to be submitted 73

[Explanatory Notes]
Unless otherwise provided, the terms used in this manual have the meanings as defined in the Act and the Regulation. The abbreviations as used in this manual have the following meanings:

Act:　　　　Act on Launching of Spacecraft, etc. and Control of Spacecraft (Act No. 76 of 2016)

Regulation:　Regulation for Enforcement of the Act on Launching of Spacecraft, etc. and Control of Spacecraft (Cabinet Office Order No. 50 of 2017)

ANNEX 3

1. Introduction
A person who intends to obtain permission to launch a spacecraft, etc., a person who intends to obtain a type certification for a launch vehicle, a person who intends to obtain a compliance certification for a launch site, and a person who intends to obtain a license for control of a spacecraft is required to submit an application to the Prime Minister pursuant to the Act and Regulation.

This manual provides guidance on the matters necessary for these applications.

2. Applications
2.1. Types of applications
Applications relating to the launching of spacecraft, etc. and the control of spacecraft are divided into the following four categories:
- Permission related to the launching of spacecraft, etc.
- Type certification for a launch vehicle
- Compliance certification for a launch site
- License related to control of spacecraft

2.2. Types of application processes
2.2.1. Application processes for the launching of spacecraft, etc.
For the processes of application for permission related to the launching of a spacecraft, etc., Figure 1 shows a flowchart for the application based on whether the applicant has a type certification or compliance certification, the standard period of time for process as well as the documents necessary for the application. In addition, Figures 2 through 4 show a flowchart for applications for permission related to the launching of a spacecraft, etc., a flowchart for applications related to a type certification for a launch vehicle, and a flowchart for applications related to a compliance certification for a launch site.

A person who intends to implement the launching of a spacecraft, etc. using a launch site located in Japan is required to obtain permission for each instance of launching. This application does not cover the launching of a spacecraft, etc. completed before the full enforcement of the Act and the launching of a sub-orbital rocket not loaded with any spacecraft.

An applicant for permission related to the launching of a spacecraft, etc., a type certification for a launch vehicle, and a compliance certification for a launch site may be the same or different persons. However, as a compliance certification for a launch site is to be applied in relation to the launch vehicle for which a type certification has been obtained, an applicant who has not obtained a certification for a launch vehicle is not eligible to apply for a compliance certification for a launch site.

2.2.2. Application processes related to the control of spacecraft

Figure 5 is a flowchart of application procedures for obtaining license related to the control of a spacecraft.

A person who intends to implement the control of a spacecraft using a spacecraft control facility located in Japan is required to obtain a license related to the control of a spacecraft, even if the spacecraft is launched outside Japan. In addition, even if a spacecraft control facility for normal operation is located outside Japan, if the control of a spacecraft is to be implemented using a spacecraft control facility located in Japan for a certain period, for example, for the purpose of initial operation, it is necessary to obtain the license.

This application does not cover the control of a spacecraft currently conducted at the time of the full enforcement of the Act. Here, "control of a spacecraft currently conducted" generally means that the control has been already commenced on the orbit (including the case of commencement of the control of a spacecraft only using a spacecraft control facility located outside Japan).

This application must be submitted by a person who intends to implement the control of a spacecraft, separately from an application for permission related to the launching of a spacecraft, etc.

ANNEX 3

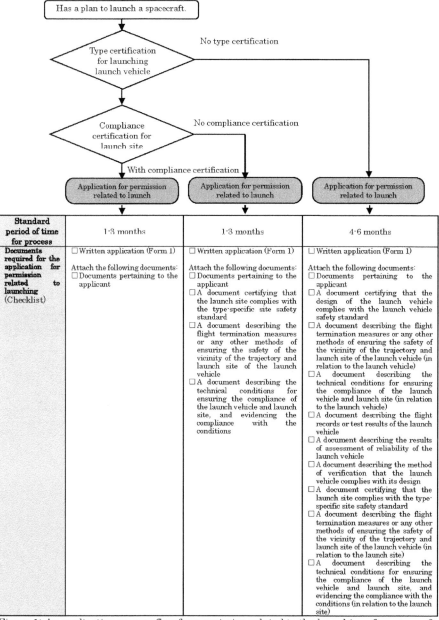

Figure 1: An application process flow for permission related to the launching of a spacecraft, etc., standard period of time for process and documents necessary for the application (depending on whether the applicant has obtained a certification)

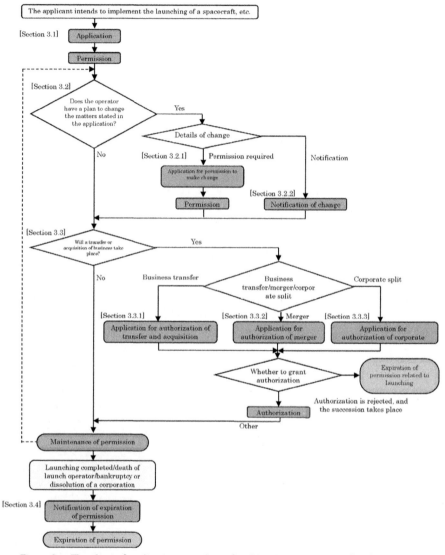

Figure 2 Flowchart of application procedures for obtaining permission related to the launching of a spacecraft, etc.

ANNEX 3

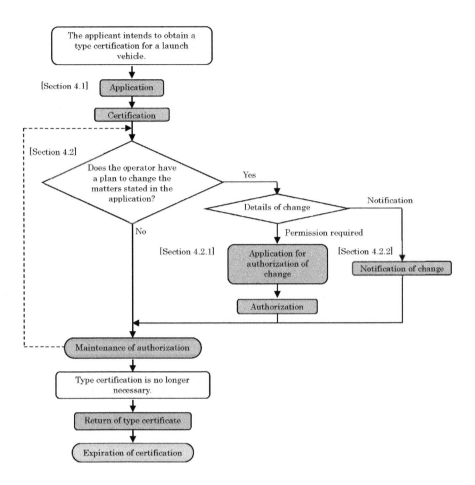

Figure 3: Flowchart of procedures for application of type certification for a launch vehicle

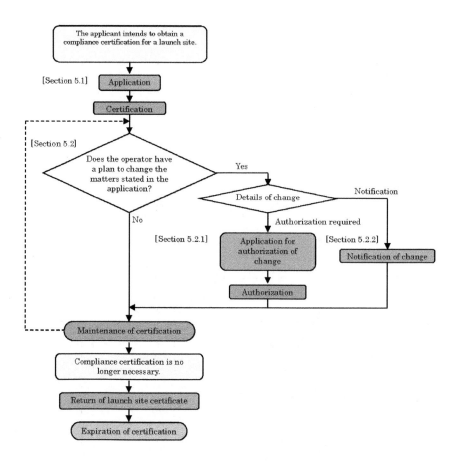

Figure 4: Flowchart of procedures for application of compliance certification for a launch site

ANNEX 3

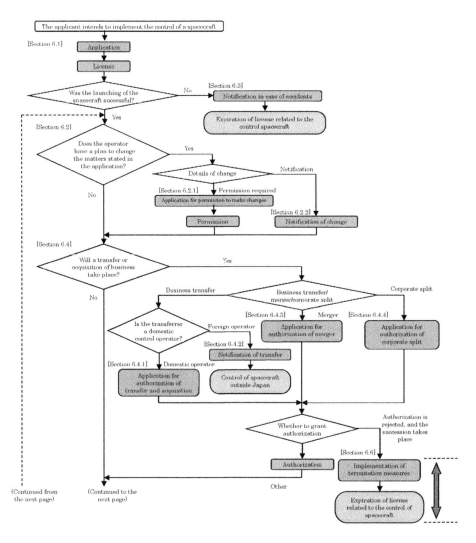

Figure 5 Flowchart of application procedures for obtaining license related to the control of a spacecraft (1/2)

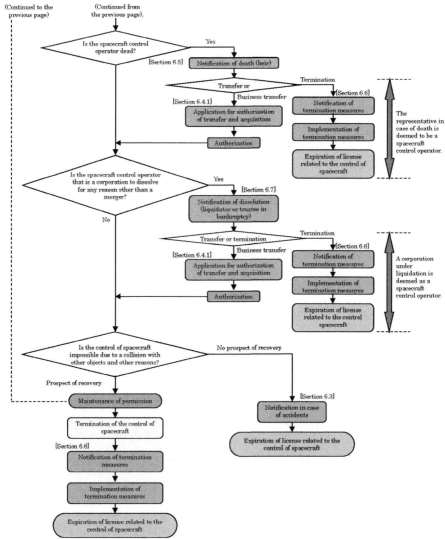

Figure 5   Flowchart of application procedures for obtaining license related to the control of a spacecraft (2/2)

2.3. Language to be used for written applications, etc.

> [Language to be used for written applications, etc.]
>
> Article 37 of the Regulation (Terms of Documents, etc.)
>
> (1) Written applications and written notifications as provided in this Cabinet Office Order must be prepared in Japanese; provided, however, that addresses, names and contact information may be written in a foreign language.
>
> (2) Documents to be attached to written applications and written notifications as provided in this Cabinet Office Order must be prepared in Japanese or English; provided, however, that Japanese translations must be submitted for any documents written in English.
>
> (3) If a person is unable to submit the documents under the preceding paragraph in a language provided in that paragraph due to special circumstances, notwithstanding the provisions of that paragraph, the person may submit the document together with its Japanese translations.

A person who intends to obtain permission to launch spacecraft, etc., a person who intends to obtain a type certification for a launch vehicle, a person who intends to obtain a compliance certification for a launch site, and a person who intends to obtain a license related to control of spacecraft is required to submit written applications in Form 1, Form 9, Form 13 and Form 17, respectively, attaching necessary documents. Information items to be stated in an application form and guidelines for preparation thereof is shown in 3.1.1, 4.1.1, 5.1.1 and 6.1.1, and the necessary attached documents are shown in 3.1.2, 4.1.2, 5.1.2 and 6.1.2. For a detailed example, please see Chapter 7.

An application form must be written in Japanese; however, an address, name and contact information may be written in a foreign language. Documents to be attached to the application form must be written in Japanese or English. For English documents, Japanese translations must be attached. If, due to any special circumstance, the applicant cannot submit attached documents in Japanese or English, documents in another language, together with the Japanese translation thereof, may be submitted.

3. Permission relating to launching of spacecraft, etc.

> [Permission relating to launching of spacecraft, etc.]
> Article 4 of the Act (Permission)
> (1) A person who intends to implement the launching of spacecraft, etc. using a launch site located in Japan or onboard a ship or aircraft with Japanese nationality must obtain permission from the Prime Minister for each instance of launching.
> (2) A person who intends to obtain permission under the preceding paragraph must submit a written application to the Prime Minister, pursuant to the provisions of Cabinet Office Order, specifying the following information, attaching the documents specified by Cabinet Office Order:
>
> Article 5 of the Regulation (Application, etc. for Permission Related to Launching of Spacecraft, etc.)
> (1) A person who intends to obtain the permission under Article 4, paragraph (1) of the Act must submit a written application using Form 1 to the Prime Minister.
> (2) The following documents must be attached to the written application under the preceding paragraph:

If the permission related to the launching of spacecraft, etc. is granted, a notice to that effect is provided, and a permission certificate for the launching of a spacecraft, etc. is issued. Refrain from making the permission certificate available to the public by such way as posting it on a website, so as to prevent forgery, etc.

Even the permission is granted, the operator may not implement the launching of a spacecraft, etc. unless the operator has taken security measures for compensation for damages.

3.1. Written application for permission

3.1.1. Information items to be entered in the application form and guides for preparation

A person who intends to obtain permission related to the launching of a spacecraft, etc. needs to submit a written application (Form 1) containing the following information.

If any document, etc. already submitted to another organization contains the following information, such document can be used as an attachment by indicating the relevant portion.

ANNEX 3

(i) the address, name and contact information
(ii) the design or type certification number of the launch vehicle
(iii) the place, design and facility of launch site or compliance certification number
(iv) the launch plan
(v) the type, name of vehicle and serial number of the launch vehicle
(iv) number of spacecraft to be loaded on the launch vehicle, as well as the names, purposes and methods of use of the respective spacecraft
(vii) names of officers or employees in charge of business with respect to the launching of spacecraft, etc.
(viii) whether the applicant falls under any of the grounds for disqualification under Article 5 of the Act

The following is the guideline on these information items.

(i) the address, name and contact information

○ If the person who intends to implement the launching of a spacecraft, etc. is an individual:
- Write the name and address as stated in the residence record.
- If the applicant is a foreign national, write the address and name as stated in a document issued by the relevant foreign government or a document equivalent thereto.

○ If the person who intends to implement the launching of a spacecraft, etc. is a corporation:
- State the corporation name and address as stated in the certificate of registered information.
- If the applicant is a foreign corporation, write the name and head office or principal office of the corporation as stated in a document issued by the relevant foreign government or competent international organization or a document equivalent thereto.

For contact information, state the address, name, corporation name, section in charge, person in charge, etc. to enable receiving of mail.

(ii) the design or type certification number of the launch vehicle

If the launch vehicle has not obtained a type certification for a launch vehicle, describe the design of the launch vehicle in Attachment 1 to Form 1.

If a launch vehicle which has obtained a type certification is to be used, write the type certification number. In this case, it is not necessary to submit Attachment 1 to Form 1.

(iii) the place, design and facility of launch site or compliance certification number

If the launch site has not obtained a compliance certification for a launch site, describe the location, design and facility of the launch site in Attachment 2 to Form 1.

If a launch site which has obtained a type certification is to be used, write the compliance certification number. In this case, it is not necessary to submit Attachment 2 to Form 1.

(iv) the launch plan

Fill in the necessary information in Attachment 3 to Form 1, based on Chapter 6 of the Guidelines on Permission Relating to Launching of Spacecraft, etc.

(v) the type, name of vehicle and serial number of the launch vehicle

- Type of launch vehicle
  "Type of launch vehicle" means the name indicating the type of launch vehicle regardless of the configuration of vehicles (e.g. whether the vehicle has an auxiliary booster).
  For example, for existing launch vehicles, the types of launch vehicle are "H-IIA launch vehicle," "Epsilon launch vehicle," etc.
  No name which is offensive or would constitute a trademark infringement is to be used.

- Name of vehicle
  "Name of vehicle" means the name which differs depending on the configuration of vehicles.
  For example, for an H-IIA launch vehicle, the name of vehicle is H2A202 Type, H2A204 Type, etc.

ANNEX 3

No name which is offensive or which would constitute a trademark infringement is to be used.

➢ Serial number

For the serial number, assign a number, alphabetical character, etc. to the same type of launch vehicle so as to avoid any overlapping.

> (vi) **number of spacecraft to be loaded on the launch vehicle, as well as the names, purposes and methods of use of the respective spacecraft**

➢ Number of spacecraft

State the number of spacecraft to be loaded on the launch vehicle.

➢ Name of spacecraft

State the names of spacecraft to be loaded on the launch vehicle.

No name which is offensive or which would constitute the trademark infringement is to be used.

➢ Purposes and methods of use

Refer to 6.1.1(v) of this Manual to describe the purpose and methods of use of spacecraft to be loaded on the launch vehicle.

If, at the time of the application, the applicant intends to replace the payload with dummy mass, the applicant is required to state the intention and make a notification upon the replacement. If the applicant replaces the payload with dummy mass without this statement, the applicant needs to obtain permission to make changes.

> (vii) **names of officers or employees in charge of business with respect to the launching of spacecraft, etc.**

State the names and addresses of officers or employees as stated in their residence certificates.

If the applicant is a corporation, state the names and addresses of officers or employees as stated in their residence records.

Here, "employees" means employees of an applicant having authority and responsibilities for the applicant's business with respect to the launching of a spacecraft, etc. (Article 6 of the Regulation), for example, the chief of the section in charge of the business of launching.

> **(viii) whether the applicant falls under any of the grounds for disqualification under Article 5 of the Act**

A person who falls under any of the following grounds is not eligible to obtain permission related to the launching of a spacecraft, etc. Check the box to indicate whether the applicant falls under any of these grounds.

> Article 5 of the Act (Grounds for Disqualification)
> A person who falls under any of the following items may not obtain permission under paragraph (1) of the preceding Article:
> (i) a person who has violated the provisions of this Act or an order based on this Act or the laws and regulations of a foreign country (meaning countries or regions outside Japan; the same applies hereinafter) equivalent thereto, and has been sentenced to a fine or severer punishment (including a punishment under the laws and regulations of a foreign country equivalent thereto), and for whom three years have not elapsed since the date on which execution of the sentence was completed or since the date on which that person ceased to be subject to the execution of the sentence;
> (ii) a person whose permission has been rescinded pursuant to Article 12, and for whom three years have not elapsed since the date of that rescission;
> (iii) an adult ward or a person who is treated in the same manner under the laws and regulations of a foreign country;
> (iv) a corporation whose officers engaged in the business thereof or employees specified by Cabinet Office Order fall under any of the preceding three items; and
> (v) an individual whose employees specified by Cabinet Office Order fall under any of items (i) through (iii).

3.1.2. Attachments

A person who intends to obtain permission related to the launching of a spacecraft, etc. is required to submit a written application referred to in 3.1.1, together with the following documents (Article 5, paragraph (2) of the Regulation).

ANNEX 3

○ With both type certification and compliance certification
   I. Documents pertaining to the applicant

○ With type certification, but without compliance certification
   I. Documents pertaining to the applicant
   II. Documents necessary in relation to a launch site without compliance certification

○ Without type certification or compliance certification
   I. Documents pertaining to the applicant
   II. Documents necessary in relation to a launch site without compliance certification
   III. Documents necessary in relation to a launch vehicle without type certification

---

The following is the guideline on these documents.

I. Documents pertaining to the applicant

○ If the applicant is an individual, the following documents:
1) A copy of the resident record or a document in lieu thereof
   This is limited to a certificate containing the registered domicile (or, in the case of a foreign national, the person's nationality, etc. provided in Article 30-45 of the Residential Basic Book Act (Act No. 81 of 1967)).
2) The following documents related to employees:
   a copy of the resident record or a document in lieu thereof

○ If the applicant is a corporation, the following documents:
1) Its articles of incorporation and certificate of registered information, or a document equivalent thereto;
   If the applicant is a foreign corporation, submit a document issued by the relevant foreign government or competent international organization or a document equivalent thereto that contains the name and head office or principal office of the corporation.
2) The following documents related to the officers and employees as provided in Article 5, item (iv) of the Act:
   a copy of the resident record or a document in lieu thereof

II. Documents necessary in relation to launch site without compliance certification

1) A document certifying that the launch site complies with the type-specific site safety standard
   See 5.1.2 (ii) of the Manual.
2) A document describing the flight termination measures or any other methods of ensuring the safety of the vicinity of the trajectory and launch site of the launch vehicle (in relation to the launch site).
   See 5.1.1 (vi) of the Manual.
3) A document describing the technical conditions for ensuring the compliance of the launch vehicle and launch site, and evidencing the compliance with the conditions (in relation to the launch site).
   See 5.1.2 (iii) of the Manual.

III. Documents necessary in relation to a launch vehicle without type certification

1) A document certifying that the design of the launch vehicle complies with the launch vehicle safety standard
   See 4.1.2 (ii) of the Manual.
2) A document describing the flight termination measures or any other methods of ensuring the safety of the vicinity of the trajectory and launch site of the launch vehicle (in relation to the launch vehicle)
   See 4.1.1 (iii) of the Manual.
3) A document describing the technical conditions for ensuring the compliance of the launch vehicle and launch site (in relation to the launch vehicle)
   See 4.1.4 (iv) of the Manual.
4) A document describing the flight records or test results of the launch vehicle
   See 4.1.2 (iii) of the Manual.
5) A document describing the results of assessment of reliability of the launch vehicle
   See 4.1.2 (iv) of the Manual.
6) A document describing the method of verification that the launch vehicle complies with its design
   See 4.1.2 (v) of the Manual.

3.2. Change related to permission

[Permission relating to the launching of spacecraft, etc.]

ANNEX 3

Article 7 of the Act (Permission, etc. Related to Change)

(1) When a person who obtained the permission under Article 4, paragraph (1) (hereinafter referred to as a "launch operator") intends to change any matter set forth in items (ii) through (v) of paragraph (2) of that Article (including when a change has been made to the launch vehicle safety standard and the design of the launch vehicle for which the permission was granted no longer satisfies the launch vehicle safety standard, and when a change has been made to the type-specific site safety standard and the launch site for which the permission was granted no longer satisfies the type-specific site safety standard), the person must obtain permission from the Prime Minister pursuant to the provisions of Cabinet Office Order; provided, however, that this does not apply to minor changes specified by Cabinet Office Order.

(2) When there has been a change to any of the items set forth in Article 4, paragraph (2), item (i) or (vi), or any minor change specified by Cabinet Office Order as referred to in the proviso to the preceding paragraph, a launch operator must make a notification to the Prime Minister to that effect without delay.

(3) The provisions of the preceding Article apply mutatis mutandis to the permission under paragraph (1).

Article 9 of the Regulation (Application, etc. for Permission Related to Change)

(1) When a launch operator intends to make any change to the matters set forth in Article 4, paragraph (2), items (ii) through (v) of the Act, the launch operator must obtain permission for the change from the Prime Minister, by submitting a written application using Form 3 attaching a document relating to the changed items contained in the documents set forth in Article 5, paragraph (2), item (ii) and (iii) and a copy of the permission certificate under paragraph (4) of that Article pertaining to the launching of the spacecraft, etc.

(2) When the Prime Minister grants the permission to make any changes under Article 7, paragraph (1) of the Act, the Prime Minister is to notify the launch operator to that effect, order the launch operator to return the permission certificate under Article 5, paragraph (4) pertaining to the launching of the spacecraft, etc. and reissue the permission certificate using Form 2.

(3) The minor changes specified by Cabinet Office Order, as referred to in the proviso to Article 7, paragraph (1) of the Act, are changes that would not result in a substantial change in the matters set forth in Article 4, paragraph (2), items (ii) through (v) of the Act.

> (4) When a launch operator intends to make a notification under Article 7, paragraph (2) of the Act, the launch operator must submit to the Prime Minister a written notification using Form 4, attaching a document pertaining to the changed matters and a copy of the permission certificate referred to in Article 5, paragraph (4) pertaining to the launching of the spacecraft, etc.

If any change occurs to the information stated in the application documents, it is necessary to submit an application for permission or a notification of change as follows, depending on the items to be changed and the nature of change. For information on the matters requiring an application for permission or a notification as well as specific examples thereof, see Chapter 7 of Guidelines on Permission Related to Launching of Spacecraft, etc.

3.2.1. Application for permission related to change

An operator that intends to make any change to any of the following matters is required to submit an application for permission to make a change, unless the change would not result in any substantial change.

- the design or type certification number of the launch vehicle
- the place, design and facility of the launch site or compliance certification number
- the launch plan
- the number of spacecraft to be loaded on the launch vehicles, as well as the purposes and methods of use of the respective spacecraft

An applicant for permission to make change is required to submit the following documents (Article 9, paragraph (1) of the Regulation).

> 1) a written application for permission to make change
> 2) documents related to the changed matters
> 3) a copy of the permission certificate

1) a written application for permission to make change
   Submit a written application stating the matters pertaining to the change (Form 3).
2) documents related to the changed matters
   Submit the documents relating to the changed matters contained in the documents specified in II. and III. of 3.1.2.
3) a copy of the permission certificate

Submit a copy of the permission certificate for the launching of a spacecraft, etc. issued.

If the permission is granted, a notice to that effect will be provided. In this case, the applicant will receive a new permission certificate, so the applicant is requested to return the existing permission certificate.

3.2.2. Notification of change

An operator that intends to make any change to any of the following matters is required to submit a notification of change.
- name and address of the launch operator
- the type, name of vehicle and serial number of the launch vehicle
- the names of spacecraft to be loaded on the launch vehicle
- names of officers or employees in charge of business with respect to the launching of a spacecraft, etc.
- whether the operator falls under any of the disqualification grounds under Article 5 of the Act
- the matters which require the application for permission under 3.2.1, that do not involve substantial changes

A notifier of change is required to submit the following documents (Article 9, paragraph (4) of the Regulation).

1) notification of change
2) documents related to the changed matters
3) a copy of the permission certificate

1) notification of change
   Submit a written notification stating the matters related to change (Form 4).
2) documents related to the changed matters
   Submit the documents relating to the changed matters contained in the documents specified in I, II and III of 3.1.2.
3) a copy of the permission certificate
   Submit a copy of the permission certificate for the launching of a spacecraft, etc. issued.

In the case of a notification of change, a new permission certificate will not be issued, and it is not necessary to return the permission certificate already issued.

3.3. Succession

[Permission relating to the launching of a spacecraft, etc.]

Article 10 of the Act (Succession)

(1) When a launch operator transfers the business with respect to the launching of spacecraft, etc. permitted under Article 4, paragraph (1), if the transferrer and the transferee have obtained authorization on that transfer from the Prime Minister in advance pursuant to the provisions of Cabinet Office Order, the transferee succeeds to the status of the launch operator under the provisions of this Act.

(2) When a corporation that is a launch operator is to be extinguished by a merger and if the merger has been authorized in advance by the Prime Minister pursuant to the provisions of Cabinet Office Order, a corporation surviving the merger or a corporation established as a result of the merger succeeds to the status of the launch operator under the provisions of this Act.

(3) When a corporation that is a launch operator has the business with respect to the launching of spacecraft, etc. permitted under Article 4, paragraph (1) succeeded to by corporate split, and that corporate split has been authorized in advance by the Prime Minister pursuant to the provisions of Cabinet Office Order, a corporation which succeeded to the business as a result of the corporate split succeeds to the status of the launch operator under the provisions of this Act.

(4) Articles 5 and 6 (limited to the part concerning item (iii) (limited to the part concerning the ability to execute the launch plan)) apply mutatis mutandis to the authorization under the preceding three paragraphs.

(5) When a launch operator transfers its business with respect to the launching of spacecraft, etc. permitted under Article 4, paragraph (1), or when a corporation that is a launch operator is extinguished by a merger or has the business succeeded to through a corporate split, if a disposition to refuse the authorization under paragraphs (1) through (3) is rendered (if an application for the authorization is not submitted, when the transfer of business, merger or corporate split takes place), the permission under Article 4, paragraph (1) ceases to be effective.

Article 10 of the Regulation (Application for Authorization on Succession of Status of Launch Operator)

(1) A person who intends to obtain authorization under Article 10, paragraph (1) of the Act must submit to the Prime Minister a written application using Form 5,

> attaching the following documents and a copy of the permission certificate under Article 5, paragraph (4) pertaining to the transferrer:
>
> (2) A person who intends to obtain authorization under Article 10, paragraph (2) of the Act must submit to the Prime Minister a written application using Form 6, attaching the following documents and a copy of the permission certificate under Article 5, paragraph (4) for the corporation whose business was succeeded to:
>
> (3) A person who intends to obtain authorization under Article 10, paragraph (3) of the Act must submit to the Prime Minister a written application using Form 7, attaching the following documents and a copy of the permission certificate under Article 5, paragraph (4) for the corporation whose business was succeeded to:

If the succession is authorized and any change occurs to the launch plan (excluding organizational structure for the execution of launch plan), an operator is required to submit an application for permission or a notification of change.

3.3.1. Business transfer

A launch operator that intends to transfer its business with respect to the launching of spacecraft, etc. is required to submit the following documents (Article 10, paragraph (1) of the Regulation).

> 1) a written application for authorization (Form 5)
> 2) a document set forth in Article 5, paragraph (2), item (i) of the Regulation in relation to the transferee (a document related to the applicant)
> 3) a document evidencing that the transferee has a sufficient ability to execute the launch plan (a document related to the organizational structure for executing the launch plan)
> 4) a copy of the contract for the transfer and acquisition
> 5) if the transferrer or the transferee is a corporation, the minutes of resolution of a general meeting of shareholders or general meeting of members or a written consent of members with unlimited liability or all members on the transfer or acquisition, or a document evidencing the decision on the transfer or acquisition
> 6) a copy of the permission certificate

3.3.2. Merger

If a corporation that is a launch operator is to be extinguished by a merger, and if it intends to have a corporation surviving the merger or a corporation to be incorporated in the merger succeed to the business with respect to the launching of spacecraft, etc., the operator is required to submit the following documents (Article 10, paragraph (2) of the Regulation).

---

1) a written application for authorization (Form 6)
2) a document stating the method and conditions of the merger
3) a document set forth in Article 5, paragraph (2), item (i)(b) pertaining to the corporation surviving the merger or corporation to be incorporated in the merger (a document related to the applicant)
4) a document evidencing that the corporation surviving the merger or corporation to be incorporated in the merger has a sufficient ability to execute the launch plan (a document related to the organizational structure for executing the launch plan)
5) a copy of the merger contract and a statement explaining the merger ratio
6) the minutes of resolution of a general meeting of shareholders or general meeting of members or a written consent of members with unlimited liability or all members on the merger, or a document certifying the decision on the merger
7) a copy of the permission certificate

3.3.3. Corporate split

If a corporation that is a launch operator intends to have its business with respect to the launching of spacecraft, etc. succeeded to by a corporate split, it is required to submit the following documents (Article 10, paragraph (3) of the Regulation).

---

1) a written application for authorization (Form 7)
2) a document stating the method and conditions of the corporate split
3) a document set forth in Article 5, paragraph (2), item (i)(b) pertaining to the corporation succeeding to the business with respect to the launching of spacecraft, etc. by the corporate split (a document related to the applicant)
4) a document evidencing that the corporation succeeding to the business with respect to the launching of spacecraft, etc. by the corporate split has a sufficient ability to execute the launch plan (a document related to the organizational structure for executing the launch plan)
5) a copy of a corporate split contract (for the incorporation-type corporate split, a corporate split plan) and a statement explaining a split ratio
6) the minutes of resolution of a general meeting of shareholders or general meeting of members or a written consent of members with unlimited liability or all members

ANNEX 3

> on the corporate split, or a document evidencing the decision on the corporate split
> 7) a copy of the permission certificate

3.4. Notification of expiration of permission

> [Permission relating to the launching of a spacecraft, etc.]
>
> Article 11 of the Act (Expiration of Permission Due to Death or Other Reasons)
>
> In addition to the case provided in paragraph (5) of the preceding Article, if a launch operator falls under any of the following items, the permission under Article 4, paragraph (1) ceases to be effective; in this case, the persons respectively specified in these items must notify the Prime Minister to that effect within thirty days from the date when they fall under the relevant item:
>
> (i) when the person has deceased: the heir;
> (ii) when the corporation has dissolved due to an order commencing bankruptcy proceedings: its bankruptcy trustee;
> (iii) when the corporation has dissolved for reasons other than a merger or an order commencing bankruptcy proceedings: its liquidator; and
> (iv) when the launching of spacecraft, etc. has been completed: an individual who was formerly the launch operator, or an officer representing the corporation which was formerly the launch operator.
>
> Article 11 of the Regulation (Notification of Death or Other Reasons)
>
> (1) When the person specified by the items of Article 11 of the Act makes a notification under that Article, the person must submit a written notification using Form 8 to the Prime Minister.
> (2) If any of items (i) through (iii) of Article 11 of the Act is applicable, a permission certificate under Article 5, paragraph (4) pertaining to the launching of the spacecraft, etc. must be attached to the written notification under the preceding paragraph.

When the launching of a spacecraft, etc. is completed, the permission related to the launching of the spacecraft, etc. ceases to be effective. The former launch operator is required to submit the necessary documents within thirty days from the day of completion of the launching.

> ○ The operator has completed the launching of a spacecraft, etc.
>   Notifier: an individual who was a launch operator or an officer representing the corporation which was a launch operator
>   Documents to be submitted: a written notification (Form 8)

When a launch operator falls under any of the following matters, the permission related to the launching of the spacecraft, etc. ceases to be effective. The persons specified in the respective items are required to submit the necessary documents within thirty days from the day when the person falls under the relevant item.

> ○ If an individual that was a launch operator is dead
>   Notifier: the heir
>   Documents to be submitted: a written notification (Form 8) and an original of the permission certificate
>
> ○ If a corporation that was formerly a launch operator dissolves due to an order commencing bankruptcy proceedings
>   Notifier: the trustee in bankruptcy
>   Documents to be submitted: a written notification (Form 8) and an original of the permission certificate
>
> ○ If a corporation that was formerly a launch operator dissolves due to any reason other than a merger or order commencing bankruptcy proceedings
>   Notifier: the liquidator
>   Documents to be submitted: a written notification (Form 8) and an original of the permission certificate

ANNEX 3

4. Type certification for launch vehicle

> [Type certification for launch vehicle]
>
> Article 13 of the Act (Type Certification)
>
> (1) The Prime Minister grants a type certification for the design of a launch vehicle upon application.
>
> (2) A person who intends to obtain a type certification referred to in the preceding paragraph must submit a written application to the Prime Minister, pursuant to the provisions of Cabinet Office Order, specifying the following information, attaching a document certifying that the design of the launch vehicle complies with the launch vehicle safety standard and other documents specified by Cabinet Office Order.
>
> Article 13 of the Regulation (Application, etc. for Type Certification for Design of Launch vehicle)
>
> (1) A person who intends to obtain a type certification under Article 13, paragraph (1) of the Act must submit a written application using Form 9 to the Prime Minister.
>
> (2) The following documents must be attached to the written application under the preceding paragraph:

If the type certification for a launch vehicle is granted, a notice to that effect is provided, and a type certificate is issued. Refrain from making the type certificate available to the public by such way as posting it on a website, so as to prevent forgery, etc.

4. Application for type certification

4.1.1. Information items to be entered in application form and guide for preparation

A person who intends to obtain a type certification for a launch vehicle needs to submit a written application (Form 9) containing the following information.

If any document, etc. already submitted to another organization contains the following information, it can be used as an attachment by indicating the relevant portion.

> (i) the address, name and contact information
>
> (ii) design of the launch vehicle
>
> (iii) the flight termination measures or other means of ensuring the safety of the vicinity of the trajectory and launch site of the launch vehicle

> (iv) the technical conditions for ensuring the compliance of the launch vehicle and launch site

The following is the guideline on these information items.

(i) the address, name and contact information

○ If a person who intends to obtain a type certification for a launch vehicle is an individual:
- Write the name and address as stated in the residence record.
- If the applicant is a foreign national, write the address and name as stated in a document issued by the relevant foreign government or a document equivalent thereto.

○ If a person who intends to obtain a type certification for a launch vehicle is a corporation:
- State the corporation name and address as stated in the certificate of registered information.
- If the applicant is a foreign corporation, write the name and head office or principal office of the corporation as stated in a document issued by the relevant foreign government or competent international organization or a document equivalent thereto.

For contact information, state the address, name, corporation name, section in charge, person in charge, etc. to enable the receiving of mail.

(ii) design of the launch vehicle

Provide an overview of the system and launching capability of launch vehicle, and an overview of the functions related to the flight safety operation, using Attachment 1 to Form 1.

An applicant may submit a single application for a type certification for a launch vehicle in relation to two or more configurations of vehicles (e.g. whether the launch vehicle has an auxiliary booster). In this case, if the outcomes of design differ depending on the configurations of vehicles, such as the flight capability, each of these designs need to conform to the launch vehicle safety standard.

Determine and describe the payload mass, orbit and effect of season of launching considering the anticipated range.

> **(iii) the flight termination measures or other means of ensuring the safety of the vicinity of the trajectory and launch site of the launch vehicle**

Describe the approach, specific methods and verification methods for ensuring the safety of the vicinity of the launch site from the lift-off until putting the spacecraft into orbit, in relation to its normal operation, contingent situations and pre-launching. Add attachments to provide additional, detailed explanation if necessary. For flight termination measures, describe the means to terminate the flight and the entire configuration.

> **(iv) the technical conditions for ensuring the compliance of the launch vehicle and launch site**

Describe the technical conditions relating to the interface between the launch vehicle and launch site as well as the method for verifying their compatibility. Add attachments to provide additional, detailed explanation if necessary.

The technical conditions for the launch vehicle and launch site differ depending on their combinations. The following is an example:

➢ Specification of the physical and electronic interface between the launch site and launch vehicle, as well as the interface drawing.
➢ Specification of the launcher of the launch site for the determination of the trajectory at the initial stage of lift-off
➢ Mechanisms for the lift-off count down sequence and emergency suspension of a launch vehicle

4.1.2. Attachments
A person who intends to obtain a type certification for a launch vehicle is required to submit a written application referred to in 4.1.1, together with the following documents (Article 13, paragraph (2) of the Regulation).

(i) documents pertaining to the applicant
(ii) a document certifying that the design of the launch vehicle complies with the launch vehicle safety standard
(iii) a document describing the flight records or test results of the launch vehicle
(iv) a document describing the results of assessment of reliability of the launch vehicle
(v) a document describing the method of verification that the launch vehicle complies with its design

The following is the guideline on these documents.

(i) documents pertaining to the applicant

○ If the applicant is an individual, a copy of the resident record or a document in lieu thereof

This is limited to a certificate containing the registered domicile (or, in the case of a foreign national, the person's nationality, etc. provided in Article 30-45 of the Residential Basic Book Act (Act No. 81 of 1967)).

○ If the applicant is a corporation, its articles of incorporation and certificate of registered information, or a document equivalent thereto

If the applicant is a foreign corporation, submit a document issued by the foreign government or competent international organization or a document equivalent thereto that contains the name and head office or principal office of the corporation.

(ii) a document certifying that the design of the launch vehicle complies with the launch vehicle safety standard

Refer to Chapter 6 of the Guidelines on Type Certification for Launch Vehicles and indicate a design drawing, system block drawing, analytical findings, test results, etc. evidencing the conformity of the design of the launch vehicle with the launch vehicle safety standard (Article 7 of the Regulation). The following are examples. The applicant may use existing documents, such as design drawings, by making references to the relevant part.

➢ Flight capability
- flight plan

- the orbit to be inserted, the spacecraft to be loaded
- system configuration of the launch vehicle and allocation of propellant
- sequence of events for the flight
- the profiles of nominal trajectory and dispersed trajectory
- flight safety operation
- also consider the feasibility of flight safety, including the position of the planned impact area for separated objects
- conditions for the calculation of the nominal trajectory and dispersed trajectory
  - including the conditions of inputting information on anticipated launch pad (e.g. positional coordinate, altitude), vehicle (e.g. mass, propulsion capacity, thrust pattern), environmental conditions (e.g. wind, atmospheric density), as well as explanation of the relevant data.
  - Nominal trajectory and dispersed trajectory, the source of dispersion of data (dispersion of the conditions including thrust and wind)
  - data evidencing that the flight will be controlled in an appropriate way
  - in case of an application for a type certification for two or more configurations of vehicles (e.g. whether the launch vehicle has an auxiliary booster), the results of design for each configuration of vehicle

➢ Measure for preventing the failure and malfunction of ignition device, etc.
- Outline of system of ignition device, etc. of the launch vehicle
- Failure tolerant design
  - block diagram *including information on power system and control line
  - result of verification test of failure tolerant design *including the conditions for the test
- Shield design of pyrotechnics
  - specifications of pyrotechnics and components such as shields
  - description of pyrotechnic igniting mechanisms
  - block diagram *including information on power system and control line
  - results of verification of tests
  - results of verification of electromagnetic compatibility *Also consider the margin for maximum non-ignition energy

➢ Function for flight safety operation
- System overview of data measurement and transmission for flight safety operation
- information on functions, main specifications and performances of each

component (e.g. frequency band, transmission power, transmission cycle, modulation methods and error budget)
- block diagram *including power system
- positions and instrumentation information on equipment onboard the vehicle *all related equipment, including a transceiver and antenna
- specifications of data interface with the ground station (e.g. data format, frequency, transmission cycle, delay)
- results of function test and RF link test

➢ Function for flight termination
- overview of flight termination functions system (e.g. command receiver and equipment for flight termination) *including a description of flight termination means
- information on functions, main specifications and performances of each component (e.g. frequency band, transmission power, transmission cycle, modulation methods and error budget)
- description of flight termination mechanism *status, transition of status and flow of data (control signals) for each stage of receiving a command, activation of flight termination mechanisms, and termination of the flight.
- description of the destruction mechanism for inadvertent separation *a description of a mechanism for an automatic termination of the flight in case of the inadvertent separation of a stage loaded with a command receiver when the flight termination is to be executed by way of receiving a signal from the ground, or of a stage other than the stage loaded with a device for determination when the flight termination is to be executed by the launching vehicle.
- block diagram *including power system
- positions and instrumentation information on equipment onboard the vehicle *all related equipment, including a receiver, antenna and flight termination mechanisms
- failure tolerant design *when it is used as a protection measure against an unexpected activation
  - block diagram *including information on control line
  - results of verification test of failure tolerant design *including the conditions for the test
- shield design of pyrotechnics *in cases of a flight termination system using pyrotechnics

ANNEX 3

- block diagram
- results of verification of electromagnetic compatibility *Also consider the margin for maximum non-ignition energy
  - specifications of data interface with the ground station (e.g. data format, encryption, health check, delay)
  - results of function test and RF link test
  - results of analysis of shape of fragments caused by the termination of flight (e.g. status of vehicle and fragments after the termination of flight)
  - result of analysis of expected casualties (Ec)
➤ Reliability and redundancy of safety-critical systems, etc.
  - results of reliability analysis and tests
  - block diagram *including power system, and can be omitted in case of duplication with the equipment indicated above.
  - positions and instrumentation information on equipment onboard the vehicle *can be omitted in case of duplication with the equipment indicated above.
  - results of function test and RF link test
➤ Mitigation of the generation of orbital debris relating to the separation of spacecraft, etc.
  - design to mitigate the generation of debris
    - overview of the system
    - block diagram
    - test results
➤ Mitigation of the generation of orbital debris relating to the orbital stage
    - design to mitigate the generation of debris
    - overview of the system
    - block diagram
    - test results
    - operation plan
  - in cases of reentry after the completion of the mission, the results of performance test, assessment of ground risks and operation plan.

(iii) a document describing the flight records or test results of the launch vehicle

For the same type of launch vehicle as one which was successfully launched in the past, submit a document containing the following information on the past flight

record. In addition, submit test results, etc. evidencing the performance of a solid rocket engine and liquid rocket motor.
- total number of occasions of launch
- date of launch, launch site, information on payload (name, mass and orbit), orbit insertion error, whether the vehicle was successfully launched.
- whether there was any material change in the safety-critical system, etc. since the commencement of the launch of the same type of launch vehicle, and if so, the outline and reasons for the change.
- other matters for special attention (e.g. result of investigation of the cause of critical failure and measures taken to prevent recurrence)

For a launch vehicle which has no record of launching, describe the conditions and results of an engine combustion test and flight test for a similar launch vehicle (a non-orbital flight is also acceptable).

(iv) a document describing the results of assessment of reliability of the launch vehicle

Analyze the reliability of the launch vehicle using appropriate evaluation methods, in relation to various fields including avionics, propulsion system and structural body, and describe the probability of an event requiring the termination of flight of launch vehicle for each launch phase.
If the flight termination involves the destruction of a liquid rocket engine, solid rocket motor, etc., describe the size and dispersion of fragments after the destruction which would be necessary for the calculation of expected casualties (Ec).

(v) a document describing the method of verification that the launch vehicle complies with its design

Explain the means for verification that the launch vehicle actually manufactured complies with its design, including an inspection manual.

4.2. Change of certification

[Type certification for a launch vehicle]
Article 14 of the Act (Change of Design, etc.)
(1) When a person who obtained a type certification under paragraph (1) of the

preceding Article intends to change any matter set forth in item (ii) of paragraph (2) of that Article (including when a change has been made to the launch vehicle safety standard and the design of the launch vehicle for which the type certification was granted no longer satisfies the launch vehicle safety standard), the person must obtain authorization from the Prime Minister pursuant to the provisions of Cabinet Office Order; provided, however, that this does not apply to minor changes specified by Cabinet Office Order.
(2) When there has been a change to any of the items set forth in Article 13, paragraph (2), item (i) or (iii), or any minor change specified by Cabinet Office Order as referred to in the proviso to the preceding paragraph, the person who obtained a type certification under Article 13, paragraph (1) must make a notification to that effect to the Prime Minister without delay.
(3) The provisions of paragraph (3) of the preceding Article apply mutatis mutandis to the approval under paragraph (1).

Article 14 of the Regulation (Application, etc. for Change of Designs, etc.)
(1) When a person who obtained a type certification under Article 13, paragraph (1) of the Act intends to make any changes to the matters set forth in item (ii) of paragraph (2) of that Article, the person must obtain authorization for the change from the Prime Minister, by submitting a written application using Form 11 attaching the following documents:
  (i)   a document relating to the changed items contained in the documents set forth in items (ii) through (iv) of paragraph (2) of the preceding Article;
  (ii) a document certifying that the changed design of the launch vehicle satisfies the launch vehicle safety standard provided in Article 7; and
  (iii) a copy of the type certificate under Article 13, paragraph (4) of the Act.
(2) When the Prime Minister grants authorization to make any changes under Article 14, paragraph (1) of the Act, the Prime Minister is to notify the person who obtained the type certification under Article 13, paragraph (1) of the Act to that effect, order the person to return the type certificate under paragraph (4) of that Article pertaining to the type certification for the design of the launch vehicle and reissue a type certification using Form 10.
(3) The minor changes specified by Cabinet Office Order, as referred to in the proviso to Article 14, paragraph (1) of the Act, are changes that would not result in a substantial change in the matters set forth in Article 13, paragraph (2), item (ii) of the Act.

> (4) When a person who obtained a type certification under Article 13, paragraph (1) of the Act intends to make a notification under Article 14, paragraph (2) of the Act, the person must submit to the Prime Minister a written notification using Form 12, attaching a document pertaining to the changed matters and a copy of the type certificate referred to in Article 13, paragraph (4) of the Act.

If any change occurs to the information stated in the application documents, it is necessary to submit an application for authorization or a notification of change as follows, depending on the items to be changed and the nature of change. For information on the matters requiring an application for authorization or a notification as well as specific examples thereof, see Chapter 7 of Guidelines on Type Certification for Launch Vehicles.

4.2.1. Application for authorization of change

An operator that intends to make any change to any of the following matters is required to submit an application for authorization related to change, unless the change would not result in any substantial change.
- design of the launch vehicle

An applicant for authorization related to change is required to submit the following documents (Article 14, paragraph (1) of the Regulation).

> 1) a written application for permission to make change
> 2) documents related to the changed matters
> 3) a document certifying that the changed design of the launch vehicle satisfies the launch vehicle safety standard
> 4) a copy of the type certificate

1) a written application for permission to make change
   Submit a written application stating the matters pertaining to the change (Form 11).
2) documents related to the changed matters
   Submit the documents relating to the changed matters contained in the documents specified in (iii) through (v) of 4.1.2.
3) a document certifying that the design of the launch vehicle complies with the launch vehicle safety standard
   Submit a document referred to in 4.1.2(ii) after the change.

4) a copy of the type certificate

Submit a copy of the type certificate issued.

If the authorization is granted, a notice to that effect will be provided. In this case, the applicant will receive a new type certificate, so the applicant is requested to return the existing type certificate.

4.2.2. Notification of change

An operator that intends to make any change to any of the following matters is required to submit a notification of change.

- name of person who obtained the type certification
- the flight termination measures or other means of ensuring the safety of the vicinity of the trajectory and launch site of the launch vehicle
- the technical conditions for ensuring the compliance of the launch vehicle and launch site
- the matters which require the application for authorization under 4.2.1 and do not involve substantial changes

The operator is required to submit the following documents (Article 14, paragraph (4) of the Regulation).

1) a written notification of change
2) documents related to the changed matters
3) a copy of the type certificate

1) a written notification of change

Submit a written notification stating the matters related to change (Form 12).

2) documents related to the changed matters

Submit the documents relating to the changed matters contained in the documents specified in 4.1.2.

3) a copy of the type certificate

Submit a copy of the type certificate issued.

In the case of a notification of change, a new type certificate will not be issued, and it is not necessary to return the type certificate already issued.

5. Compliance certification for launch site

[Compliance certification for a launch site]

Article 16 of the Act (Compliance Certification)

(1) The Prime Minister grants a compliance certification with respect to a launch site located in Japan or onboard a ship or aircraft with Japanese nationality, according to the type of launch vehicle related to the launching of spacecraft, etc. to be implemented using the relevant launch site (limited to a type of launch vehicle whose design has obtained a type certification under Article 13, paragraph (1) or foreign certification), upon application.

(2) A person who intends to obtain a compliance certification under the preceding paragraph must submit a written application to the Prime Minister, pursuant to the provisions of Cabinet Office Order, specifying the following information, attaching a document certifying that the launch site complies with the type-specific site safety standard and other documents specified by Cabinet Office Order.

Article 16 of the Regulation (Application, etc. for Compliance Certification for Launch Site)

(1) A person who intends to obtain a compliance certification under Article 16, paragraph (1) of the Act must submit a written application using Form 13 to the Prime Minister.

(2) The following documents must be attached to the written application under the preceding paragraph:

If the compliance certification for a launch site is granted, a notice to that effect is provided, and a launch site certificate is issued. Refrain from making the launch site certificate available to the public by such way as posting it on a website, so as to prevent forgery, etc.

5.1. Application for compliance certification
5.1.1. Information items to be entered in application form and guide for preparation

A person who intends to obtain a compliance certification for a launch site needs to submit a written application (Form 13) containing the following information.

If any document, etc. already submitted to another organization contains the following information, such document can be used as an attachment by indicating the relevant portion.

ANNEX 3

> (i) the address, name and contact information
> (ii) the place, design and facility of launch site
> (iii) the type certification number
> (iv) type
> (v) date of type certification
> (vi) the flight termination measures or other means of ensuring the safety of the vicinity of the trajectory and launch site of the launch vehicle

The following is the guideline on these information items.

(i) the address, name and contact information

○ If a person who intends to obtain a compliance certification for a launch site is an individual:
- Write the name and address as stated in the residence record.
- If the applicant is a foreign national, write the address and name as stated in a document issued by the relevant foreign government or a document equivalent thereto.

○ If a person who intends to obtain a compliance certification for a launch site is a corporation:
- State the corporation name and address as stated in the certificate of registered information.
- If the applicant is a foreign corporation, write the name and head office or principal office of the corporation as stated in a document issued by the relevant foreign government or competent international organization or a document equivalent thereto.

For contact information, state the address, name, corporation name, section in charge, person in charge, etc. to enable receiving of mail.

(ii) the place, design and facility of launch site

➢ "Place"
State the address of the launch site (meaning a facility equipped with a function of lifting off a launch vehicle). If facilities necessary for the launching, including radio

equipment, are located in different places, state the respective addresses.

➤ "Design and facility"

Describe the constituent elements of the launch site and show a layout drawing. Although the design and composition of a launch site differs, the following composition may be possible, for example:
- place of storage of hazardous materials including explosives
- buildings for assembling launch vehicles and spacecraft
- vicinity of launch pad
- buildings for flight safety operation

In the description, the applicant is to show that it is possible to ensure the safety of the vicinity of the launch site. For this purpose, the boundary with outside and facilities to prevent the entry of third parties are to be included in the layout drawing.

(iii) the type certification number

The compliance certification for a launch site is required for each type of launch vehicle.
State the type certification number of the launch vehicle to be launched at the launch site.

(iv) type

State the type of the launch vehicle subject to the type certification that is to be launched at the launch site.

(v) date of type certification

State the date of the type certification for the launch vehicle subject to the type certification that is to be launched at the launch site.

(vi) the flight termination measures or other means of ensuring the safety of the vicinity of the trajectory and launch site of the launch vehicle

Describe the approach, specific methods and verification methods for ensuring the safety of the vicinity of the launch site from the launching until putting the spacecraft into orbit, in relation to its normal operation, contingent situations, and pre-launching. Add attachments to provide additional, detailed explanation.

For flight termination measures, describe the means to terminate the flight and the entire configuration.

5.1.2. Attachments

A person who intends to obtain a compliance certification for a launch site is required to submit a written application referred to in 5.1.1, together with the following documents (Article 16, paragraph (2) of the Regulation).

(i) documents pertaining to the applicant

(ii) a document certifying that the launch site complies with the type-specific site safety standard

(iii) a document describing the technical conditions for ensuring the compliance of the launch vehicle and launch site, and evidencing the compliance with the conditions

---

The following is the guideline on these documents.

(i) documents pertaining to the applicant

○ if the applicant is an individual, a copy of the resident record or a document in lieu thereof

This is limited to a certificate containing the registered domicile (or, in the case of a foreign national, the person's nationality, etc. provided in Article 30-45 of the Residential Basic Book Act (Act No. 81 of 1967)).

○ if the applicant is a corporation, its articles of incorporation and certificate of registered information, or a document equivalent thereto

If the applicant is a foreign corporation, submit a document issued by the foreign government or competent international organization or a document equivalent thereto that contains the name and head office or principal office of the corporation.

(ii) a document certifying that the launch site complies with the type-specific site safety standard

Refer to Chapter 6 of the Guidelines on Compliance Certification for Launching Sites and indicate a design drawing, system block drawing, analytical findings, test results, etc. evidencing the conformity of the launch site with the type-specific site safety standard (Article 8 of the Regulation). The following are examples. The applicant may use existing documents, such as design drawings, by making references to the relevant part.

- Securing restricted areas and measures to prevent entry of third parties
    - Securing of restricted areas
        - An explanation that restricted areas can be secured.
    - Measures to prevent entry of third parties
        - Explanation of measures to prevent entry of third parties, including fences, entry and leaving management system, surveillance by guards
- Launcher
    - System overview of launcher
    - Means to ensure safety and design information
- Measure for preventing the failure and malfunction of ignition device, etc.
    - System overview of ignition device, etc. of the launch vehicle system including the launch site
    - Failure tolerant design
        - block diagram *including information on power system and control line
        - result of verification test of failure tolerant design *including the conditions for the test
    - Shield design of pyrotechnics
        - specifications of pyrotechnics and components such as shields
        - description of pyrotechnics igniting mechanisms
        - block diagram *including information on power system and control line
        - result of verification of tests
        - results of verification of electromagnetic compatibility *Also consider the margin for maximum non-ignition energy
- Function for flight safety operation
    - System overview of data measurement and transmission for the flight safety operation
    - Information on functions, main specifications and performances of each component (e.g. frequency band, transmission power, transmission cycle, modulation methods and error budget)

- Block diagram *including power system
- Specifications of data interface with the launch vehicle (e.g. data format, frequency, transmission cycle, delay)
- Results of function test and RF link test

➢ Function for flight termination (explanation of functions including the facilities and launch vehicle body)
- Outline of flight termination functions system (e.g. command transmitter) *including a description of flight termination means
- Information on functions, main specifications and performances of each component (e.g. frequency band, transmission power, transmission cycle, modulation methods and error budget)
- a connection between ground systems and airborne flight termination functions *for example, method of verification and answer back for system operations of the vehicle including the stages of command transmission, activation of the flight termination mechanisms and the completion of the flight termination as well as the ground systems.
- block diagram *including power system
- failure tolerant design *in case where it is used as a protection against an unexpected activation
    - block diagram *including control line
    - result of verification test of failure tolerant design *including the conditions for the test
- consideration for the prevention of unexpected activation of airborne pyrotechnics
    - measures for the prevention of unexpected RF radiation
    - appropriate layout of RF link equipment.
    - ensuring a time period during which electromagnetic radiation is prohibited
- specifications of data interface with the launch vehicle (e.g. data format, encryption, health check, delay)
- results of function test and RF link test
- function for data collection after the termination of flight

➢ Reliability and redundancy of safety-critical systems, etc.
- results of reliability analysis and tests
- block diagram *including power system, and can be omitted in case of duplication with the equipment indicated above.

· results of function test and RF link test

> **(iii) a document describing the technical conditions for ensuring the compliance of the launch vehicle and launch site, and evidencing the compliance with the conditions**

Describe the technical conditions relating to the interface between the launch vehicle and launch site as well as the method for verifying their compatibility. Add attachments to provide additional, detailed explanation.

The technical conditions for the launch vehicle and launch site differ depending on their combinations. The following is an example:

- Specification of the physical and electronic interface between the launch site and launch vehicle, as well as the interface drawing.
- Specification and performance of the launcher of the launch site for the determination of the trajectory at the initial stage of lift-off
- Mechanisms for the lift-off count down sequence and emergency suspension of a launch vehicle

5.2. Change of certification

> [Compliance certification for launch site]
>
> Article 17 of the Act (Change of Location, etc. of Launch Site)
>
> (1) When a person who obtained a compliance certification under paragraph (1) of the preceding Article intends to change any matter set forth in item (ii) or (iv) of paragraph (2) of that Article (including the case when a change has been made to the type-specific site safety standard and the launch site for which the compliance certification was granted no longer satisfies the type-specific site safety standard), the person must obtain authorization from the Prime Minister pursuant to the provisions of Cabinet Office Order; provided, however, that this does not apply to minor changes specified by Cabinet Office Order.
>
> (2) When there has been a change to any of the matters set forth in Article 16, paragraph (2), item (i) or (v), or any minor changes specified by Cabinet Office Order as referred to in the proviso to the preceding paragraph, the person who obtained a compliance certification under Article 16, paragraph (1) must make a notification to that effect to the Prime Minister without delay.

(3) The provisions of paragraph (3) of the preceding Article apply mutatis mutandis to the approval under paragraph (1).

Article 17 of the Regulation (Application, etc. for Change of Location of Launch Site and Other Matters)

(1) When a person who obtained a compliance certification under Article 16, paragraph (1) of the Act intends to make any changes to the matters set forth in item (ii) or (iv) of paragraph (2) of that Article, the person must obtain authorization on the change from the Prime Minister, by submitting a written application using Form 15 attaching the following documents:
 (i) the document relating to the changed items set forth in the documents provided in item (ii) of paragraph (2) of the preceding Article;
 (ii) a document certifying that the launch site after the change satisfies the type-specific site safety standard provided in Article 8; and
 (iii) a copy of the launch site certificate under Article 16, paragraph (4) of the Act.
(2) When the Prime Minister grants the authorization to make any changes under Article 17, paragraph (1) of the Act, the Prime Minister is to notify the person who obtained the compliance certification under Article 16, paragraph (1) to that effect, order the person to return the launch site certificate under paragraph (4) of that Article pertaining to the compliance certification for the launch site and reissue a launch site certificate using Form 14.
(3) The minor changes specified by Cabinet Office Order, as referred to in the proviso to Article 17, paragraph (1) of the Act, are changes that would not result in a substantial change in the matters set forth in Article 16, paragraph (2), items (ii) or (iv) of the Act.
(4) When a person who obtained a compliance certification under Article 16, paragraph (1) of the Act intends to make a notification under Article 17, paragraph (2) of the Act, the person must submit to the Prime Minister a written notification using Form 16, attaching a document pertaining to the changed matters and a copy of the launch site certificate referred to in Article 16, paragraph (4) of the Act.

If any change occurs to the information stated in the application documents, it is necessary to submit an application for authorization or a notification of change as follows, depending on the items to be changed and the nature of change. For information on the matters requiring an application for authorization or a notification as well as specific examples thereof, see Chapter 7 of Guidelines on

Compliance Certification for Launching Sites.

5.2.1. Application for authorization of change
An operator that intends to make any change to any of the following matters is required to submit an application for authorization related to change, unless the change would not result in any substantial change.
- the place, design and facility of the launch site
- the flight termination measures or other means of ensuring the safety of the vicinity of the trajectory and launch site of the launch vehicle

An applicant for authorization related to change is required to submit the following documents (Article 17, paragraph (1) of the Regulation).

---
1) a written application for authorization to make change
2) documents related to the changed matters
3) a document certifying that the launch site after the change satisfies the type-specific site safety standard
4) a copy of the launch site certificate

---

1) a written application for authorization to make change
   Submit a written application stating the matters pertaining to the change (Form 15).
2) documents related to the changed matters
   Submit a document referred to in 5.1.2(iii) related to the change.
3) a document certifying that the launch site after the change satisfies the type-specific site safety standard
   Submit a document referred to in 5.1.2(ii) after the change.
4) a copy of the launch site certificate
   Submit a copy of the launch site certificate issued.

If the authorization is granted, a notice to that effect will be provided. In this case, the applicant will receive a new launch site certificate, so the applicant is requested to return the existing launch site certificate.

5.2.2. Notification of change
An operator that intends to make any change to any of the following matters is required to submit a notification of change.

- name of person who obtained the compliance certification
- the type of the launch vehicle
- the date of type certification for the launch vehicle
- the matters which require the application for authorization under 5.2.1, that do not involve substantial changes

A notifier of change is required to submit the following documents (Article 17, paragraph (4) of the Regulation).

> 1) a written notification of change
> 2) documents related to the changed matters
> 3) a copy of the launch site certificate

1) a written notification of change
   Submit a written notification stating the matters related to change (Form 16).
2) documents related to the changed matters
   Submit the documents relating to the changed matters contained in the documents specified in 5.1.2.
3) a copy of the launch site certificate
   Submit a copy of the launch site certificate issued.

In the case of a notification of change, a new launch site certificate will not be issued, and it is not necessary to return the launch site certificate already issued.

6. License related to control of spacecraft

> [License related to control of spacecraft]
> Article 20 of the Act (License)
> (1) A person who intends to implement the control of a spacecraft using a spacecraft control facility located in Japan must obtain a license from the Prime Minister for each of the spacecraft.
> (2) A person who intends to obtain the license under the preceding paragraph must submit a written application to the Prime Minister, pursuant to the provisions of Cabinet Office Order, specifying the following information, attaching the documents specified by Cabinet Office Order:
>
> Article 20 of the Regulation (Application, etc. for License Related to Control of Spacecraft)
> (1) A person who intends to obtain the license under Article 20, paragraph (1) of the Act must submit a written application using Form 17 to the Prime Minister.
> (2) The following documents must be attached to the written application under the preceding paragraph.

The specific examples of cases requiring or not requiring license related to the control of a spacecraft are as follows. If an applicant is unable to precisely determine whether the application for license is required, consult with the National Space Policy Secretariat, Cabinet Office (hereinafter the "NSPS") in advance.

○ Cases requiring license related to the control of a spacecraft
- a case where the applicant intends to control a spacecraft launched by a launch vehicle in Japan or abroad by using a spacecraft control facility located in Japan.
- a case where the applicant intends to control the spacecraft by generating a signal for identifying and controlling the position of a spacecraft, in whole or part, at an operational site in Japan (i.e. the place where the computer is located), and transmitting the signal from a ground station in a foreign country through a network, etc. without using any ground station located in Japan.
- a case where the applicant intends to implement the initial operation of a spacecraft to be delivered to a foreign country using a spacecraft control facility located in Japan.
- a case where the applicant intends to start the control of a spacecraft which was controlled by using only a spacecraft control facility located in a foreign country

before the termination of the control using the former facility, using a spacecraft control facility located in Japan.
- a case where the applicant intends to start the control of a spacecraft which was transported to a space station as a cargo of a space station cargo vehicle and then deployed by the space station, using a spacecraft control facility located in Japan.
- a case where the applicant intends to control an object separated from a spacecraft, using a spacecraft control facility located in Japan, after the separation thereof.

○ Cases not requiring license related to the control of a spacecraft
- a case where the spacecraft is to be launched using a launch vehicle in Japan but not to be controlled using a spacecraft control facility located in Japan.
- a case where the applicant intends to control a spacecraft by generating a signal for identifying and controlling the position of spacecraft at an operational site outside Japan, and transmitting the signal from a ground station in Japan through a network, etc.
- a case where the applicant intends to monitor the position, attitude and condition of the spacecraft using a spacecraft control facility but does not intend to control all of these (a case where the applicants does not intend to conduct any bus control at all).
- a case where the applicant intends to operate the spacecraft in an integral manner by transporting it to a space station as a cargo of a space station cargo vehicle and then placing it inside or outside the space station.
- a case where the applicant does not intend to control any objects separated from a spacecraft, after the separation thereof (a case where the applicant does not intend to monitor the position, attitude or condition of the spacecraft or where the applicant does not intend to control all of them).

If the license related to the control of a spacecraft is granted, a notice to that effect is provided, and a license certificate for control of a spacecraft is issued. Refrain from making the license certificate for control of a spacecraft available to the public by such way as posting it on a website, so as to prevent forgery, etc.

6.1. Written application for license
6.1.1. Information items to be entered in the application form and guides for

preparation

A person who intends to obtain a license related to the control of a spacecraft needs to submit a written application (Form 17) containing the following information.

If any document, etc. already submitted to another organization contains the following information, it can be used as an attachment by indicating the relevant portion.

(i)	the address, name and contact information
(ii)	the name of spacecraft
(iii)	the location of the spacecraft control facility
(iv)	the orbit of spacecraft
(v)	purposes and methods of use of spacecraft
(vi)	the configuration of spacecraft
(vii)	content of the termination measures specified in Article 22, item (iv) of the Act
(viii)	control plan
(ix)	name and address of the representative in case of death
(x)	name of officers or employees in charge of business of control of spacecraft, etc.
(xi)	whether the applicant falls under any of the grounds for disqualification under Article 21 of the Act

The following is the guideline on these information items.

(i) the address, name and contact information

○ If the person who intends to obtain a license to control spacecraft is an individual:
- Write the name and address as stated in the residence record.
- If the applicant is a foreign national, write the address and name as stated in a document issued by the relevant foreign government or a document equivalent thereto.

○ If the person who intends to obtain a license to control spacecraft is a corporation:
- State the corporation name and address as stated in the certificate of registered information.
- If the applicant is a foreign corporation, write the name and head office or principal office of the corporation as stated in a document issued by the foreign government or competent international organization or a document equivalent

thereto.

For contact information, state the address, name, corporation name, section in charge, person in charge, etc. to enable the receiving of mail.

(ii) the name of spacecraft

State the name of the spacecraft for the control.
No name which is offensive or which would constitute trademark infringement is to be used.

(iii) the location of the spacecraft control facility

State the address of the spacecraft control facility.
If the operational site and ground station are located in different places, state the place of the operational site in this item, and the place of the ground station in "(viii) Control plan." If the operational sites are located in two or more places, also state the addresses of these sites. In addition, if the applicant is not able to precisely determine the operational sites required to be stated depending on the manner of the control of a spacecraft, consult with the NSPS in advance.

(iv) the orbit of spacecraft

Describe the orbit of the spacecraft expected at the time of the application, together with the type (i.e. earth orbit and planetary transfer orbit, etc.) For example, in the case of an earth orbit, state information such as semi-major axis, orbital eccentricity, inclination, right ascension of the ascending node, argument of perigee and time of perigee passage. For the specific values of these items, an applicant is permitted to state the values allowing some realistic margin so as to eliminate the need of making repeated applications or notifications of change.

(v) purposes and methods of use of spacecraft

In relation to the purposes of uses of spacecraft, describe the relevant field and purpose, such as telecommunication, scientific observation and Earth observation, and the relevant method, such as commercial use, use for research and development and academic study.

If the spacecraft covers two or more fields, describe information for all of the applicable items.

(vi) the configuration of spacecraft

State the dimension (at the time of operation), total mass, design life, power system, attitude control system, propulsion system, type of propellant, mass of propellant, major structural material and main onboard devices in Attachment 1·1 of Form 17. For the specific values of these items, including a value to be determined immediately before the time of launch such as the mass of propellant, an applicant is permitted to state the values allowing some realistic margin so as to eliminate the need for making repeated applications or notifications of change. In addition, include an overview and a spacecraft system diagram of the spacecraft in Attachment 1·2 of Form 17.

Describe other design measures for the configuration of spacecraft in Attachments 1·3 through 1·5 according to 6.2 of the Guidelines on License Related to Control of Spacecraft.

(vii) content of the termination measures specified in Article 22, item (iv) of the Act

Measures for termination of a spacecraft include a measure to lower the altitude of the spacecraft (including a measure to lower the orbit by natural decay) and burn it in the atmosphere, a measure to elevate the orbit so as to avoid the risk of interfering with the control of other spacecraft, and a measure to guide the spacecraft to be put into the orbit of or fall to a celestial body other than the Earth.

Indicate the measures specified in Article 22, item (iv) of the Act to be adopted, as well as the details thereof.

If the applicant intends to use two or more termination measures at the time of the application, describe all of them.

(viii) control plan

For the control of a spacecraft, an applicant needs to devise plans for measures required for avoiding a collision with another spacecraft and other measures for the

ANNEX 3

prevention of harmful contamination, etc. of outer space. In addition, an applicant needs to prove that it has sufficient ability for the execution of the control plan. If the applicant is an individual, the applicant also needs to prove that the representative in case of death also has the ability.

For this purpose, describe the following information in Attachment 2 to Form 17, in accordance with 6.3 of the Guidelines on License Related to Control of Spacecraft.

- Overview of spacecraft control facility
  - Describe the operational site for the control of spacecraft (i.e. the place where the computer is located) and ground station.
- Method of control of spacecraft
  - Describe the outline of the communication route for receiving and transmitting signals between the operational site and the spacecraft, including a ground station.
  - Describe the details of operation for each operational phase.
  - If the spacecraft is to constitute a constellation, and if the control of another spacecraft has already commenced, describe the information identifying the spacecraft, including its name. If a license has been obtained for the control of the spacecraft, describe the name and license number of the spacecraft.
- Prevention of interference with the control of other spacecraft upon separation or docking
  - If any components, etc. of the spacecraft is to be separated or docked, describe the procedures for control so as to avoid any adverse effect on the control of other spacecraft.
- Prevention of break-up upon the occurrence of anomalies
  - state the fact that a measure to prevent the break-up of the spacecraft or termination measures will be implemented in the case of detection of any anomalies in the position, attitude and condition of the spacecraft.
- Prevention of collision with other spacecraft, etc.
  - In the case of a spacecraft capable of transferring to another obit, describe the method of obtaining information on the probability of a collision with other spacecraft, etc., and the condition for determining whether to take measures to avoid collision if the relevant information is obtained.
- Termination measures
  - Describe the details of measures to be taken as the termination measures of the spacecraft.

- Establishment of organizational structure for the implementation of the control of spacecraft
  - state the fact that appropriate organizational structures will be established, including organization and business relating to control, response to anomalies, establishment of security measures and personnel training.

(ix) name and address of the representative in case of death

This information is required when the person who intends to obtain a license related to the control of a spacecraft is an individual.

If the representative in case of death is an individual, state the name and address as indicated in the resident record, or if the representative in case of death is a corporation, state the corporate name and address as indicated in its certificate of registered information.

(x) name of officers or employees in charge of business of control of spacecraft, etc.

State the names and addresses of officers or employees as stated in their residence certificates.

If the applicant is a corporation, state the names and addresses of officers or employees as stated in their residence records.

Here, "employees" means employees of an applicant having authority and responsibilities for the applicant's business with respect to the control of spacecraft (Article 21 of the Regulation), for example, the chief of the section in charge of business of control.

(x) whether the applicant falls under any of the grounds for disqualification under Article 21 of the Act

A person who falls under any of the following grounds is not eligible to obtain a license related to the control of a spacecraft. Check the box to indicate whether the applicant falls under any of these grounds.

Article 21 of the Act (Grounds for Disqualification)
A person who falls under any of the following items may not obtain a license under

paragraph (1) of the preceding Article:

(i) a person who has violated the provisions of this Act or orders based on this Act or the laws and regulations of a foreign country equivalent thereto, and has been sentenced to a fine or severer punishment (including a punishment under the laws and regulations of a foreign country equivalent thereto), and for whom three years have not elapsed since the date on which execution of the sentence was completed or since the date on which that person ceased to be subject to the execution of the sentence;

(ii) a person whose license has been rescinded pursuant to Article 30, paragraph (1), and for whom three years have not elapsed since the date of that rescission;

(iii) an adult ward or a person who is treated in the same manner under the laws and regulations of a foreign country;

(iv) a corporation whose officers engaged in the business thereof or employees specified by Cabinet Office Order fall under any of the preceding three items;

(v) an individual whose employees specified by Cabinet Office Order fall under any of items (i) through (iii); and

(vi) an individual whose representative in case of death falls under any of the preceding items.

6.1.2. Attachments

A person who intends to obtain a license related to the control of a spacecraft is required to submit a written application referred to in 6.1.1, together with the following documents (Article 20, paragraph (2) of the Regulation).

(i) documents pertaining to the applicant

(ii) a document certifying that the configuration of the spacecraft satisfies the standard specified by Article 22 of the Regulation

The following is the guideline on these documents.

(i) documents pertaining to the applicant

○ if the applicant is an individual, the following documents:

1) a copy of the resident record or a document in lieu thereof

This is limited to a certificate containing the registered domicile (or, in the case of a foreign national, the person's nationality, etc. provided in Article 30-45 of the

Residential Basic Book Act (Act No. 81 of 1967)).
2) the following documents related to employees and representative in case of death: a copy of the resident record or a document in lieu thereof

○ if the applicant is a corporation, the following documents:
1) its articles of incorporation and certificate of registered information, or a document equivalent thereto;
If the applicant is a foreign corporation, submit a document issued by the relevant foreign government or competent international organization or a document equivalent thereto that contains the name and head office or principal office of the corporation.
2) The following documents related to the officers and employees as provided in Article 21, item (v) of the Act
a copy of the resident record or a document in lieu thereof

(ii) a document certifying that the configuration of the spacecraft satisfies the standard specified by Article 22 of the Regulation

---

Refer to 6.2 of the Guidelines on License Related to Control of Spacecraft and indicate a design drawing, system block drawing, analytical findings, test results, etc. evidencing the conformity of the configuration of the spacecraft with the standard related to the configuration of spacecraft (Article 22 of the Regulation). The following are examples. The applicant may use existing documents such as design drawings by making references to the relevant part.

➢ Mechanism for the prevention of unintended release of objects
- Indicate that the results of the structural analysis or environmental test have shown that the components of the spacecraft will not easily come off or scatter. Generally, the most severe environmental conditions of a spacecraft are during the period between the lift-off of the launch vehicle and the separation of the spacecraft.
- If the spacecraft has a separable or deployable system including an antenna, solar battery panel, etc. on the orbit, indicate that it has a mechanism to prevent the equipment, etc. from easily scattering during the sequence.
- Indicate that the release of combustion products is minimized.

➢ Mechanism for separation or docking, if applicable

- If the components, etc. of the spacecraft is to be separated or deployed (including the case of separation of a slave spacecraft from a master spacecraft), indicate that it has a mechanism to prevent the equipment, etc. from easily scattering during the sequence. In addition, indicate that the equipment, etc. deployed can be placed on an appropriate orbit so as to avoid a severe adverse effect on the control of a manned spacecraft, etc. and other spacecraft.
- If the docking with other spacecraft, etc. or debris capturing are to be implemented, indicate that the spacecraft has a mechanism to prevent the generation of debris, etc. due to the collision in the course of docking or capturing activities.

➢ Mechanism to prevent break-up
- Indicate that the spacecraft is equipped with a function for transmitting the position, attitude and condition of the spacecraft to the spacecraft control facility, either directly or via other radio equipment, so as to detect any anomalies that may lead to break-up.
- Indicate that the spacecraft is equipped with a function of removing or making safe all remaining energy in the case of anomalies, so as to prevent the spacecraft from breaking up and generating debris.

➢ For spacecraft for reentry into the Earth, a mechanism for the ensuring of public safety
- Indicate that a spacecraft or its components, etc. which are to fall to Earth within or after the period of control of the spacecraft has a mechanism to be completely combusted in the atmosphere, or, if they are not to be completely combusted, that they have a configuration so that the risk to the estimated point of landing or water landing will not exceed the standards stipulated in the international standards or standards provided by space agency of each state.

➢ In the case of retrieving substances derived from another celestial body by guiding them to fall to Earth, a mechanism for the prevention of the deterioration of the environment of the Earth
- In the case of retrieving a spacecraft or its components and parts which were put into orbit around a celestial body other than the Earth or which fell to the celestial body, by guiding them to fall to Earth (including substances obtained from other celestial body or attached substances derived from the other celestial body), indicate that measures in compliance with the

Planetary Protection Policy stipulated by the Committee on Space Research (COSPAR) have been implemented.
- In the case of putting the spacecraft into the orbit around a celestial body other than the Earth or guiding the spacecraft to fall on the celestial body, a mechanism to prevent the harmful contamination of the celestial body
  - In the case of a spacecraft or its components, etc. which are put into orbit around a celestial body other than the Earth or which are to be guided to fall to the celestial body, indicate that measures in compliance with the Planetary Protection Policy stipulated by the Committee on Space Research (COSPAR) have been implemented.

6.2. Change related to permission

[License related to control of spacecraft]

Article 23 of the Act (Permission, etc. Related to Change)

(1) When a person who obtained the license under Article 20, paragraph (1) (hereinafter referred to as a "spacecraft control operator") intends to change any matter set forth in items (iv) through (viii) of paragraph (2) of that Article, the person must obtain a license from the Prime Minister pursuant to the provisions of Cabinet Office Order; provided, however, that this does not apply to minor changes specified by Cabinet Office Order.

(2) When there has been a change to any of the matters set forth in Article 20, paragraph (2), items (i) through (iii) or item (ix), or any minor changes specified by Cabinet Office Order as referred to in the proviso to the preceding paragraph, a spacecraft control operator must make a notification to the Prime Minister to that effect without delay.

(3) The provisions of the preceding Article apply mutatis mutandis to the license under paragraph (1).

Article 25 of the Regulation (Application, etc. for Permission Related to Change)

(1) When a spacecraft control operator intends to make any changes to the matters set forth in Article 20, paragraph (2), items (iv) through (viii) of the Act, the spacecraft control operator must obtain permission from the Prime Minister, by submitting a written application using Form 19 attaching a document relating to the changed items contained in the documents set forth in Article 20, paragraph (2), item (ii) and a copy of the license certificate under paragraph (4) of that Article pertaining to the control of the spacecraft.

ANNEX 3

> (2) When the Prime Minister grants the permission to make any changes under Article 23, paragraph (1) of the Act, the Prime Minister is to notify the spacecraft control operator to that effect, order the spacecraft control operator to return the license certificate under Article 20, paragraph (4) pertaining to the control of the spacecraft and reissue the license certificate using Form 18.
> (3) The minor changes specified by Cabinet Office Order, as referred to in the proviso to Article 23, paragraph (1) of the Act, are changes that would not result in a substantial change in the matters set forth in Article 20, paragraph (2), items (iv) through (viii) of the Act.
> (4) When a spacecraft control operator intends to make a notification under Article 23, paragraph (2) of the Act, the spacecraft control operator must submit to the Prime Minister a written notification using Form 20, attaching a document pertaining to the changed matters and a copy of the license certificate under Article 20, paragraph (4) pertaining to the control of the spacecraft.

If any change occurs to the information stated in the application documents, it is necessary to submit an application for permission or a notification of change as follows, depending on the items to be changed and the nature of change. For information on the matters requiring an application for permission or a notification as well as specific examples thereof, see Chapter 7 of Guidelines on License Related to Control of Spacecraft.

6.2.1. Application for permission related to change

An operator that intends to make any change to any of the following matters is required to submit an application for permission to make change, unless the change would not result in any substantial change.
- purpose and method of use of the spacecraft
- the configuration of the spacecraft
- the details of termination measures
- the control plan
- name and address of the representative in case of death

An applicant for permission to make change is required to submit the following documents (Article 25, paragraph (1) of the Regulation).

1) a written application for permission to make change
2) documents related to the changed matters

3) a copy of the permission certificate

1) a written application for permission to make change
    Submit a written application stating the matters pertaining to the change (Form 19).
2) documents related to the changed matters
    Submit the documents relating to the changed matters contained in the documents specified in 6.1.2.
3) a copy of the license certificate
    Submit a copy of the license certificate of the control of spacecraft issued.

If the permission to make change is granted, a notice to that effect will be provided. In this case, the applicant will receive a new license certificate, so the applicant is requested to return the existing license certificate.

6.2.2. Notification of change

An operator that intends to make any change to any of the following matters is required to submit a notification of change.
- Name and address of the spacecraft control operator
- Place of the spacecraft control facility:
- Orbit of the spacecraft
- Name of officers or employees in charge of business of control of spacecraft, etc.
- Whether the applicant falls under any of the disqualification grounds under Article 21 of the Act
- The matters which require the application for permission under 6.2.1, that do not involve substantial changes

A notifier of change is required to submit the following documents (Article 25, paragraph (4) of the Regulation).

1) a written notification of change
2) documents related to the changed matters
3) a copy of the permission certificate

1) a written notification of change
    Submit a written notification stating the matters related to change (Form 20).
2) documents related to the changed matters

Submit the documents relating to the changed matters contained in the documents specified in 6.1.2.

3) a copy of the license certificate

Submit a copy of the license certificate of the control of spacecraft issued.

In the case of a notification of change, a new license certificate will not be issued, and it is not necessary to return the license certificate already issued.

6.3. Notification in case of accidents

[License related to control of spacecraft]

Article 25 of the Act (Measures in Case of Accident)

If a spacecraft control operator becomes unable to implement the control of the spacecraft without taking any termination measures pertaining to the license under Article 20, paragraph (1) due to the collision of a spacecraft pertaining to the license under that paragraph with another object or any other accidents, and if there is no prospect of recovery, the spacecraft control operator must promptly make a notification to the Prime Minister to that effect, the circumstances of the accident and the matters specified by Cabinet Office Order which assist with the identification of the position of the spacecraft after the occurrence of the accident, pursuant to the provisions of Cabinet Office Order. In this case, the license under that paragraph ceases to be effective.

Article 26 of the Regulation (Notification in Case of Accident)

(1) When a spacecraft control operator intends to make a notification under Article 25 of the Act, the spacecraft control operator must submit a written notification using Form 21 to the Prime Minister.
(2) The matters specified by Cabinet Office Order, as referred to in Article 25 of the Act, are as follows:
   (i) the date, time and location of the accident; and
   (ii) the orbit of the spacecraft after the occurrence of the accident.

If a spacecraft control operator is unable to implement the control of the spacecraft without taking any termination measures due to the collision of the spacecraft and another object or any other accidents, and if there is no prospect of recovery, the spacecraft control operator must submit a notification of accident (Form 21).
State the following information in the notification.

- situation of the accident
- date and place of occurrence of the accident
- the orbit of the spacecraft after the occurrence of the accident

Submit this notification even in the case where the operator is not able to commence the control of spacecraft due to a failure of launch and other reasons.

6.4. Succession

[License related to control of spacecraft]

Article 26 of the Act (Succession)

(1) When a spacecraft control operator intends to transfer its business with respect to the control of spacecraft licensed under Article 20, paragraph (1) to a person who intends to implement the control of spacecraft using a spacecraft control facility located in Japan, if the transferrer and transferee have obtained authorization for that transfer and acquisition from the Prime Minister in advance pursuant to the provisions of Cabinet Office Order, the transferee succeeds to the status of the spacecraft control operator under the provisions of this Act.

(2) When a spacecraft control operator intends to transfer the business with respect to the control of spacecraft licensed under Article 20, paragraph (1) to a person who intends to implement the control of spacecraft without using a spacecraft control facility located in Japan, the spacecraft control operator must make a notification to the Prime Minister in advance to that effect pursuant to the provisions of Cabinet Office Order.

(3) When a corporation that is a spacecraft control operator is to be extinguished by merger and that merger has been authorized in advance by the Prime Minister pursuant to the provisions of Cabinet Office Order, a corporation surviving the merger or a corporation established as a result of the merger succeeds to the status of the spacecraft control operator under this Act.

(4) When a corporation that is a spacecraft control operator has the business with respect to the control of spacecraft licensed under Article 20, paragraph (1) succeeded to by corporate split, and that corporate split has been authorized in advance by the Prime Minister pursuant to the provisions of Cabinet Office Order, a corporation which succeeded to the business as a result of the corporate split succeeds to the status of the spacecraft control operator under the provisions of this Act.

(5) Articles 21 and 22 (limited to the part concerning item (iii) (limited to the part

concerning ability to implement the control plan)) apply mutatis mutandis to the authorization under paragraph (1) and the preceding two paragraphs.

(6) When a spacecraft control operator transfers its business with respect to the control of spacecraft licensed under Article 20, paragraph (1), or when a corporation that is a spacecraft control operator is extinguished by a merger or has the business succeeded to through a corporate split, if a disposition to refuse the authorization under paragraphs (1), (3) or (4) is rendered (if an application for the authorization is not submitted, when the transfer of business, merger or corporate split takes place), the license under Article 20, paragraph (1) ceases to be effective, and its transferee (except for the transferee with respect to business transfer provided in paragraph (2)), a corporation surviving the merger or a corporation established as a result of the merger, or a corporation which succeeded to that business by the corporate split, must take termination measures for which the license under paragraph (1) of that Article was granted, within 120 days from the day of the disposition (if an application for the authorization is not submitted, the day of the transfer of business, merger or corporate split). In this case, until the termination measures are completed (in the case provided in the preceding Article, until the notification under that Article is submitted), the provisions of Article 24, the first sentence of the preceding Article, Article 31, Article 32 and Article 33, paragraph (3) (including penal provisions relating to these provisions) apply by deeming these persons as spacecraft control operators.

Article 27 of the Regulation (Application, etc. for Authorization on Succession of Status of Spacecraft Control Operator)

(1) A person who intends to obtain authorization under Article 26, paragraph (1) of the Act must submit to the Prime Minister a written application using Form 22, attaching the following documents and a copy of the license certificate under Article 20, paragraph (4) pertaining to the transferrer.

(2) When a spacecraft control operator intends to make a notification under Article 26, paragraph (2) of the Act, the spacecraft control operator must submit to the Prime Minister a written notification using Form 23, attaching a document set forth in the items of the preceding paragraph and a copy of the license certificate under Article 20, paragraph (4) pertaining to the transferrer.

(3) A person who intends to obtain authorization under Article 26, paragraph (3) of the Act must submit to the Prime Minister a written application using Form 24, attaching the following documents and a copy of the license certificate under

> Article 20, paragraph (4) for the corporation whose business was succeeded to.
> (4) A person who intends to obtain authorization under Article 26, paragraph (4) of the Act must submit to the Prime Minister a written application using Form 25, attaching the following documents and a copy of the license certificate under Article 20, paragraph (4) for the corporation whose business was succeeded to:

If the succession is authorized and any change occurs to the control plan (excluding the organizational structure for the execution of the control plan), an operator is required to submit an application for permission or a notification of change.

6.4.1. Business transfer

A spacecraft control operator that intends to transfer its business with respect to the control of spacecraft is required to submit the following documents (Article 27, paragraph (1) of the Regulation).

> 1) a written application for authorization (Form 22)
> 2) a document set forth in Article 20, paragraph (2), item (i) of the Regulation in relation to the transferee (a document related to the applicant)
> 3) a document evidencing that the transferee has a sufficient ability to execute the control plan (a document related to the organizational structure for executing the control plan)
> 4) a copy of the contract for the transfer and acquisition
> 5) if the transferrer or the transferee is a corporation, the minutes of resolution of a general meeting of shareholders or general meeting of members or a written consent of members with unlimited liability or all members on the transfer or acquisition, or a document evidencing the decision on the transfer or acquisition.
> 6) a copy of the permission certificate

6.4.2. Notification of transfer of business to person who intends to implement the control of spacecraft without using a spacecraft control facility located in Japan

A spacecraft control operator that intends to transfer its business with respect to the control of spacecraft to a person who intends to implement the control of spacecraft without using a spacecraft control facility located in Japan is required to submit the following documents (Article 27, paragraph (2) of the Regulation).

> 1) a written notification (Form 23)

2) a document set forth in Article 20, paragraph (2), item (i) of the Regulation in relation to the transferee (a document related to the applicant)
3) a document evidencing that the transferee has a sufficient ability to execute the control plan (a document related to the organizational structure for executing the control plan)
4) a copy of the contract for the transfer and acquisition
5) if the transferrer or the transferee is a corporation, the minutes of resolution of a general meeting of shareholders or general meeting of members or a written consent of members with unlimited liability or all members on the transfer or acquisition, or a document evidencing the decision on the transfer or acquisition.
6) a copy of the permission certificate

6.4.3. Merger

If a corporation that is a spacecraft control operator is to be extinguished by a merger, and if it intends to have a corporation surviving the merger or a corporation to be incorporated in the merger succeed to the business with respect to the control of spacecraft, the operator is required to submit the following documents (Article 27, paragraph (3) of the Regulation).

1) a written application for authorization (Form 26)
2) a document stating the method and conditions of the merger
3) a document set forth in Article 20, paragraph (2), item (i)(b) pertaining to the corporation surviving the merger or corporation to be incorporated in the merger (a document related to the applicant)
4) a document evidencing that the corporation surviving the merger or corporation to be incorporated in the merger has a sufficient ability to execute the control plan (a document related to the organizational structure for executing the control plan)
5) a copy of the merger contract and a statement explaining the merger ratio
6) the minutes of resolution of a general meeting of shareholders or general meeting of members or a written consent of members with unlimited liability or all members on the merger, or a document certifying the decision on the merger
7) a copy of the permission certificate

6.4.4. Corporate split

If a corporation that is a spacecraft control operator intends to have its business with respect to the control of spacecraft succeeded to by a corporate split, it is required to submit the following documents (Article 27, paragraph (4) of the Regulation).

1) a written application for authorization (Form 25)
2) a document stating the method and conditions of the corporate split
3) a document set forth in Article 20, paragraph (2), item (i)(b) pertaining to the corporation succeeding to the business with respect to the control of spacecraft by the corporate split (a document related to the applicant)
4) a document evidencing that the corporation succeeding to the business with respect to the launching of spacecraft, etc. by the corporate split has a sufficient ability to execute the launch plan (a document related to the organizational structure for executing the control plan)
5) a copy of a corporate split contract (for the incorporation-type corporate split, a corporate split plan) and a statement explaining a split ratio
6) the minutes of resolution of a general meeting of shareholders or general meeting of members or a written consent of members with unlimited liability or all members on the corporate split, or a document evidencing the decision on the corporate split
7) a copy of the permission certificate

6.5. Notification of death

[License related to control of spacecraft]

Article 27 of the Act (Notification and Other Procedures Related to Death)

(1) When a spacecraft control operator has deceased, the heir must make a notification to the Prime Minister to that effect without delay.
(2) When a spacecraft control operator has deceased, the license under Article 20, paragraph (1) ceases to be effective, and the representative in case of death must take termination measures pertaining to the license under Article 20, paragraph (1), within 120 days from the date of death, unless the transfer of business with respect to the control of the spacecraft has been authorized under paragraph (1) of the preceding Article. In this case, until the implementation of the business transfer or the completion of the termination measures (in the case provided in Article 25, until the notification provided therein is submitted), provisions of Article 24, the first sentence of Article 25, Article 26, paragraphs (1) and (5), Article 31, Article 32 and Article 33, paragraph (3) (including penal provisions relating to these provisions) apply by deeming the representative in case of death as a spacecraft control operator.

ANNEX 3

> Article 28 of the Regulation (Notification of Death)
> When an heir makes a notification under Article 27, paragraph (1) of the Act, the heir must submit a written notification using Form 26 to the Prime Minister.

When an individual who is a spacecraft control operator has deceased, the heir is required to submit a notification (Form 26).

A representative in case of death is required to transfer the business with respect to the control of the spacecraft or take termination measures, within 120 days from the date of death of the spacecraft control operator.

6.6. Termination measures

> [License related to control of spacecraft]
> Article 28 of the Act (Termination Measures)
> (1) When a spacecraft control operator intends to terminate the control of a spacecraft pursuant to the control plan pertaining to the license under Article 20, paragraph (1), the spacecraft control operator must make a notification to the Prime Minister to that effect in advance and take the termination measures pertaining to the license under that paragraph, pursuant to the provisions of Cabinet Office Order.
> (2) When the termination measures are taken pursuant to the preceding paragraph, the license under Article 20, paragraph (1) ceases to be effective.
>
> Article 29 of the Regulation (Notification of Termination Measures)
> When a spacecraft control operator makes a notification under Article 28, paragraph (1) of the Act, the spacecraft control operator must submit a written notification using Form 27 to the Prime Minister.

When a spacecraft control operator intends to terminate the control of a spacecraft pursuant to its control plan, the operator is required to submit a written notification (Form 27) in advance.

The license related to the control of a spacecraft ceases to be effective upon the implementation of the termination measures specified in the written notification.

6.7. Notification of dissolution

> [License related to control of spacecraft]
> Article 29 of the Act (Notification and Other Procedures Related to Dissolution)
> (1) When a corporation that is a spacecraft control operator dissolves for reasons other

> than a merger, a liquidator or a bankruptcy trustee must make a notification to the Prime Minister to that effect without delay.
> (2) When a corporation that is a spacecraft control operator dissolves for reasons other than a merger, the license under Article 20, paragraph (1) ceases to be effective, and the corporation under liquidation (meaning a corporation under liquidation or special liquidation, or a corporation against which bankruptcy proceedings have been commenced; hereinafter the same applies in this paragraph) must take termination measures pertaining to the license under Article 20, paragraph (1) within 120 days from the date of dissolution, unless the transfer of business with respect to the control of spacecraft has been authorized under Article 26, paragraph (1). In this case, until the implementation of the business transfer or the completion of the termination measures (in the case provided in Article 25, until the notification provided therein is submitted), provisions of Article 24, the first sentence of Article 25, Article 26, paragraphs (1) and (5), Article 31, Article 32 and Article 33, paragraph (3) (including penal provision relating to these provisions) apply by deeming the corporation under liquidation as a spacecraft control operator.
>
> Article 30 of the Regulation (Notification of Dissolution)
> When a liquidator or bankruptcy trustee makes a notification under Article 29, paragraph (1) of the Act, the liquidator or bankruptcy trustee must submit a written notification using Form 28 to the Prime Minister.

When a corporation that is a spacecraft control operator is to be dissolved due to grounds other than a merger, its liquidator or bankruptcy trustee is required to submit a written notification (Form 28).

A corporation under liquidation (meaning a corporation under liquidation or special liquidation, or a corporation against which bankruptcy proceedings have been commenced) is required to transfer the business with respect to the control of the spacecraft or take termination measures, within 120 days from the date of dissolution.

ANNEX 3

7. Scope of applicants

7.1. Application for permission related to launching of spacecraft, etc.

A person who intends to implement the launching of a spacecraft, etc. using a launch site located in Japan is required to submit an application.

7.2. Application for license related to control of spacecraft

The control of a spacecraft often involves two or more parties, including when the manufacturer (manufacturing business) and operator (operating business) of the spacecraft are different.

The following shows the criteria to be applied to the major cases involving two or more parties.

○ A case where the operator of a spacecraft bus and the operator of missions are different

In this case, the operator of a spacecraft bus is generally required to submit an application, as the person engaged in the detection and control of position, etc. of a spacecraft using a spacecraft control facility is generally the operator of the spacecraft bus.

○ A case where the manufacturer of a spacecraft conducts an initial operation thereof on an orbit and delivers the spacecraft to the operator on the orbit

<Case where the initial operation is to be conducted by the manufacturer of spacecraft>

(i) A case where the spacecraft manufacturer proactively engages in regular maintenance works or emergency responses even after the delivery of the spacecraft

In this case, as both businesses can be considered to be engaged in the control of the same spacecraft, it is recommended that these two businesses jointly submit an application for license related to the control of spacecraft. In this case, describe the involvement of these two businesses in the control plan.

(ii) A case where the spacecraft manufacturer is not involved in the control of a spacecraft after the delivery, or even in a case where the spacecraft manufacturer is involved, it is to only conduct ancillary works under the instruction of the operator of the spacecraft

In this case, it is recommended that the spacecraft manufacturer submit an application for license related to the control of a spacecraft, and take procedures to transfer the business to the operator of the spacecraft (see 6.4) before the spacecraft is delivered on the orbit.

<Case where the initial operation is to be conducted by the operator of spacecraft>
During the initial operation, if the operator conducts the control of a spacecraft by giving instructions to the manufacturer from time to time, and the manufacturer is not involved in the control of the spacecraft after the initial operation at its discretion, including regular maintenance works or emergency responses, it is recommended that only the operator submit an application for the license.

8. Example of preparation of application forms

Attachment 1 shows an example of a form for the application of permission related to the launching of a spacecraft, etc., and Attachment 2 an example of a form for the application of license related to the control of a spacecraft.

ANNEX 3

9. List of main agencies and sections in charge

9.1. List of main agencies and sections relating to launching of a spacecraft, etc.

Table 1 shows the list of main agencies and sections related to the laws and regulations applicable to the launching of a spacecraft, etc.

Table 1 List of main agencies and sections relating to the launching of a spacecraft, etc.

Acts and provisions	Contact	Example of case where procedures, statutory qualifications, etc. are required
Article 4 of the Radio Act	Ministry of Internal Affairs and Communications, Telecommunications Bureau, Radio Department	
	· Fixed and Satellite Radio Communications Division, Fixed Radio Communications Office TEL: 03-5253-5886	· Case requiring an application for and acquisition of a license for a radio station related to a ground meteorological radar
	· Land Mobile Communications Division TEL: 03-5253-5895	· Cases requiring an application for and acquisition of a license for a radio station related to a launch vehicle
Fire Service Act	Fire and Disaster Management Agency, Dangerous Goods Safety Office TEL: 03-5253-7524 (Application) Municipal firefighting headquarters having jurisdiction over the location of the manufacturing site, etc.	· Case of manufacturing or storage of dangerous goods · Case requiring facilities, etc. for fire prevention
Poisonous and Deleterious Substances Control Act	Ministry of Health, Labour and Welfare, Pharmaceutical Safety and Environmental Health Bureau, Pharmaceutical Evaluation Division, Chemical Safety Office TEL: 03-3595-2298	In cases where a propellant includes deleterious chemical substances such as hydrazine
Industrial Safety and Health Act	(Outline of system) Ministry of Health, Labour and Welfare, Labour Standards Bureau, Industrial Safety and Health Department, Safety Division TEL: 03-3595-3225 (Notification, etc.) Labor standards office having jurisdiction over the location of the workplace	· In cases where a notification of installing a crane, etc. or a completion inspection is necessary. · In cases where business involving the operation of machine equipment, etc. requiring qualifications is to be conducted, such as a crane operator's license and the completion of a skill training course for forklift operation.
Explosives Control Act	(Outline of system) Ministry of Trade, Industry and Economy, Industrial and Product Safety Policy Group, Assistant to Director, Mine Safety and Explosive Control Division TEL: 03-3501-1870 (Application) Prefectures, etc. having jurisdiction over the location of workplaces, etc.	In cases of manufacturing, sale, storage, consumption or other handling of explosives
High Pressure Gas Safety Act	(Outline of system) Ministry of Trade, Industry and Economy, Industrial and Product Safety Policy Group, High Pressure Gas Safety Office TEL: 03-3501-1706 (Application) Prefectures, etc. having jurisdiction over the location of workplaces, etc.	In cases of manufacturing, storage, sale, transportation or other handling or consumption of high pressure gases, and cases of manufacturing and handling containers for high pressure gases
Article 99-2 of the Civil Aeronautics Act	Ministry of Land, Infrastructure, Transport and Tourism, Civil Aviation Bureau, Aviation Safety and Security Department, Flight Standards Division TEL: 03-5253-8737	Case of launching a launch vehicle
Procedures related to notices to mariners, etc.	Japan Coast Guard, Hydrographic and Oceanographic Department, Chart and Navigational Information Division, Notices to Mariners Office TEL: 03-3595-3615	In cases where any object is to fall on the sea upon the launching of a spacecraft, etc. and reentry of the orbital stage

9.2. List of main agencies and sections relating to control of spacecraft

Table 2 shows the list of main agencies and sections related to the laws and regulations applicable to the control of a spacecraft.

Table 2 List of main agencies and sections relating to control of spacecraft

Acts and provisions	Contact	Example of case where procedures, statutory qualifications, etc. are required
Article 4 of the Radio Act	Ministry of Internal Affairs and Communications, Telecommunications Bureau, Radio Department	
	· Radio Policy Division, International Frequency Policy Office TEL: 03-5253-5878	· Cases requiring international frequency coordination for spacecraft
	· Fixed and Satellite Radio Communications Division TEL: 03-5253-5816	· Cases requiring application and acquisition of radio station licenses related to spacecraft and Earth stations
	· Fixed and Satellite Radio Communications Division, Fixed Radio Communications Office TEL: 03-5253-5886	· Case requiring an application for and acquisition of a license for a radio station related to a ground meteorological radar
	· Land Mobile Communications Division TEL: 03-5253-5895	· Cases requiring an application for and acquisition of radio station licenses for amateur stations to be established on spacecraft and amateur stations for remote control of the radio equipment
	· Regional Bureaus of Telecommunications and Okinawa Office of Telecommunications	· Cases requiring application and acquisition of radio station licenses related to small spacecraft and Earth stations
Fire Service Act	Fire and Disaster Management Agency, Dangerous Goods Safety Office TEL: 03-5253-7524 (Application) Municipal firefighting headquarters having jurisdiction over the location of manufacturing site, etc.	· Case of manufacturing or storage of dangerous goods · Case requiring facilities, etc. for fire prevention
Poisonous and Deleterious Substances Control Act	Ministry of Health, Labour and Welfare, Pharmaceutical Safety and Environmental Health Bureau, Pharmaceutical Evaluation Division, Chemical Safety Office TEL: 03-3595-2298	In cases where a propellant includes deleterious chemical substances such as hydrazine
Industrial Safety and Health Act	(Outline of system) Ministry of Health, Labour and Welfare, Labour Standards Bureau, Industrial Safety and Health Department, Safety Division TEL: 03-3595-3225 (Notification, etc.) Labor standards office having jurisdiction over the location of the workplace	· In cases where a notification of installing a crane, etc. or a completion inspection is necessary. · In cases where business involving the operation of machine equipment, etc. requiring qualifications is to be conducted, such as a crane operator's license and the completion of a skill training course for forklift operation.
Explosives Control Act	(Outline of system) Ministry of Trade, Industry and Economy, Industrial and Product Safety Policy Group, Assistant to Director, Mine Safety and Explosive Control Division TEL: 03-3501-1870 (Application) Prefectures, etc. having jurisdiction over the location of the workplace, etc.	In cases of manufacturing, sale, storage, consumption or other handling of explosives
High Pressure Gas Safety Act	(Outline of system) Ministry of Trade, Industry and Economy, Industrial and Product Safety Policy Group, High Pressure Gas Safety Office TEL: 03-3501-1706 (Application) Prefectures, etc. having jurisdiction over the location of the workplace, etc.	In cases of manufacturing, storage, sale, transportation or other handling or consumption of high pressure gases, and cases of manufacturing and handling containers for high pressure gases
Article 52 of the Foreign Exchange and Foreign Trade Act	Ministry of Trade, Industry and Economy, Security Export Licensing Division TEL: 03-3501-2801	Case of exporting a spacecraft
Procedures related to notices to mariners, etc.	Japan Coast Guard, Hydrographic and Oceanographic Department, Chart and Navigational Information Division, Notices to Mariners Office TEL: 03-3595-3615	In cases where any object is to fall on the sea upon the reentry of a spacecraft

ANNEX 3

10. Checklist of documents to be submitted

The following are checklists for the documents to be submitted for applications.

- Checklist for application of permission related to the launching of a spacecraft, etc.
- Checklist for application of type certification for a launch vehicle
- Checklist for application of compliance certification for a launch site
- Checklist for application of a license related to the control of spacecraft

■ Checklist for application for permission related to the launching of a spacecraft, etc.

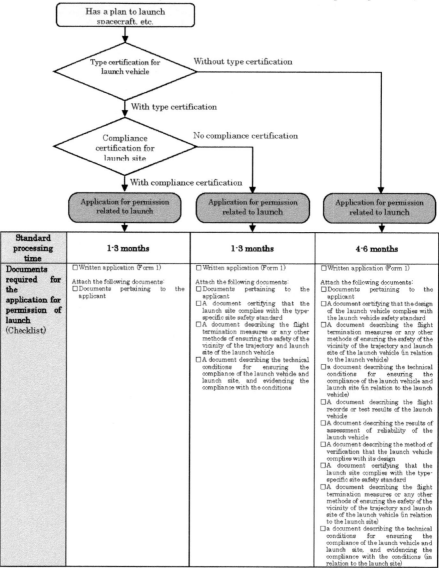

Standard processing time	1-3 months	1-3 months	4-6 months
Documents required for the application for permission of launch (Checklist)	☐ Written application (Form 1) Attach the following documents: ☐ Documents pertaining to the applicant	☐ Written application (Form 1) Attach the following documents: ☐ Documents pertaining to the applicant ☐ A document certifying that the launch site complies with the type-specific site safety standard ☐ A document describing the flight termination measures or any other methods of ensuring the safety of the vicinity of the trajectory and launch site of the launch vehicle ☐ A document describing the technical conditions for ensuring the compliance of the launch vehicle and launch site, and evidencing the compliance with the conditions	☐ Written application (Form 1) Attach the following documents: ☐ Documents pertaining to the applicant ☐ A document certifying that the design of the launch vehicle complies with the launch vehicle safety standard ☐ A document describing the flight termination measures or any other methods of ensuring the safety of the vicinity of the trajectory and launch site of the launch vehicle (in relation to the launch vehicle) ☐ a document describing the technical conditions for ensuring the compliance of the launch vehicle and launch site (in relation to the launch vehicle) ☐ A document describing the flight records or test results of the launch vehicle ☐ A document describing the results of assessment of reliability of the launch vehicle ☐ A document describing the method of verification that the launch vehicle complies with its design ☐ A document certifying that the launch site complies with the type-specific site safety standard ☐ A document describing the flight termination measures or any other methods of ensuring the safety of the vicinity of the trajectory and launch site of the launch vehicle (in relation to the launch site) ☐ a document describing the technical conditions for ensuring the compliance of the launch vehicle and launch site, and evidencing the compliance with the conditions (in relation to the launch site)

ANNEX 3

- ■ Checklist for application of type certification for a launch vehicle
- ☐ Written application (Form 9)

Attach the following documents:
- ☐ Documents pertaining to the applicant
- ☐ A document certifying that the design of the launch vehicle complies with the launch vehicle safety standard
- ☐ A document describing the flight records or test results of the launch vehicle
- ☐ A document describing the results of assessment of reliability of the launch vehicle
- ☐ A document describing the method of verification that the launch vehicle complies with its design

- ■ Checklist for application of compliance certification for launch site
- ☐ Written application (Form 13)

Attach the following documents:
- ☐ Documents pertaining to the applicant
- ☐ A document certifying that the launch site complies with the type-specific site safety standard
- ☐ A document describing the technical conditions for ensuring the compliance of the launch vehicle and launch site, and evidencing the compliance with the conditions

ANNEX 3

- ■ Checklist for application of license related to the control of spacecraft
- ☐ Written application (Form 17)

Attach the following documents:
- ☐ Documents pertaining to the applicant
- ☐ A document certifying that the configuration of the spacecraft satisfies the launch vehicle safety standard provided in Article 22

Annex 4 – Act on Ensuring Appropriate Handling of Satellite Remote Sensing Data (Act No. 77 of November 16, 2016)

Chapter I General Provisions

(Purpose)
Article 1
In line with the basic principles provided for in the Space Basic Act (Act No. 43 of 2008), to ensure the appropriate handling of Satellite Remote Sensing Data in Japan, Act provides responsibilities of the national government, establishes a licensing system for the Use of Satellite Remote Sensing Instruments, and provides obligations of a Satellite Remote Sensing Data Holder, certification of a person handling Satellite Remote Sensing Data, supervision by the Prime Minister, and other necessary matters concerning handling of Satellite Remote Sensing Data.

(Definitions)
Article 2
In this Act, the meanings of the terms set forth in the following items are as provided respectively in those items:
(i) "Satellite" means an artificial object which is used by launched into Earth orbit or beyond or placed on a celestial body other than the Earth;
(ii) "Satellite Remote Sensing Instruments" means equipment onboard a satellite which is used by launched into the Earth orbit (hereinafter referred to as "Earth Orbiting Satellite"), which detects electromagnetic waves emitted or reflected from objects existing on the ground surface or water surface (including underground or underwater which is near the surface) or in the air (hereinafter referred to as "Ground Emitted Electromagnetic Waves, etc."), and which records information concerning intensity, frequency and phase of Ground Emitted Electromagnetic Waves, etc. as well as information on the position and other status of the Earth Orbiting Satellite when that information was detected (referred to as "detected information" in the following item) in electronic or magnetic records (meaning records created electronically, magnetically and any other form which cannot be recognized through human senses and which is used for information processing in computers; the same applies hereinafter), and has the function of sending such information to the ground; when this function is operated under appropriate

conditions, the strength of those electronic or magnetic records received on the ground and made recognizable through visualization on a computer screen (hereinafter referred to as "distinguishing accuracy of target" in this Article and Article 21, paragraph (1)) fulfills the criteria specified by Cabinet Office Order which determines the thresholds as capable of discerning the movement of vehicles, ships, aircraft and other moving facilities; and which has a ground radio station capable of sending or receiving a signal necessary to activate or deactivate these functions and those electronic or magnetic records to other ground radio stations (meaning electrical equipment for sending or receiving codes using electromagnetic waves, and a computer connected to such equipment via telecommunication line; the same applies hereinafter) by the use of electromagnetic waves.

(iii) "Ground Radio Station for Command and Control" means a ground radio station which has the function of transmitting signals necessary for the operation of Satellite Remote Sensing Instruments to that Satellite Remote Sensing Instruments, directly or via other ground radio stations using electromagnetic waves; and for which signals include the time at which to activate the function of detecting Ground Emitted Electromagnetic Waves, etc. of the Satellite Remote Sensing Instruments, the time at which to send electronic or magnetic records recording detected information (hereinafter referred to as "Electromagnetic Data of Detected Information") to the ground, communication method used for sending such information, and the determination and change of distinguishing accuracy of target;

(iv) "Use of Satellite Remote Sensing Instruments" means to operate Satellite Remote Sensing Instruments onboard an Earth Orbiting Satellite by the use of Ground Radio Station for Command and Control operating on its own or managed by another person, and to send Electromagnetic Data of Detected Information to the ground after setting methods for sending necessary signals to that Satellite Remote Sensing Instruments from that Ground Radio Station for Command and Control.

(v) "Specified User Organization" means a national governmental organization or local governmental organization prescribed by Cabinet Order as an entity capable of performing an appropriate Use of Satellite Remote Sensing Instruments.

(vi) "Satellite Remote Sensing Data" means Electromagnetic Data of Detected Information which is sent to the ground by a person other than Specified User Organization by Use of Satellite Remote Sensing Instruments through the Ground Radio Station for Command and Control located in Japan, and electronic or magnetic records processed from that Electromagnetic Data of Detected Information; and when such information and processed data fall under the criteria of information specified by Cabinet Office Order where the use of such information is likely to cause adverse effect on ensuring the peace and security of the

Annex 4 – Act on Ensuring Appropriate Handling of Satellite Remote Sensing Data (Act No. 77 of November 16, 2016)

international community and the national security of Japan provided in Article 14 of the Space Basic Act (hereinafter referred to as "Ensuring Peace of the International Community, etc.") in view of their distinguishing accuracy of target, the extent and degree of modification of Electromagnetic Data of Detected Information through processing, the elapsed time since that Electromagnetic Data of Detected Information was recorded, and other circumstances; and copies of that information on electromagnetic recording medium (means recording medium of electronic or magnetic records).

(vii) "Specified Data Handling Organization" means Specified User Organization and a national or local governmental organization in Japan or a governmental organization of a foreign country (meaning a country or region outside of Japan; the same applies hereinafter) prescribed by Cabinet Order as an entity capable of performing an appropriate handling of Satellite Remote Sensing Data.

(viii) "Satellite Remote Sensing Data Holder" means a person possessing Satellite Remote Sensing Data (other than Specified Data Handling Organization).

(Responsibilities of the National Government)
Article 3

(1) As a part of measures with respect to space development and use in order to contribute to the Ensuring Peace of the International Community, etc., the national government is responsible for taking necessary measures to ensure that persons conducting the Use of Satellite Remote Sensing Instruments and Satellite Remote Sensing Data Holders will comply with their obligations pursuant to the provisions of this Act.

(2) In taking measures under the preceding paragraph, the national government must give due consideration to ensure the sound development of various activities leveraging the values created by the Use of Satellite Remote Sensing Instruments.

Chapter 2 License pertaining to Use of Satellite Remote Sensing Instruments

(License)
Article 4

(1) A person who intends to conduct the Use of Satellite Remote Sensing Instruments by use of a Ground Radio Station for Command and Control located in Japan (excluding Specified User Organization) must obtain a license from the Prime Minister per Satellite Remote Sensing Instruments.

(2) A person who intends to obtain a license under the preceding paragraph must submit a written application to the Prime Minister, pursuant to the provisions of Cabinet Office Order, specifying the following information, attaching the documents specified by Cabinet Office Order.

(i) the name and address;
(ii) the type, structure and capability of the Satellite Remote Sensing Instruments;
(iii) the orbit of the Earth Orbiting Satellite equipped with the Satellite Remote Sensing Instruments;
(iv) the location, structure and capability of the Ground Radio Station for Command and Control and the other ground radio stations which are used as a passing point when transmitting necessary signals to operate the Satellite Remote Sensing Instruments (referred to as a "ground radio station for command and control" in Article 6, item (i)) and the methods of management thereof;
(v) the location, structure and capability of the ground radio stations necessary to receive Electromagnetic Data of Detected Information transmitted from the Satellite Remote Sensing Instruments (including those used as a passing point for receipt of information; hereinafter referred to as "Receiving Station") and the methods of management thereof;
(vi) the methods of management of Satellite Remote Sensing Data;
(vii) if the applicant is an individual, the name and address of the person who will conduct the Use of the Satellite Remote Sensing Instruments upon death of the applicant on behalf of the applicant (hereinafter referred to as "Successor"); and
(viii) other matters specified by Cabinet Office Order.

(Grounds for Disqualification)
Article 5
A person who falls under any of the following items may not obtain a license under paragraph (1) of the preceding Article:
(i) a person who has violated the provisions of this Act or other Acts specified by Cabinet Order relating to the regulations of conducts which are likely to cause adverse effect on Ensuring Peace of the International Community, etc., or orders based on these Acts or the laws and regulations of a foreign country equivalent thereto, and has been sentenced to a fine or severer punishment (including a punishment under the laws and regulations of a foreign country equivalent thereto), and for whom five (5) years have not elapsed since the date on which execution of the sentence was completed or since the date on which that person ceased to be subject to the execution of the sentence;
(ii) a person whose license has been rescinded pursuant to the provisions of Article 17, paragraph (1) or whose certification has been rescinded pursuant to the provisions of Article 25, paragraph (1) or Article 26, paragraph (1), and for whom three (3) years have not elapsed since the date of that rescission;
(iii) a person for which a public notice has been made pursuant to the provisions of Article 3, paragraph (1) of the Act on Special Measures Concerning Asset Freezing, etc. of International Terrorists Conducted by Japan Taking into Consideration

United Nations Security Council Resolution 1267, etc. (Act No. 124 of 2014) (limited to a person currently specified in the list provided in that paragraph), or a person who has been designated under Article 4, paragraph (1) of that Act (referred to as an "international terrorist" in Article 21, paragraph (3), item (i)(c));

(iv) an adult ward or a person who is treated in the same manner under the laws and regulations of a foreign country;

(v) a corporation whose officers engaged in the business thereof or employees specified by Cabinet Office Order falls under any of the preceding items;

(vi) an individual whose employees specified by Cabinet Office Order falls under any of the items (i) through (iv) above; or

(vii) an individual whose Successor falls under any of the preceding items.

(Requirements for License)
Article 6

The Prime Minister may not grant the license under Article 4, paragraph (1), unless the Prime Minister finds that the application for the license under that paragraph meets all of the following requirements:

(i) for the structure and capability of the Satellite Remote Sensing Instruments, the orbit of the Earth Orbiting Satellite on which the Satellite Remote Sensing Instruments are installed, the location, structure and capability of the ground radio station for command and control, etc., and Receiving Stations as well as methods of management thereof, the necessary and appropriate measures have been taken to prevent persons other than the applicant from the Use of Satellite Remote Sensing Instruments, or other criteria specified by Cabinet Office Order as those not being likely to cause adverse effect on Ensuring the Peace of the International Community, etc. have been met;

(ii) measures for prevention of divulgence, loss or damage of Satellite Remote Sensing Data and any other necessary and appropriate measures to be specified by Cabinet Office Order for the safety management of the Satellite Remote Sensing Data have been taken;

(iii) the applicant (in the case of an individual, including his/her Successor) has the ability to properly implement measures to prevent persons other than the applicant provided in item (i) from the Use of Satellite Remote Sensing Instruments and measures for safety management of the Satellite Remote Sensing Data provided in the preceding item; and

(iv) it is found that that Use of Satellite Remote Sensing Instruments is unlikely to cause adverse effect on Ensuring Peace of the International Community, etc.

(Change of License)
Article 7

(1) When a person who received a license under Article 4, paragraph (1) (hereinafter referred to as a "Satellite Remote Sensing Instruments User") intends to change any matter set forth in items (ii) to (viii) of paragraph (2) of that Article, such person must obtain a license from the Prime Minister pursuant to the provisions of Cabinet Office Order; provided, however, that this does not apply to minor changes that are specified by Cabinet Office Order.

(2) When there has been a change to any of the items set forth in Article 4, paragraph (2), item (i) or any minor changes specified by Cabinet Office Order as referred to in the proviso to the preceding paragraph, a Satellite Remote Sensing Instruments User must make a notification to the Prime Minister to that effect without delay.

(3) The provisions of the preceding Article apply mutatis mutandis to the license under paragraph (1).

(Measures to Prevent Unauthorized Use of Satellite Remote Sensing Instruments)
Article 8

(1) For signals which are necessary to operate Satellite Remote Sensing Instruments and are provided for the use for information processing by computer, a Satellite Remote Sensing Instruments User must perform a conversion process through the use of computer and conversion codes (meaning codes used for a signal conversion process; hereinafter the same applies in this Article), so that it cannot be reconstructed without using conversion codes corresponding to the conversion codes used in the conversion process (referred to as "corresponding conversion code" in paragraph (5)), and take other necessary and appropriate measures specified by Cabinet Office Order to prevent the Use of Satellite Remote Sensing Instruments by a person other than the Satellite Remote Sensing Instruments User.

(2) For Electromagnetic Data of Detected Information transmitted from the Satellite Remote Sensing Instruments, a Satellite Remote Sensing Instruments User must perform a conversion process through the use of a computer and data conversion codes (meaning codes used for a conversion process of electronic or magnetic records; the same applies hereinafter), so that it cannot be reconstructed without using data conversion codes corresponding to the data conversion codes used in the conversion process (referred to as "corresponding data conversion code" in paragraphs (4) and (5)), or take other necessary and appropriate measures specified by Cabinet Office Order to prevent Electromagnetic Data of Detected Information sent from the Satellite Remote Sensing Instruments from being received by a Receiving Station other than the Receiving Station covered by license under Article 4, paragraph (1) and used as Satellite Remote Sensing Data.

(3) A Satellite Remote Sensing Instruments User must not provide a conversion code to any other person (if a person managing a Ground Radio Station for Command and Control is different from the Satellite Remote Sensing Instruments User, any person other than the person managing the Ground Radio Station for Command and Control).
(4) A Satellite Remote Sensing Instruments User must not provide a corresponding data conversion Code to any other person (if a person managing a Receiving Station is different from the Satellite Remote Sensing Instruments User, any person other than the person managing the Receiving Station).
(5) A Satellite Remote Sensing Instruments User must take measures for the prevention of divulgence, loss or damage of conversion codes, corresponding conversion codes, data conversion codes and corresponding data conversion codes (hereinafter referred to as "conversion codes, etc." in this paragraph), and any other necessary and appropriate measures specified by Cabinet Office Order for the safety management of the conversion codes, etc.

(Suspension of Function Other than Orbit Pertaining to Application)
Article 9
If the Earth Orbiting Satellite installed with Satellite Remote Sensing Instruments that has been licensed under Article 4, paragraph (1) does not stay in the orbit licensed under that paragraph, a Satellite Remote Sensing Instruments User must immediately send a signal to the Satellite Remote Sensing Instruments from the Ground Radio Station for Command and Control to stop its function of detecting Ground Emitted Electromagnetic Waves, etc., and keep such function stopped until that Earth Orbiting Satellite is placed into the orbit licensed under that paragraph.

(Receiving Station Used for Receiving Electromagnetic Data of Detected Information)
Article 10
(1) When receiving Electromagnetic Data of Detected Information sent from the Satellite Remote Sensing Instruments, a Satellite Remote Sensing Instruments User may not use any Receiving Stations other than those licensed under Article 4, paragraph (1) which is managed by the Satellite Remote Sensing Instruments User, Specified Data Handling Organization, or a person certified under Article 21, paragraph (1).
(2) When receiving Electromagnetic Data of Detected Information sent from the Satellite Remote Sensing Instruments, if a Satellite Remote Sensing Instruments User uses a Receiving Station which is managed by a person certified under in Article 21, paragraph (1), and if such certification is rescinded pursuant to the provisions of Article 25, paragraph (1) or Article 26, paragraph (1), the Prime Minister must promptly notify the Satellite Remote Sensing Instruments User to that effect.

(3) Upon receipt of the notice referred to in the preceding paragraph, the Satellite Remote Sensing Instruments User must take measures to ensure that Electromagnetic Data of Detected Information will not be sent to the Receiving Station from the Satellite Remote Sensing Instruments if receiving by the Receiving Station provided in that paragraph is possible and to ensure change the data conversion code, and take any other necessary and appropriate measures specified by Cabinet Office Order to prevent the Electromagnetic Data of Detected Information sent from Satellite Remote Sensing Instruments from being received by the Receiving Station and used as Satellite Remote Sensing Data.

(Measures to be Taken in Case of Fault)
Article 11
If a Satellite Remote Sensing Instruments User is unable to conduct the Use of Satellite Remote Sensing Instruments without taking any termination measures (meaning the Termination Measures provided in Article 15, paragraph (2); the same applies in Article 13, paragraph (6) and Article 14, paragraph (2)) due to the fault of the Satellite Remote Sensing Instruments or the Earth Orbiting Satellite installed with such Satellite Remote Sensing Instruments or any other reasons, and if there is no prospect of recovery, the Satellite Remote Sensing Instruments User must promptly make a notification to the Prime Minister to that effect pursuant to the provisions of Cabinet Office Order. In this case, the license under Article 4, paragraph (1) ceases to be effective.

(Log)
Article 12
(1) A Satellite Remote Sensing Instruments User must keep a log pursuant to the provisions of Cabinet Office Order (including electronic or magnetic records, if the electronic or magnetic records have been prepared in lieu thereof; the same applies hereinafter), and specify matters specified by Cabinet Office Order concerning the status of the Use of Satellite Remote Sensing Instruments in the log.
(2) The log referred to in the preceding paragraph must be preserved pursuant to the provisions of Cabinet Office Order.

(Succession)
Article 13
(1) When a Satellite Remote Sensing Instruments User intends to transfer its business in respect of the Use of Satellite Remote Sensing Instruments licensed under Article 4, paragraph (1) to a person who intends to conduct the Use of Satellite Remote Sensing Instruments using a Ground Radio Station for Command and Control located in Japan, and when the transferor and transferee have obtained an authorization on that transfer from the Prime Minister in advance pursuant to the provisions of

Cabinet Office Order, the transferee succeeds to the status of the Satellite Remote Sensing Instruments User under the provisions of this Act.

(2) When a Satellite Remote Sensing Instruments User intends to transfer the business in respect of the Use of Satellite Remote Sensing Instruments licensed under Article 4, paragraph (1) to a person who intends to conduct the Use of Satellite Remote Sensing Instruments without using a Ground Radio Station for Command and Control located in Japan, that Satellite Remote Sensing Instruments User must make a notification to the Prime Minister in advance to that effect pursuant to the provisions of Cabinet Office Order.

(3) When a corporation that is a Satellite Remote Sensing Instruments User is to be extinguished by merger and that merger has been authorized in advance by the Prime Minister pursuant to the provisions of Cabinet Office Order, a corporation surviving the merger or a corporation established as a result of the merger succeeds to the status of the Satellite Remote Sensing Instruments User under the provisions of this Act.

(4) When a corporation that is a Satellite Remote Sensing Instruments User transfers the business in respect of the Use of Satellite Remote Sensing Instruments licensed under Article 4, paragraph (1) by corporate split, and that corporate split has been authorized in advance by the Prime Minister pursuant to the provisions of Cabinet Office Order, a corporation which succeeded to the business as a result of the corporate split succeeds to the status of the Satellite Remote Sensing Instruments User under this Act.

(5) Article 5 and 6 (limited to the part concerning item (iii)) apply mutatis mutandis to the authorization under paragraphs (1), (3), and (4).

(6) When a Satellite Remote Sensing Instruments User transfers its business in respect to the Use of Satellite Remote Sensing Instruments licensed under Article 4, paragraph (1), or when a corporation that is a Satellite Remote Sensing Instruments User is extinguished by merger or transfers such business through corporate split, if a disposition to refuse the authorization under paragraphs (1), (3) or (4) is rendered (in the case where an application for such authorization is not submitted, when the transfer, merger or corporate split takes place), the license under Article 4, paragraph (1) ceases to be effective, and its transferee (except for the transferee in respect of business transfer provided in paragraph (2)), a corporation surviving the merger or a corporation established as a result of the merger, or a corporation which succeeded to that business by corporate split, must take termination measures within 120 days from the day of such disposition (in the case where an application for such authorization is not submitted, the day of the transfer, merger or corporate split). In this case, until the termination measures are completed (in the case provided in Article 11, until the notification under that Article is submitted), such persons are deemed as Satellite Remote Sensing Instruments Users, and the provisions of

Articles 8 through 10, the first sentence of Article 11, Article 12, Article 27, Article 28 and Article 29, paragraph (1) (including penal provisions relating to these provisions) apply.

(Reporting of Death)
Article 14
(1) When a Satellite Remote Sensing Instruments User has deceased, the heir must make a notification to the Prime Minister to that effect without delay.
(2) When a Satellite Remote Sensing Instruments User has deceased, the license under Article 4, paragraph (1) ceases to be effective, and the Successor must take termination measures within 120 days from the date of death, unless the transfer of business in respect to the Use of Satellite Remote Sensing Instruments has been authorized under Article 13, paragraph (1). In this case, until the implementation of the business transfer or the completion of the termination measures (in the case provided in Article 11, until the notification provided therein is submitted), the Successor is deemed as a Satellite Remote Sensing Instruments User, and Articles 8 through 10, the first sentence of Article 11, Article 12, Article 13, paragraphs (1) and (5), Article 27, Article 28 and Article 29, paragraph (1) (including penal provisions relating to these provisions) apply.

(Termination Measures)
Article 15
(1) A Satellite Remote Sensing Instruments User may terminate the Use of Satellite Remote Sensing Instruments at any time, in addition to the cases referred to in Article 13, paragraph (6), Article 14, paragraph (2), Article 16, paragraph (2) or Article 17, paragraph (2).
(2) When a Satellite Remote Sensing Instruments User terminates the Use of Satellite Remote Sensing Instruments, it must take any of the following measures (hereinafter referred to as "Termination Measures"), pursuant to the provisions of Cabinet Office Order, and make a notification to the Prime Minister of the contents of the measure taken without delay.
 (i) measure to send a signal from the Ground Radio Station for Command and Control to the Satellite Remote Sensing Instruments to stop its function of detecting Ground Emitted Electromagnetic Waves, etc., or any other necessary measures specified by Cabinet Office Order to stop such function completely; or
 (ii) measure to ensure that a signal is sent from the Ground Radio Station for Command and Control to the Satellite Remote Sensing Instruments to stop its function until a restart signal (meaning a signal necessary to recover the function of detecting Ground Emitted Electromagnetic Waves, etc. if the function has been suspended; the same applies hereinafter) is received and that a notification of

information on the restart signal and the method of creation thereof is made to the Prime Minister, and any other necessary measures specified by Cabinet Office Order in order to ensure that the function cannot be restored unless the restart signal is received by Satellite Remote Sensing Instruments;
(3) When the Termination Measures are taken pursuant to the provisions of the preceding paragraph, the license under Article 4, paragraph (1) ceases to be effective.
(4) A person who has taken the Termination Measures set forth in item (ii) of paragraph (2) must not provide information on the restart signal under that item and the method of creating said signal to any persons other than Specified User Organizations or a person who has been newly licensed under Article 4, paragraph (1) to conduct the Use of Satellite Remote Sensing Instruments pertaining to the Termination Measures.

(Notification of Dissolution)
Article 16
(1) When a corporation that is a Satellite Remote Sensing Instruments User has dissolved for reasons other than a merger, a liquidator or a bankruptcy trustee must make a notification to the Prime Minister to that effect without delay.
(2) When a corporation that is a Satellite Remote Sensing Instruments User has dissolved for reasons other than a merger, the license under Article 4, paragraph (1) ceases to be effective, and the corporation under liquidation (meaning a corporation in liquidation or special liquidation, or a corporation against which bankruptcy proceedings have been commenced; hereinafter the same applies in this paragraph) must take termination measures within 120 days from the dissolution, unless the transfer of business in respect to the Use of Satellite Remote Sensing Instruments has been authorized under Article 13, paragraph (1). In this case, until the implementation of the business transfer or the completion of the termination measures (in the case provided in Article 11, until the notification provided therein is submitted), the corporation under liquidation is deemed as a Satellite Remote Sensing Instruments User, and Articles 8 through 10, the first sentence of Article 11, Article 12, Article 13, paragraphs (1) and (5), Article 27, Article 28 and Article 29 paragraph (1) (including penal provision relating to these provisions) apply.

(Rescission of License)
Article 17
(1) If a Satellite Remote Sensing Instruments User falls under any of the following items, the Prime Minister may rescind the license under Article 4, paragraph (1) or order the suspension of the Use of Satellite Remote Sensing Instruments for a specified period not exceeding one (1) year:

(i) the Satellite Remote Sensing Instruments User has received the license under Article 4, paragraph (1) or Article 7, paragraph (1) or the authorization under Article 13, paragraph (1), (3) or (4) by deception or other wrongful means;

(ii) the Satellite Remote Sensing Instruments User has come to fall under any of the items of Article 5;

(iii) the Satellite Remote Sensing Instruments User has ceased to conform to any of the items of Article 6;

(iv) the Satellite Remote Sensing Instruments User has changed a matter for which a license must be obtained pursuant to the provisions of Article 7, paragraph (1) without obtaining the license under that paragraph;

(v) the Satellite Remote Sensing Instruments User has received Electromagnetic Data of Detected Information which was transmitted from a Satellite Remote Sensing Instruments in violation of the provisions of Article 10, paragraph (1);

(vi) the Satellite Remote Sensing Instruments User has violated the order made under this paragraph, Article 19, paragraph (1), Article 29, paragraph (1) or (2);

(vii) the Satellite Remote Sensing Instruments User has provided Satellite Remote Sensing Data in violation of the provisions of Article 18, paragraph (3); or

(viii) the Satellite Remote Sensing Instruments User has violated the conditions attached to the license under Article 4, paragraph (1) or Article 7, paragraph (1), or the authorization under Article 13, paragraph (1), (3) or (4), pursuant to the provisions of Article 30, paragraph (1).

(2) If the license of a Satellite Remote Sensing Instruments User under Article 4, paragraph (1) is rescinded pursuant to the provisions of the preceding paragraph, that Satellite Remote Sensing Instruments User must take termination measures within 120 days from the date of such rescission unless transfer of business in respect to the Use of Satellite Remote Sensing Instruments has been authorized under Article 13, paragraph (1). In this case, until the implementation of the business transfer or the completion of the termination measures (in the case provided in Article 11, until the notification provided therein is submitted), the person is deemed as a Satellite Remote Sensing Instruments User, and Articles 8 through 10, the first sentence of Article 11, Article 12, Article 13, paragraphs (1) and (5), Article 27, Article 28 and Article 29, paragraph (1) (including penal regulations relating to these provisions) apply.

Chapter 3 Regulations Concerning Handling of Satellite Remote Sensing Data

(Restriction on Provision of Satellite Remote Sensing Data)
Article 18
(1) When providing the Satellite Remote Sensing Data to a person who obtained a certification under Article 21, paragraph (1) for the handling of Satellite Remote

Sensing Data, a Satellite Remote Sensing Data Holder must do so after verifying that the recipient of such provision is a person who has obtained that certification, by requiring such recipient to present a certificate under paragraph (4) of that Article, clearly indicating the categories of the Satellite Remote Sensing Data specified by Cabinet Office Order as referred to in paragraph (1) of that Article, and provide the information using cryptography or any other method of transmission whereby it is not easy to restore the contents thereof or any other method specified by Cabinet Office Order as necessary and appropriate for prevention of acquisition and use of Satellite Remote Sensing Data by any person other than the recipient of the Satellite Remote Sensing Data, pursuant to the provisions of Cabinet Office Order.

(2) When providing the Satellite Remote Sensing Data to a Satellite Remote Sensing Instruments User (limited to those who have obtained a license under Article 4, paragraph (1) for the Use of Satellite Remote Sensing Instruments pertaining to the relevant Satellite Remote Sensing Data) or Specified Data Handling Organization, a Satellite Remote Sensing Data Holder must provide the data by clearly indicating to the recipient the categories of the Satellite Remote Sensing Data specified by Cabinet Office Order as referred to in Article 21, paragraph (1) and by the method specified by Cabinet Office Order referred to in the preceding paragraph, pursuant to the provisions of Cabinet Office Order.

(3) A Satellite Remote Sensing Data Holder may not provide the Satellite Remote Sensing Data except when such provision is made pursuant to the provisions of the preceding two paragraphs for the examination or research conducted by each House or by a committee of each House or research committee of the House of Councilors pursuant to the provisions of Article 104, paragraph (1) of the Diet Act (Act No. 79 of 1947) (including the as applied mutatis mutandis pursuant to Article 54-4, paragraph (1) of the same Act) or Article 1 of the Act on Witnesses' Oath, Testimony, etc. Before Both Houses of the Diet (Act No. 225 of 1947), litigation proceedings or any other court proceedings, an execution of judicial decisions, an investigation of criminal cases, or the audit by the Board of Audit, or in any other case equivalent thereto where such provision is necessary for the public interest as specified by a Cabinet Order, or such provision is carried out in an urgent situation when measures must be taken to rescue human life, for disaster relief or for other emergencies.

(Order Prohibiting Provision of Satellite Remote Sensing Data)
Article 19

(1) If the Prime Minister believes on the sufficient ground that the use of Satellite Remote Sensing Data is likely to cause adverse effect on Ensuring of Peace of the International Community, etc., the Prime Minister may issue an order to a Satellite Remote Sensing Data Holder (excluding a natural person who has neither domicile nor residence in

Japan or a corporation or any other organization which does not have a principal office in Japan that handles Satellite Remote Sensing Data in a foreign country (hereinafter referred to as a "Foreign Handler")) to prohibit provision of the Satellite Remote Sensing Data designating the scope and time period.

(2) The prohibition order under the preceding paragraph must be limited to the minimum extent required for the Ensuring of Peace of the International Community, etc.

(3) The provisions of the preceding two paragraphs apply mutatis mutandis to a Satellite Remote Sensing Data Holder (limited to a Foreign Handler). In this case, the term "to prohibit provision" in paragraph (1) is deemed to be replaced with "not to provide", and the term "issue an order to" in the preceding paragraph is deemed to be replaced with "request."

(Safety Management Measures for Satellite Remote Sensing Data)
Article 20
A Satellite Remote Sensing Data Holder must take measures for prevention of divulgence, loss or damage of Satellite Remote Sensing Data and any other necessary and appropriate measures specified by Cabinet Office Order for the safety management of the relevant Satellite Remote Sensing Data.

Chapter 4 Certification of Persons Handling Satellite Remote Sensing Data

(Certification)
Article 21
(1) A person handling Satellite Remote Sensing Data (excluding Specified Data Handling Organization) may, upon application, obtain a certification from the Prime Minister that states that person is found to be capable of properly handling Satellite Remote Sensing Data, according to the categories of Satellite Remote Sensing Data specified by Cabinet Office Order having regard to circumstances such as Distinguishing Accuracy of Target, the scope and degree of information changed as a result of processing of Electromagnetic Data of Detected Information, or the time elapsed since the relevant Electromagnetic Data of Detected Information was recorded.

(2) A person who intends to obtain a certification referred to in the preceding paragraph must submit a written application to the Prime Minister, pursuant to the provisions of Cabinet Office Order, specifying the following information, attaching the documents evidencing that the certification standards set forth in each item of the following paragraph are met and other documents specified by Cabinet Office Order.

(i) the name and address;
(ii) the category of the Satellite Remote Sensing Data;
(iii) the purpose and method of usage of the Satellite Remote Sensing Data;
(iv) the methods of management of the Satellite Remote Sensing Data;
(v) if the applicant receives the Satellite Remote Sensing Data at a Receiving Station, the place of such station; and
(vi) other matters prescribed by Cabinet Office Order.

(3) If the Prime Minister finds that the application for the certification under paragraph (1) conforms to the criteria set forth in the following, the Prime Minister must grant the certification under that paragraph:
(i) the applicant does not fall under any of the following:
   (a) a person who has violated the provisions of this Act or other Acts specified by Cabinet Order relating to the regulations of conducts which are likely to cause adverse effect on Ensuring Peace of the International Community, etc., or orders based on these Acts or the laws and regulations of a foreign country equivalent thereto, and has been sentenced to a fine or severer punishment (including a punishment under the laws and regulations of a foreign country equivalent thereto), and for whom five (5) years have not elapsed since date on which execution of the sentence was completed or since the date on which that person ceased to be subject to the execution of the sentence;
   (b) a person whose license has been rescinded pursuant to the provisions of Article 17, paragraph (1) or whose certification has been rescinded pursuant to the provisions of Article 25, paragraph (1) or Article 26, paragraph (1) and for whom three (3) years have not elapsed since the date of the that rescission;
   (c) an international terrorist;
   (d) an adult ward or a person who is treated in the same manner under the laws and regulations of a foreign country;
   (e) a corporation whose officers engaged in the business thereof or employees specified by a Cabinet Office Order fall under any of (a) to (d) above; or
   (f) an individual whose employees specified by a Cabinet Office Order fall under any of (a) to (d) above:
(ii) the handling by the applicant of Satellite Remote Sensing Data which belongs to the types pertaining to that application conforms to the criteria specified by Cabinet Office Order as being those that are not likely to cause adverse effect on Ensuring Peace of the International Community, etc. by taking into consideration the purpose and methods of use of Satellite Remote Sensing Data by the applicant, the capability to carry out analysis or processing of Satellite Remote Sensing Data, measures to ensure the safety management of Satellite Remote Sensing Data and other circumstances.

(4) When the Prime Minister granted the certification under paragraph (1), the Prime Minister must make a notification to that effect to the applicant and promptly issue a certificate.
(5) In the case where a certificate is lost or destroyed, the person who has received the issuance of the certificate must promptly make a notification to that effect to the Prime Minister and receive the reissuance of the certificate.

(Change of Certification)
Article 22
(1) When a person who received the certification under Article 21, paragraph (1) intends to change any matter set forth in paragraph (2), items (iii) to (vi) of that Article, such person must receive a certification from the Prime Minister pursuant to the provisions of Cabinet Office Order; provided, however, that this does not apply to minor changes specified by Cabinet Office Order.
(2) When there has been a change to any of the items set forth in Article 21, paragraph (2), item (i), or any minor changes specified by Cabinet Office Order referred to in the proviso to the preceding paragraph, the person who received the certification under Article 21, paragraph (1) must make a notification to that effect to the Prime Minister without delay.
(3) The provisions in Article 21, paragraph (3) (limited to the part pertaining to item (ii)) apply mutatis mutandis to the certification under paragraph (1) .

(Log)
Article 23
(1) A person who obtained a certification under Article 21, paragraph (1) of the Act must keep a log and record the matters specified by Cabinet Office Order in relation to the status of handling of Satellite Remote Sensing Data, pursuant to the provisions of Cabinet Office Order.
(2) The log under the preceding paragraph must be preserved pursuant to the provisions of Cabinet Office Order.

(Return of Certificate)
Article 24
(1) If a person who received the certificate has come to fall under any of the following items, such person must return the certificate (in the case provided in item (ii), the certificate which was discovered or restored) to the Prime Minister without delay:
   (i) when the certification under Article 21, paragraph (1) has been rescinded; or
   (ii) in the case of having received the reissuance of the certificate, when the lost certificate was discovered or restored.

(2) If a person who received the issuance of the certificate has come to fall under any of the cases set forth in the following items, the person respectively specified in that items must return the certificate to the Prime Minister without delay:
   (i) when the person has deceased: a relative living together with that person or the statutory agent(s);
   (ii) when the corporation has dissolved for reasons other than a merger: the liquidator or the bankruptcy trustee or a person who assumes the obligations equivalent thereto; or
   (iii) when the corporation has been extinguished due to a merger: the representative of the corporation surviving the merger or the corporation established by the merger.

(Rescission of Certification)
Article 25
(1) If a person who received the certification under Article 21, paragraph (1) (excluding Foreign Handlers) falls under any of the following items, the Prime Minister may rescind the certification or suspend its effect for a specified period not exceeding one (1) year:
   (i) the person has provided Satellite Remote Sensing Data in violation of the provision of Article 18, paragraph (3);
   (ii) the person has violated an order under Article 19, paragraph (1) or Article 29, paragraph (2);
   (iii) the person has received the certification under Article 21, paragraph (1) or Article 22, paragraph (1) by deception or other wrongful means;
   (iv) the person has ceased to conform to the criteria set forth in any of the items of Article 21, paragraph (3);
   (v) the person has changed a matter for which a certification must be obtained pursuant to the provisions of Article 22, paragraph (1) without obtaining certification under that paragraph; or
   (vi) has violated the conditions attached to the certification under Article 21, paragraph (1) or Article 22, paragraph (1) pursuant to the provisions of Article 30, paragraph (1).
(2) A person who has become subject to the suspension of effect of certification under the preceding paragraph must promptly submit the certificate to the Prime Minister.
(3) When the period of the suspension of effect of certification under paragraph (1) expires, and, if the person who has submitted a certificate pursuant to the provisions of the preceding paragraph requests the return of that certificate, the Prime Minister must immediately return the certificate.

Article 26

(1) If a person who received the certification under Article 21, paragraph (1) (limited to Foreign Handlers; the same applies in item (iii)) falls under any of the following items, the Prime Minister may rescind the certification or suspend its effect for a specified period not exceeding one (1) year:
   (i) the person has failed to respond to a request under Article 19, paragraph (1) as applied mutatis mutandis by replacing certain terms in Article 19, paragraph (3), or a request under Article 29, paragraph (2) as applied mutatis mutandis by replacing certain terms in Article 29, paragraph (3);
   (ii) the person has fallen under any of item (i) or items (3) to (6) of paragraph (1) the preceding Article; or
   (iii) when the Prime Minister requested the person who received the certification under Article 21, paragraph (1) to provide necessary reports or tried to have the officials enter such person's office or any other places of business to inspect books, documents or other items or to question relevant persons to the extent necessary for the enforcement of this Act, no reports or a false reports have been made, or that inspection has been refused, interrupted or avoided, or no answer or a false answer has been given to such questions.
(2) The provisions of Article 25, paragraphs (2) and (3) apply mutatis mutandis to the suspension of effect of certification under the preceding paragraph.

Chapter 5 Supervision by the Prime Minister

(On-site Inspection)
Article 27
(1) The Prime Minister may, to the extent necessary for the enforcement of this Act, request a Satellite Remote Sensing Instruments User or Satellite Remote Sensing Data Holder (excluding Foreign Handlers) to provide necessary reports or have Cabinet Office officials enter its office or any other places of business to inspect books, documents or other items or to question relevant persons.
(2) An official who conducts an on-site inspection under the preceding paragraph must carry an identification card and present this at the request of relevant persons.
(3) The authority to conduct an on-site inspection under paragraph (1) shall not be construed as being granted for a criminal investigation.

(Guidance)
Article 28
The Prime Minister may give necessary guidance, advice and recommendations to Satellite Remote Sensing Instruments Users or Satellite Remote Sensing Data Holders in order to ensure the proper handling of Satellite Remote Sensing Data in Japan.

(Correction Order)

Article 29

(1) When the Prime Minister finds that a Satellite Remote Sensing Instruments User is in violation of the provisions of Article 8, Article 9 or Article 10, paragraph (3) or that a Satellite Remote Sensing Instruments User has failed to take termination measures in violation of the provisions of Article 13, paragraph (6), Article 14, paragraph (2), Article 15, paragraph (2), Article 16, paragraph (2) or Article 17, paragraph (2), the Prime Minister may order such person to take necessary measures for correcting that violation.

(2) When the Prime Minister finds that a Satellite Remote Sensing Data Holder (excluding Foreign Handlers) is in violation of the provisions of Article 18, paragraph (1) or (2) or Article 20, the Prime Minister may order such person to take necessary measures for correcting that violation.

(3) The provisions of the preceding paragraph apply mutatis mutandis to Satellite Remote Sensing Data Holders (limited to Foreign Handlers). In this case, the term "order" in that paragraph is deemed to be replaced with "request."

(Conditions of the Licenses)

Article 30

(1) Conditions may be imposed on the license under Article 4, paragraph (1) or Article 7, paragraph (1), the authorization under Article 13, paragraph (1), (3) or (4), or the certification under Article 21, paragraph (1) or Article 22, paragraph (1) (hereinafter referred to as "license, etc." in the following paragraph), and changes may be made to such conditions.

(2) The conditions referred to in the preceding paragraph must be limited to the minimum extent required for securing the proper implementation of the matters pertaining to the licenses and must not result in imposing unreasonable duties on the person receiving the license, etc.

Chapter 6 Miscellaneous Provisions

(Transitional Measures)

Article 31

Where an order is established, revised or abolished pursuant to the provisions of this Act, the necessary transitional measures (including transitional measures relating to penal provisions) may be stipulated in the order to the extent considered reasonably necessary for the establishment, revision or abolition.

(Delegation to Cabinet Office Order)
Article 32
In addition to what is provided for in this Act, the procedures for the enforcement of this Act and any other matters necessary for the enforcement of this Act are to be specified by Cabinet Office Order.

Chapter 7 Penal Provisions

Article 33
A person falling under any of the following items is punished by imprisonment of not more than three (3) years or a fine of not more than 1,000,000 yen, or both:
(i) a person who has conducted the Use of Satellite Remote Sensing Instruments in violation of the provision of Article 4, paragraph (1);
(ii) a person who has received the license under Article 4, paragraph (1) or Article 7, paragraph (1), the authorization under Article 13, paragraph (1), (3) or (4), or the certification under Article 21, paragraph (1) or Article 22, paragraph (1) by deception or other wrongful means;
(iii) a person who has changed the matters set forth in Article 4, paragraph (2), items (ii) to (viii) in violation of the provision of Article 7, paragraph (1);
(iv) a person who has received Electromagnetic Data of Detected Information transmitted from Satellite Remote Sensing Instruments in violation of the provision of Article 10, paragraph (1);
(v) a person who has provided information on the restart code or the method of creating such code in violation of the provisions of Article 15, paragraph (4);
(vi) a person who has violated the order under Article 17, paragraph (1), Article 19, paragraph (1) or Article 29, paragraph (1) or (2);
(vii) a person who has provided Satellite Remote Sensing Data in violation of the provisions of Article 18, paragraph (3); or
(viii) a person who has changed the matters set forth in Article 21, paragraph (2), items (iii) to (vi) in violation of the provisions of Article 22, paragraph (1).

Article 34
A person who failed to make a report under Article 27, paragraph (1) or made a false report, or refused, obstructed or avoided the inspection under that paragraph, or failed to answer or gave a false answer to the questions under that paragraph is to be punished by imprisonment of not more than one (1) year or a fine of not more than 500,000 yen, or both.

Annex 4 – Act on Ensuring Appropriate Handling of Satellite Remote Sensing Data (Act No. 77 of November 16, 2016)

Article 35
A person falling under any of the following items is punished by a fine of not more than 500,000 yen:
(i) a person who has failed to make the notification under Article 7, paragraph (2), Article 11, Article 13, paragraph (2), Article 15, paragraph (2) or Article 22, paragraph (2) or who made a false notification;
(ii) a person who has failed to keep a log or state necessary matters in the log or made false statements in the log in violation of the provisions of Article 12, paragraph (1) or Article 23, paragraph (1), or a person who failed to preserve a log in violation of the provisions of Article 12, paragraph (2) or Article 23, paragraph (2), ;
(iii) a person who has failed to return the certificate in violation of the provision of Article 24, paragraph (1); or
(iv) a person who has failed to submit the certificate in violation of the provisions of Article 25, paragraph (2) (including the cases applied mutatis mutandis pursuant to Article 26, paragraph (2)).

Article 36
The crimes set forth in Article 33 (limited to the part pertaining to item (vi) (limited to the portion pertaining to Article 19, paragraph (1) and Article 29, paragraph (2)) and item (vii); hereinafter the same applies in this Article) also apply to persons who have committed the crimes set forth in Article 33 outside Japan.

Article 37
Where the representative of a corporation or an agent, employee or other worker of a corporation or an individual has committed any of the violations prescribed in Article 33 to Article 35 with regard to the business of the corporation or individual, not only the offender but also the corporation or individual is to be punished by the fine prescribed in the respective Articles.

Article 38
A person falling under one of the following items is to be punished by a civil fine of not more than 100,000 yen:
(i) a person who has failed to make the notification under Article 14, paragraph (1)or Article 16, paragraph (1) or has made a false notification; or
(ii) a person who has failed to return the certificate in violation of the provisions of Article 24, paragraph (2).

Supplementary Provisions

(Effective Date)
Article 1
This Act comes into effect as of the date specified by Cabinet Order within a period not exceeding one (1) year from the date of promulgation; provided, however, that the provisions set forth in the following items come into effect as of the dates specified respectively in the following items:
(i) Provisions of Article 4 of the Supplementary Provisions: The date of promulgation
(ii) Provisions of the following article: The date specified by Cabinet Order within a period not exceeding nine (9) months from the date of promulgation

(Preparatory Actions)
Article 2
A person who intends to obtain the license under Article 4, paragraph (1) or the certification set forth in Article 21, paragraph (1) may make such application prior to the enforcement of this Act pursuant to the provisions of Article 4, paragraph (2) or Article 21, paragraph (2).

(Transitional Measures)
Article 3
(1) If an application for license under Article 4, paragraph (1) is made with regard to the Use of Satellite Remote Sensing Instruments onboard satellites which are actually put into the orbit around the Earth when this Act comes into effect (including the case where the application was made pursuant to the provisions of the preceding Article prior to the enforcement of this Act), for the purpose of application of the provisions of Article 6 (including the cases applied mutatis mutandis pursuant to Article 7, paragraph (3); hereinafter the same applies in this paragraph) and Article 17, paragraph (1), item (iii) with regard to the Use of Satellite Remote Sensing Instruments, the term "the following items" in Article 6 is deemed to be replaced with "items (ii) to (iv)" and the term "the items of Article 6" in the item is deemed to be replaced with "items (ii) to (iv) of Article 6".
(2) In the case referred to in the preceding paragraph, when the Prime Minister granted the license under Article 4, paragraph (1), the provisions of Article 8, Article 9 and Article 10, paragraph (3) do not apply to the Use of Satellite Remote Sensing Instruments regarding which such license was obtained.

Annex 4 – Act on Ensuring Appropriate Handling of Satellite Remote Sensing Data (Act No. 77 of November 16, 2016)

(Delegation to Cabinet Order)
Article 4
In addition to what is provided for in the preceding two Articles, any necessary transitional measures (including transitional measures concerning penal provisions) for the enforcement of this Act are to be specified by a Cabinet Order.

(Review)
Article 5
When five (5) years have passed after this Act comes into effect, the government is to review the state of enforcement of this Act, and, when it finds it necessary, take necessary measures based on the findings of the review.

ANNEX 5

Guidelines on Measures, etc. Under Act on Ensuring Appropriate Handling of Satellite Remote Sensing Data

December 29, 2017

National Space Policy Secretariat, Cabinet Office

Table of Contents

1. Introduction 2
2. Measures, etc. Relating to Use of SRS Instruments and SRS Data 3
   2.1. Measures, etc. relating to SRS Instruments 3
      2.1.1. Measures to Prevent Unauthorized Use of SRS Instruments 3
      2.1.2. Suspension of Function on Orbit Other Than Orbit For Which Application Was Made 6
      2.1.3. Measures necessary and appropriate to prevent receiving by a Receiving Station managed by a person whose certification was rescinded and being used as SRS Data 6
      2.1.4. Measures in Case of Fault, etc. 8
      2.1.5. Preparation and Management of Logs 8
      2.1.6. Termination Measure 11
   2.2. Measures, etc. relating to Handling of SRS Data 13
      2.2.1. Restriction on Method of Provision of SRS Data 13
      2.2.2. Order of Prohibition of Provision of SRS Data 19
      2.2.3. Safety Management Measures 20
      2.2.4. Preparation and Management of Logs 36
3. Review of Guidelines 38

[Explanatory Notes]

Unless otherwise provided, the terms used in these Guidelines have the meanings as defined in the Act and Enforcement Regulation. The abbreviations as used in these Guidelines have the following meanings:

- Act: Act on Ensuring Appropriate Handling of Satellite Remote Sensing Data (Act No. 77 of 2016)
- Enforcement Regulation: Regulation for Enforcement of the Act for Ensuring Appropriate Handling of Satellite Remote Sensing Data (Cabinet Office Order No. 41 of 2017)
- Enforcement Order: Order for Enforcement of the Act for Ensuring Appropriate Handling of Satellite Remote Sensing Data (Cabinet Order No. 282 of 2017)
- SRS Instruments: Satellite Remote Sensing Instruments
- SRS Data: Satellite Remote Sensing Data

ANNEX 5

1. Introduction

Matters necessary for the measures concerning the use of SRS Instruments and handling of SRS Data are provided in the Act and Enforcement Regulation. These Guidelines are intended to provide detailed explanations on such measures, etc.

2. Measures, etc. Relating to Use of SRS Instruments and SRS Data

2.1. Measures, etc. relating to SRS Instruments

2.1.1. Measures to Prevent Unauthorized Use of SRS Instruments

ANNEX 5

Article 8 of the Act (Measures to Prevent Unauthorized Use of Satellite Remote Sensing Instruments)

(1) For signals which are necessary to operate Satellite Remote Sensing Instruments and are provided for the use for information processing by computer, a Satellite Remote Sensing Instruments User must perform a conversion process through the use of computer and conversion codes (meaning codes used for a signal conversion process; hereinafter the same applies in this Article), so that it cannot be reconstructed without using conversion codes corresponding to the conversion codes used in the conversion process (referred to as "corresponding conversion code" in paragraph (5)), and take other necessary and appropriate measures specified by Cabinet Office Order to prevent the Use of Satellite Remote Sensing Instruments by a person other than the Satellite Remote Sensing Instruments User.

(2) For Electromagnetic Data of Detected Information transmitted from the Satellite Remote Sensing Instruments, a Satellite Remote Sensing Instruments User must perform a conversion process through the use of a computer and data conversion codes (meaning codes used for a conversion process of electronic or magnetic records; the same applies hereinafter), so that it cannot be reconstructed without using data conversion codes corresponding to the data conversion codes used in the conversion process (referred to as "corresponding data conversion code" in paragraphs (4) and (5)), or take other necessary and appropriate measures specified by Cabinet Office Order to prevent Electromagnetic Data of Detected Information sent from the Satellite Remote Sensing Instruments from being received by a Receiving Station other than the Receiving Station covered by license under Article 4, paragraph (1) and used as Satellite Remote Sensing Data.

(3) A Satellite Remote Sensing Instruments User must not provide a conversion code to any other person (if a person managing a Ground Radio Station for Command and Control is different from the Satellite Remote Sensing Instruments User, any person other than the person managing the Ground Radio Station for Command and Control).

(4) A Satellite Remote Sensing Instruments User must not provide a corresponding data conversion Code to any other person (if a person managing a Receiving Station is different from the Satellite Remote Sensing Instruments User, any person other than the person managing the Receiving Station).

(5) A Satellite Remote Sensing Instruments User must take measures for the prevention of divulgence, loss or damage of conversion codes, corresponding conversion codes, data conversion codes and corresponding data conversion codes (hereinafter referred to as "conversion codes, etc." in this paragraph), and any other necessary and appropriate measures specified by Cabinet Office Order for the safety management of the conversion codes, etc.

Article 10 of the Enforcement Regulation (Measures Specified by Cabinet Office Order as Referred to

> in Article 8, paragraph (1) of the Act)
> (1) The measures specified by Cabinet Office Order, as referred to in Article 8, paragraphs (1) and (2) of the Act, are any of the measures specified in the following items:
>   (i) making it impossible to reconstruct codes without the use of corresponding conversion codes or corresponding data conversion codes;
>   (ii) obtaining two or more frequencies and making communications depending on the use; and
>   (iii) taking a measure such that only a person authorized to use the Satellite Remote Sensing Instruments can operate the Operational Radio Station.
> (2) The provisions of Article 7, paragraphs (1) and (2) apply mutatis mutandis to the measures specified by Cabinet Office Order as the measures necessary and appropriate for the safety management of conversion codes, etc. under Article 8, paragraph (5) of the Act.

■ Article 8, paragraph (1) of the Act

An SRS Instruments User is required to take measures specified in Cabinet Office Order so as to prevent any other person from using the SRS Instruments, for example, applying conversion process to signals necessary for operation of the SRS Instruments using conversion codes, and making it impossible to restore such signals without using the corresponding conversion codes.

The details of the measures are as provided in Article 10 of the Enforcement Regulation.

■ Article 8, paragraph (2) of the Act

An SRS Instruments User is required to take measures specified in Cabinet Office Order so as to prevent any Electromagnetic Data of Detected Information transmitted from such SRS Instruments from being received by any station other than the Receiving Station covered by the license of such SRS Instruments and used, for example, by applying a conversion process to such Electromagnetic Data of Detected Information using data conversion codes, and making it impossible to restore such Electromagnetic Data of Detected Information without using the corresponding data conversion codes.

The details of the measures are as provided in Article 10 of the Enforcement Regulation.

■ Article 8, paragraphs (3) to (5) of the Act

An SRS Instruments User may not provide conversion codes and corresponding data conversion codes to any other person.

In addition, an SRS Instruments User is required to take measures against divulgence, loss or damage of conversion codes, corresponding conversion codes, data conversion codes and corresponding data conversion codes. When taking such measures, the SRS

ANNEX 5

Instruments User is required to take the measures identical to the safety management measures for SRS Data pursuant to the provisions of Article 10, paragraph (2) of the Enforcement Regulation. For the details of such measures, please see the Guidelines 2.2.3, Safety Management Measures.2.2.3

2.1.2. Suspension of Function on Orbit Other Than Orbit For Which Application Was Made

---

Article 9 of the Act (Suspension of Function Other than Orbit Pertaining to Application)

If the Earth Orbiting Satellite installed with Satellite Remote Sensing Instruments that has been licensed under Article 4, paragraph (1) does not stay in the orbit licensed under that paragraph, a Satellite Remote Sensing Instruments User must immediately send a signal to the Satellite Remote Sensing Instruments from the Ground Radio Station for Command and Control to stop its function of detecting Ground Emitted Electromagnetic Waves, etc., and keep such function stopped until that Earth Orbiting Satellite is placed into the orbit licensed under that paragraph.

A Distinguishing Accuracy of Target of SRS Instruments varies to a significant degree with the orbit of an Earth Orbiting Satellite (in particular, altitude from earth's surface). If a user uses SRS Instruments on an orbit lower than that within the scope anticipated at the time of the application, the Distinguishing Accuracy of Target is improved, allowing the user to obtain data with higher accuracy than the data that was expected at the time of obtaining a license to use such SRS Instruments.

So, an SRS Instruments User must immediately send a signal to the SRS Instruments from the ground radio station for command and control when the Earth Orbiting Satellite installed with the SRS Instruments does not stay in the orbit pertaining to the license so as to stop its function of detecting Ground Emitted Electromagnetic Waves, etc., and must hold such function suspended until such Earth Orbiting Satellite is placed into the licensed orbit.

2.1.3. Measures necessary and appropriate to prevent receiving by a Receiving Station managed by a person whose certification was rescinded and being used as SRS Data

---

Article 10 of the Act (Receiving Station Used for Receiving Electromagnetic Data of Detected Information)

(1) When receiving Electromagnetic Data of Detected Information sent from the Satellite Remote

> Sensing Instruments, a Satellite Remote Sensing Instruments User may not use any Receiving Stations other than those licensed under Article 4, paragraph (1) which is managed by the Satellite Remote Sensing Instruments User, Specified Data Handling Organization, or a person certified under Article 21, paragraph (1).
> (2) When receiving Electromagnetic Data of Detected Information sent from the Satellite Remote Sensing Instruments, if a Satellite Remote Sensing Instruments User uses a Receiving Station which is managed by a person certified under in Article 21, paragraph (1), and if such certification is rescinded pursuant to the provisions of Article 25, paragraph (1) or Article 26, paragraph (1), the Prime Minister must promptly notify the Satellite Remote Sensing Instruments User to that effect.
> (3) Upon receipt of the notice referred to in the preceding paragraph, the Satellite Remote Sensing Instruments User must take measures to ensure that Electromagnetic Data of Detected Information will not be sent to the Receiving Station from the Satellite Remote Sensing Instruments if receiving by the Receiving Station provided in that paragraph is possible and to ensure change the data conversion code, and take any other necessary and appropriate measures specified by Cabinet Office Order to prevent the Electromagnetic Data of Detected Information sent from Satellite Remote Sensing Instruments from being received by the Receiving Station and used as Satellite Remote Sensing Data.
>
> Article 11 of the Enforcement Regulation (Measures Specified by Cabinet Office Order as Referred to in Article 10, paragraph (3) of the Act)
> Article 11 (1) The measures specified by Cabinet Office Order, as referred to in Article 10, paragraph (3) of the Act, are any of the measures specified in the following items:
> (i) to ensure that Electromagnetic Data of Detected Information is not transmitted to a Receiving Station provided in Article 10, paragraph (2) of the Act; and
> (ii) to change data conversion codes.

When receiving Electromagnetic Data of Detected Information transmitted from SRS Instruments, the SRS Instruments User may not use any Receiving Station other than that managed by: (i) the SRS Instruments User, (ii) Specified Data Handlers, or (iii) person certified under Article 21, paragraph (1) of the Act.

In addition, if a Receiving Station managed by (iii) a person certified under Article 21, paragraph (1) of the Act is to be used, and if such person's certification is rescinded, a measure must be taken to make it impossible to receive Electromagnetic Data of Detected Information using the Receiving Station managed by such person.

The details of the measures are as provided in Article 11 of the Enforcement Regulation.

2.1.4. Measures in Case of Fault, etc.

Article 11 of the Act (Measures to be Taken in Case of Fault)

If a Satellite Remote Sensing Instruments User is unable to conduct the Use of Satellite Remote Sensing Instruments without taking any termination measures (meaning the Termination Measures provided in Article 15, paragraph (2); the same applies in Article 13, paragraph (6) and Article 14, paragraph (2)) due to the fault of the Satellite Remote Sensing Instruments or the Earth Orbiting Satellite installed with such Satellite Remote Sensing Instruments or any other reasons, and if there is no prospect of recovery, the Satellite Remote Sensing Instruments User must promptly make a notification to the Prime Minister to that effect pursuant to the provisions of Cabinet Office Order. In this case, the license under Article 4, paragraph (1) ceases to be effective.

Article 12 of the Enforcement Regulation (Notification in Case of Fault)

When a Satellite Remote Sensing Instruments User intends to make a notification under Article 11 of the Act, it must submit to the Prime Minister a written notification in Form 5.

When it becomes impossible to use SRS Instruments without taking a Termination Measure and it is unlikely that such system will recover (see the examples shown below), an SRS Instruments User must notify the Prime Minister of information such as the year, date, month of the occurrence of such fault, etc. (or, in the case where the year, date, month of the occurrence of a fault, etc. cannot be identified, the estimated year, date, month of the occurrence of such fault, etc.)

If it is possible to take a Termination Measure at the time of the occurrence of the fault, a user is required to do so in a certain manner.

The notification must be submitted using Form 5 of the Enforcement Regulation.

[Example of case where it is unlikely that the system will recover]
① A case where it is possible to judge that the function is unlikely to recover due to a fault, etc. of SRS Instruments from telemetry, etc.
② A case where there is no means for recovery, such as communication blackout.

2.1.5. Preparation and Management of Logs

Article 12 of the Act (Log)

(1) A Satellite Remote Sensing Instruments User must keep a log pursuant to the provisions of Cabinet Office Order (including electronic or magnetic records, if the electronic or magnetic records

have been prepared in lieu thereof; the same applies hereinafter), and specify matters specified by Cabinet Office Order concerning the status of the Use of Satellite Remote Sensing Instruments in the log.

(2) The log referred to in the preceding paragraph must be preserved pursuant to the provisions of Cabinet Office Order.

Article 13 of the Enforcement Regulation (Matters to be Stated in Logs)

(1) The matters specified by Cabinet Office Order, as referred to in Article 12, paragraph (1) of the Act, are as set forth in the following items:

   (i) the date and time of sending signals for operation of a Satellite Remote Sensing Instruments, its contents and the place of the Ground Radio Station for Command and Control, etc. used for sending signals;

   (ii) the date and time of recording the Electromagnetic Data of Detected Information, the scope of coverage, and letters, numbers, signs or any other codes for identification of these information (hereinafter referred to as "Identification Codes");

   (iii) the date and time of ground transmission of Electromagnetic Data of Detected Information and the location of the Receiving Station used for receiving data;

   (iv) status of processing or deleting the Electromagnetic Data of Detected Information; and

   (v) If any Satellite Remote Sensing Data is to be provided to a third party, the Identification Code, category, date and time of provision of the Satellite Remote Sensing Data, the name of the recipient, and, the certificate number of the recipient if the recipient has obtained a certificate under Article 21, paragraph (4) of the Act.

(2) When a Satellite Remote Sensing Instruments User creates electronic or magnetic records for the logs under Article 12, paragraph (1) of the Act, it must do so by a method of recording the created electronic or magnetic records on a file stored in a computer used by it, or by a method of preparing the records using the means file which can securely record certain information by means of a magnetic disk, CD-ROM or any other means equivalent thereto (hereinafter referred to as "Magnetic Disks, etc.")

(3) A Satellite Remote Sensing Instruments User must record in the log the matters set forth in the items of paragraph (1) without delay, for each instance of transmission of signals for operation of Satellite Remote Sensing Instruments, recording of Electromagnetic Data of Detected Information, ground transmission of Electromagnetic Data of Detected Information, processing or deleting of Electromagnetic Data of Detected Information, or provision of Satellite Remote Sensing Data, for each Satellite Remote Sensing Instruments, and must keep the record for five years from the entry of the information.

An SRS Instruments User is required to keep a log to record the status of use of the SRS Instruments, and to preserve such log.

More concretely, in relation to the following matters, if there occurs any event which requires entry of information into the log, the user is required to make such entry without delay and to preserve the log for the designated period of time.

■Article 13, paragraph (1), items (i) to (iii) of the Enforcement Regulation

Use of SRS Instruments comprises: instruction for operation to the SRS Instruments (command transmission), recording by the SRS Instruments (imaging), and transmission of the records to the ground (downlink).

In order to verify that the Ground Radio Station for Command and Control used for the transmission and the Receiving Station used for receiving is appropriately used, it is necessary that these operations are accurately recorded.

To this end, a user is required to make an entry of information including the date and time of sending the signals for operation of SRS Instruments.

■Article 13, paragraph (1), item (iv) of the Enforcement Regulation

In order to make clear the status of Electromagnetic Data of Detected Information obtained through the use of SRS Instruments, it is necessary that its processing and deleting procedures are recorded in the log.

More concretely, a user is required to make entries into log the Identification Codes of the SRS Data to be processed and the details of such processing, in the case of processing; or the Identification Codes of the SRS Data to be deleted and the method and date and time of the deletion, in case of deletion.

■Article 13, paragraph (1), item (v) of the Enforcement Regulation

Provision of SRS Data to other parties must be made in accordance with the category of such data.

In order to verify that the SRS Data is provided in an appropriate manner, it is necessary that the category of data and the name of recipients, in addition to the Identification Codes of the SRS Data and date and time of its provision, are recorded in the log.

■Article 13, paragraph (2) of the Enforcement Regulation

Information on use of the SRS Instruments is generally prepared and managed on a computer. As such, preparation of such information by way of electronic or magnetic records is acceptable.

■Article 13, paragraph (3) of the Enforcement Regulation
In order to verify that the SRS Instruments User is using the SRS Instruments in an appropriate manner, it is necessary that each the activities specified in the Enforcement Regulation is recorded in the log without delay.
Further, as the license for use of SRS Instruments is granted for the individual SRS Instruments, a user of two or more SRS Instruments is required to prepare a log for each of the SRS Instruments.

2.1.6. Termination Measure

(b) Termination Measure provided in Article 15 of the Act:
(1) A Satellite Remote Sensing Instruments User may terminate the Use of Satellite Remote Sensing Instruments at any time, in addition to the cases referred to in Article 13, paragraph (6), Article 14, paragraph (2), Article 16, paragraph (2) or Article 17, paragraph (2).
(2) When a Satellite Remote Sensing Instruments User terminates the Use of Satellite Remote Sensing Instruments, it must take any of the following measures (hereinafter referred to as "Termination Measures"), pursuant to the provisions of Cabinet Office Order, and make a notification to the Prime Minister of the contents of the measure taken without delay.
   (i) measure to send a signal from the Ground Radio Station for Command and Control to the Satellite Remote Sensing Instruments to stop its function of detecting Ground Emitted Electromagnetic Waves, etc., or any other necessary measures specified by Cabinet Office Order to stop such function completely; or
   (ii) measure to ensure that a signal is sent from the Ground Radio Station for Command and Control to the Satellite Remote Sensing Instruments to stop its function until a restart signal (meaning a signal necessary to recover the function of detecting Ground Emitted Electromagnetic Waves, etc. if the function has been suspended; the same applies hereinafter) is received and that a notification of information on the restart signal and the method of creation thereof is made to the Prime Minister, and any other necessary measures specified by Cabinet Office Order in order to ensure that the function cannot be restored unless the restart signal is received by Satellite Remote Sensing Instruments;
(3) When the Termination Measures are taken pursuant to the provisions of the preceding paragraph, the license under Article 4, paragraph (1) ceases to be effective.
(4) A person who has taken the Termination Measures set forth in item (ii) of paragraph (2) must not provide information on the restart signal under that item and the method of creating said signal to

> any persons other than Specified User Organizations or a person who has been newly licensed under Article 4, paragraph (1) to conduct the Use of Satellite Remote Sensing Instruments pertaining to the Termination Measures.
>
> Article 17 of the Enforcement Regulation (Measures to be Specified by Cabinet Office Order as Referred to in Article 15, paragraph (2), item (i) of the Act)
> (1) The measures specified by Cabinet Office Order, as referred to in Article 15, paragraph (2), item (i) of the Act, are any of the measures specified in the following items:
>   (i) sending signals from a Ground Radio Station for Command and Control to the Satellite Remote Sensing Instruments subject to the measure to stop the function to detect Ground Emitted Electromagnetic Waves, etc.; or
>   (ii) sending signals from a Ground Radio Station for Command and Control to the Satellite Remote Sensing Instruments subject to the measure not to supply power.
> (2) The measure specified by Cabinet Office Order, as referred to in Article 15, paragraph (2), item (ii) of the Act, is sending signals from a Ground Radio Station for Command and Control to the Satellite Remote Sensing Instruments subject to the measure to stop the function to detect the Ground Emitted Electromagnetic Waves, etc. until a restart signal is received, and the notification of information on the restart signal and the creation method thereof to the Prime Minister.

Use of SRS Instruments may be terminated any time, except for the cases referred to in Article 13, paragraph (6) of the Act (a case where a business transfer, merger or company split was implemented but an authorization of succession was not granted), Article 14, paragraph (2) of the Act (a case where a user is deceased but an authorization of succession was not granted), Article 16 of the Act (a case where a user was dissolved but an authorization of succession was not granted) or Article 17, paragraph (2) of the Act (a case where a license was rescinded).

Meanwhile, when terminating the use of SRS Instruments, unless the SRS Instruments' function to detect Ground Emitted Electromagnetic Waves, etc. is properly suspended, a person intending to use such SRS Instruments for an inappropriate purpose might operate the uncontrolled SRS Instruments without authorization and use such SRS Instruments in an inappropriate manner. Therefore, the Act requires that a measure be taken to appropriately stop SRS Instruments' function to detect Ground Emitted Electromagnetic Waves, etc. when terminating the use of the SRS Instruments. In such case, the details of measures must be reported to the Prime Minister without delay.

The details of the measures are as provided in Article 17 of the Enforcement Regulation.

2.2. Measures, etc. relating to Handling of SRS Data

2.2.1. Restriction on Method of Provision of SRS Data

> Article 18 of the Act (Restriction on Provision of Satellite Remote Sensing Data)
>
> (1) When providing the Satellite Remote Sensing Data to a person who obtained a certification under Article 21, paragraph (1) for the handling of Satellite Remote Sensing Data, a Satellite Remote Sensing Data Holder must do so after verifying that the recipient of such provision is a person who has obtained that certification, by requiring such recipient to present a certificate under paragraph (4) of that Article, clearly indicating the categories of the Satellite Remote Sensing Data specified by Cabinet Office Order as referred to in paragraph (1) of that Article, and provide the information using cryptography or any other method of transmission whereby it is not easy to restore the contents thereof or any other method specified by Cabinet Office Order as necessary and appropriate for prevention of acquisition and use of Satellite Remote Sensing Data by any person other than the recipient of the Satellite Remote Sensing Data, pursuant to the provisions of Cabinet Office Order.
>
> (2) When providing the Satellite Remote Sensing Data to a Satellite Remote Sensing Instruments User (limited to those who have obtained a license under Article 4, paragraph (1) for the Use of Satellite Remote Sensing Instruments pertaining to the relevant Satellite Remote Sensing Data) or Specified Data Handling Organization, a Satellite Remote Sensing Data Holder must provide the data by clearly indicating to the recipient the categories of the Satellite Remote Sensing Data specified by Cabinet Office Order as referred to in Article 21, paragraph (1) and by the method specified by Cabinet Office Order referred to in the preceding paragraph, pursuant to the provisions of Cabinet Office Order.
>
> (3) A Satellite Remote Sensing Data Holder may not provide the Satellite Remote Sensing Data except when such provision is made pursuant to the provisions of the preceding two paragraphs for the examination or research conducted by each House or by a committee of each House or research committee of the House of Councilors pursuant to the provisions of Article 104, paragraph (1) of the Diet Act (Act No. 79 of 1947) (including the as applied mutatis mutandis pursuant to Article 54-4, paragraph (1) of the same Act) or Article 1 of the Act on Witnesses' Oath, Testimony, etc. Before Both Houses of the Diet (Act No. 225 of 1947), litigation proceedings or any other court proceedings, an execution of judicial decisions, an investigation of criminal cases, or the audit by the Board of Audit, or in any other case equivalent thereto where such provision is necessary for the public interest as specified by a Cabinet Order, or such provision is carried out in an urgent situation when measures must be taken to rescue human life, for disaster relief or for other emergencies.

Article 20 of the Enforcement Regulation (Method of Provision of Satellite Remote Sensing Data)
(1) The method specified by Cabinet Office Order as the method necessary and appropriate for prevention of acquisition and use of Satellite Remote Sensing Data by any person other than the recipient of the Satellite Remote Sensing Data, as referred to in Article 18, paragraph (1) of the Act, is any of the measures specified in the following items:
   (i) cryptography or any other method of transmission whereby it is not easy to restore the contents thereof; or
   (ii) a method of encrypting Satellite Remote Sensing Data and recording it on Magnetic Disks, etc., and providing it by means of the Magnetic Disks, etc.
(2) When a Satellite Remote Sensing Data Holder provides Satellite Remote Sensing Data pursuant to the provisions of Article 18, paragraph (1) of the Act, it must require the recipient to present a certificate under Article 21, paragraph (4) of the Act in advance, and must clearly indicate the category of the Satellite Remote Sensing Data provided in Article 22.
(3) When a Satellite Remote Sensing Data Holder provides the Satellite Remote Sensing Data to a Satellite Remote Sensing Instruments User provided in Article 18, paragraph (2) of the Act pursuant to the provision of that paragraph, it must confirm the name of the Satellite Remote Sensing Instruments User and the name and type of the Satellite Remote Sensing Instruments in advance, and must clearly indicate the category of Satellite Remote Sensing Data provided in Article 22.
(4) The provisions of the preceding paragraph apply mutatis mutandis to the case of provision of Satellite Remote Sensing Data to a Specified Data Handling Organization pursuant to the provisions of Article 18, paragraph (2) of the Act. In this case, the phrase "the name of the Satellite Remote Sensing Instruments User and the name and type of the Satellite Remote Sensing Instruments" is deemed to be replaced with "the name of the Satellite Remote Sensing Instruments.

Article 21 of the Enforcement Regulation (Procedures in Case of Provision of Satellite Remote Sensing Data Due to Urgent Necessity)
(1) When a Satellite Remote Sensing Data Holder provided Satellite Remote Sensing Data due to urgent necessity for rescuing human lives, disaster relief or any other response to emergent situations (including the case of response through international cooperation) if a disaster (meaning a disaster provided in Article 2, item (i) of the Basic Act on Disaster Control Measures (Act No. 223 of 1958) occurred or is likely to occur, the Satellite Remote Sensing Data Holder must submit a document stating the following matters to the Prime Minister without delay.
   (i) the details of the situation;

> (ii) the background and process of provision of the Satellite Remote Sensing Data;
> (iii) the category of the Satellite Remote Sensing Data;
> (iv) the scope and period of the Satellite Remote Sensing Data; and
> (v) the name of the recipient (including another recipients who received the data from the recipient).
>
> (2) When submitting the document set forth in the preceding paragraph, a document clearly specifying the matters set forth in items (i) and (ii) of that paragraph and any other necessary documents must be attached.

> Article 2 of the Act (Definitions)
>
> In this Act, the meanings of the terms set forth in the following items are as provided respectively in those items:
>
> (i) to (iv)   (the rest is omitted.)
>
> (v) "Specified User Organization" means a national governmental organization or local governmental organization prescribed by Cabinet Order as an entity capable of performing an appropriate Use of Satellite Remote Sensing Instruments.
>
> (vi)   (the rest is omitted.)
>
> (vii) "Specified Data Handling Organization" means Specified User Organization and a national or local governmental organization in Japan or a governmental organization of a foreign country (meaning a country or region outside of Japan; the same applies hereinafter) prescribed by Cabinet Order as an entity capable of performing an appropriate handling of Satellite Remote Sensing Data.
>
> (viii)   (the rest is omitted.)
>
>
> Article 1 of the Enforcement Order (National Governmental Organizations Specified by Cabinet Order as Referred to in Article 2, item (v) of the Act)
>
> The national governmental organization to be specified by Cabinet Order, as referred to in Article 2, item (v) of the Act on Ensuring Appropriate Handling of Satellite Remote Sensing Data (hereinafter referred to as the "Act") is the Cabinet Secretariat.
>
>
> Appended Table 1 of the Enforcement Order (Re: Article 2)
>
> Cabinet Office
>
> Fair Trade Commission
>
> National Public Safety Commission
>
> National Police Agency
>
> Financial Services Agency

ANNEX 5

Ministry of Internal Affairs and Communications
Fire and Disaster Management Agency
Ministry of Justice
Public Prosecutors Office
Public Security Examination Commission
Public Security Intelligence Agency
Ministry of Foreign Affairs
Ministry of Finance
National Tax Agency
Ministry of Education, Culture, Sports, Science and Technology
Sports Agency
Agency for Cultural Affairs
Ministry of Health, Labour and Welfare
Ministry of Agriculture, Forestry and Fisheries
Forestry Agency
Fisheries Agency
Ministry of Economy, Trade and Industry
Agency for Natural Resources and Energy
Small and Medium Enterprise Agency
Ministry of Land, Infrastructure, Transport and Tourism
Meteorological Agency
Japan Coast Guard
Ministry of the Environment
Nuclear Regulation Authority
Ministry of Defense
Acquisition, Technology and Logistics Agency
Board of Audit of Japan

■ Article 18, paragraph (1) of the Act

For handling SRS Data, appropriate management according to data categories is necessary. In addition, if the data is not provided in an appropriate manner, Ensuring of the Peace of the International Community, etc. may be harmed as a result of divulgence, etc. of SRS Data.

Therefore, for the provision of SRS Data, it is necessary to prevent divulgence to third parties by checking a certificate possessed by the recipient, clearly specifying the data of appropriate categories, and providing such data in an appropriate manner.

As SRS Data is electronic or magnetic records, provision by way of telecommunication or portable storage media is anticipated. In such case, there is a risk of divulgence to third parties if appropriate measures are not taken. Therefore, for providing SRS Data, prevention of divulgence to third parties is required by taking measures such as encryption.

The method of providing SRS Data as referred to above is set forth in Article 20, paragraph (1) of the Enforcement Regulation.

■ Article 8, paragraph (2) of the Act

For providing SRS Data to an SRS Instruments User (limited to an SRS Instruments User pertaining to such SRS Data), it is necessary to verify that the recipient is the user of such system by confirming the name of such SRS Instruments User or the name and category of such SRS Instruments.

For providing SRS Data to a Specified Data Handler, it is necessary to verify that the recipient is a Specified Data Handler by confirming the name of such Specified Data Handler.

In Article 2, paragraph (1) of the Enforcement Order, national or local public organizations listed in item (i) and (ii) of the paragraph shall take appropriate measures equivalent to that Satellite Remote Sensing Data Holder shall take the safety management measures as specified by the Enforcement Regulation, according to Article 20 of the Act.

Then Satellite Remote Sensing Data Holder is necessary to confirm regulations of the national or local public organization if the organization has measures equivalent to the safety management measures before providing SRS Data to the organization, by inquiring with Cabinet Office about the organization if necessary.

The method of providing SRS Data as referred to above is as set forth in Article 20, paragraphs (3) and (4) of the Enforcement Regulation.

■ Article 18, paragraph (3) of the Act

For the examination or research conducted by each House, court proceedings, execution of judicial decisions, investigation of criminal cases or any other case being equivalent thereto where such provision is necessary for the public interest as specified by a Cabinet Order, or for the response to an urgent situation where measures must be taken to rescue human life, disaster relief or for other emergencies, restriction on provision of SRS Data may be detrimental to the execution of duties. In these cases, provision of

SRS Data is permitted under Article 18, paragraph (3) of the Act.

In the case of the urgent necessity for responding to the rescue of human life, disaster relief or any other emergencies, SRS Data may be provided without taking measures such as confirmation of a certificate of the recipient or using a communication method incapable of recovering secret codes or any other details thereof, which are usually required when providing SRS Data, as such measures would cause difficulties for swift response to disaster, etc.

In this connection, under the Enforcement Regulation, a SRS Data Holder who provided SRS Data due to "urgent necessity for rescuing human lives, disaster relief or any other response to emergent situations (including the case of response through international cooperation)" in the case where "a disaster (meaning a disaster provided in Article 2, item (i) of the Basic Act on Disaster Control Measures (Act No. 223 of 1958) occurred or is likely to occur," is required to submit information summarizing the details of such data to the Prime Minister.

Therefore, whether the provision of SRS Data falls under the aforementioned cases is determined based on whether an "urgent necessity for rescuing human lives, disaster relief or any other response to emergent situations" exists in the case where "wind storms, tornados, torrential rains, heavy snows, floods, landslides, debris flows, storm surges, earthquakes, tsunamis, volcanic eruptions and any other abnormal natural phenomena, large-scale fire or explosion, or any other damage caused by events specified by a Cabinet Order as events with the equivalent degree of damage (e.g. emission of large amount of radioactive substance, sinking of a vessel causing distress to a large number of people, or any other large-scale accidents)" as provided in Article 1, item (i) of the Basic Act on Disaster Control Measures occurred or is likely to occur.

Such determination is made considering the individual and specific circumstances of the case. For example, the following cases would be considered to satisfy the conditions.

- a case of providing SRS Data to respond to a disaster, in the case where a forecast or warning of such disaster has been made (including the case of using SRS Data for discussion of such forecast or warning).
- a case where a disaster response headquarters relating to such disaster is established, and where SRS Data is provided for the purpose of responding to such disaster.
- a case where SRS Data is provided for the purpose of responding to a disaster within the international framework among space organizations and disaster prevention organizations relating to a disaster (e.g. the Sentinel Asia, the International Charter "Space and Major Disasters")

2.2.2. Order of Prohibition of Provision of SRS Data

Article 19 of the Act (Order Prohibiting Provision of Satellite Remote Sensing Data)

(1) If the Prime Minister believes on the sufficient ground that the use of Satellite Remote Sensing Data is likely to cause adverse effect on Ensuring of Peace of the International Community, etc., the Prime Minister may issue an order to a Satellite Remote Sensing Data Holder (excluding a natural person who has neither domicile nor residence in Japan or a corporation or any other organization which does not have a principal office in Japan that handles Satellite Remote Sensing Data in a foreign country (hereinafter referred to as a "Foreign Handler")) to prohibit provision of the Satellite Remote Sensing Data designating the scope and time period.

(2) The prohibition order under the preceding paragraph must be limited to the minimum extent required for the Ensuring of Peace of the International Community, etc.

(3) The provisions of the preceding two paragraphs apply mutatis mutandis to a Satellite Remote Sensing Data Holder (limited to a Foreign Handler). In this case, the term "to prohibit provision" in paragraph (1) is deemed to be replaced with "not to provide", and the term "issue an order to" in the preceding paragraph is deemed to be replaced with "request."

Article 3 of the Enforcement Regulation (Standards Specified by Cabinet Office Order as Referred to in Article 2, item (vi) of the Act)

(Standards to be Specified by Cabinet Office Order as Referred to in Article 2, item (vi) of the Act)

Categories	Standards
(i) Raw data	(a) for data recorded by an optical sensor, data with Distinguishing Accuracy of Target not exceeding 2 meters, which is within five years after the recording.
	(b) for data recorded by a SAR sensor, data with Distinguishing Accuracy of Target not exceeding 3 meters, which is within five years after the recording.
	(c) for data recorded by a hyperspectral sensor, data with Distinguishing Accuracy of Target not exceeding 10 meters and detectable wavelength bands exceeding 49, which is within five years after the recording.
	(d) for data recorded by a thermal infrared sensor, data with Distinguishing Accuracy of Target not exceeding 5 meters, which is within five years after the recording.

ANNEX 5

(ii) Standard data	(a) for data recorded by an optical sensor, data with Distinguishing Accuracy of Target less than 25 centimeters.
	(b) for data recorded by a SAR sensor, data with Distinguishing Accuracy of Target less than 24 centimeters.
	(c) for data recorded by a hyperspectral sensor, data with Distinguishing Accuracy of Target not exceeding 5 meters and detectable wavelength bands exceeding 49.
	(d) for data recorded by a thermal infrared sensor, data with Distinguishing Accuracy of Target not exceeding 5 meters.

(2) Notwithstanding the provisions of the preceding paragraph, the standards specified by Cabinet Office Order, as referred to in Article 2, item (vi) of the Act pertaining to Satellite Remote Sensing Data subject to an order prohibiting provision pursuant to the provisions of Article 19, paragraph (1) of the Act, are to be specified by the Prime Minister by a public notice.

Prohibition of provision of SRS Data under Article 19, paragraph (1) of the Act is ordered in case where there is a sufficient reason to believe that such provision is likely to cause harm to Ensuring of the Peace of the International Community, etc.

As the standards for the SRS Data subject to the prohibition are determined depending on the circumstances of individual and specific cases, it is difficult to set specific standards in advance.

Therefore, such standards are to be determined by the Prime Minister in a public notice depending on the circumstances of individual and specific cases.

2.2.3. Safety Management Measures

Article 6 of the Act (Requirements for License)
The Prime Minister may not grant the license under Article 4, paragraph (1), unless the Prime Minister finds that the application for the license under that paragraph meets all of the following requirements:
(ii) measures for prevention of divulgence, loss or damage of Satellite Remote Sensing Data and any other necessary and appropriate measures to be specified by Cabinet Office Order for the safety management of the Satellite Remote Sensing Data have been taken;

Article 20 of the Act (Safety Management Measures for Satellite Remote Sensing Data)
A Satellite Remote Sensing Data Holder must take measures for prevention of divulgence, loss or

damage of Satellite Remote Sensing Data and any other necessary and appropriate measures specified by Cabinet Office Order for the safety management of the relevant Satellite Remote Sensing Data.

Article 7 of the Enforcement Regulation (Measures Specified by Cabinet Office Order as Referred to in Article 6, item (ii) of the Act)

(1) The measures specified by Cabinet Office Order, as referred to in Article 6, item (ii) and Article 20 of the Act, are as specified in the lower column of the following table, in accordance with the categories of Satellite Remote Sensing Data respectively specified in the upper column of the table.

Categories of Satellite Remote Sensing Data	Measures

(i) Raw data	(a) Organizational safety management measures
	  (i) that a basic policy for safety management of the Satellite Remote Sensing Data is established.
	  (ii) that the responsibilities and authorities as well as businesses of person in charge of handling Satellite Remote Sensing Data are made clear.
	  (iii) that an organization for handling business in case of divulgence, loss or damage of Satellite Remote Sensing Data is established.
	  (iv) that regulations on safety management measures have been established and implemented, and operation of such regulations is being assessed and improved.
	(b) Human safety management measures
	  (i) that confirmation is made that the person in charge of handling Satellite Remote Sensing Data does not fall under any of Article 5, items (i) to (iv) and Article 21, paragraph (3), item (i)(a) to (d) of the Act;
	  (ii) that a person in charge of handling Satellite Remote Sensing Data has taken has taken measures to ensure that information on Satellite Remote Sensing Data handled by such person in the course of business and any other special confidential information (meaning unpublished information which such person may learn in the course of business) will not be used for any purpose other than for the ensuring of appropriate operation of such business or any other purpose found to be necessary.
	  (iii) that necessary education and training are provided to a person

	in charge of handling Satellite Remote Sensing Data.
	(c) Physical safety management measure
	   (i) that facilities for handling Satellite Remote Sensing Data are clearly distinguished.
	   (ii) that measures have been taken to restrict entry into and bringing any device into facilities for handling Satellite Remote Sensing Data.
	   (iii) that for a computer and portable memory device (meaning a portable media or device capable of being inserted into or connected to a computer or its peripheral equipment to store information; hereinafter the same applies in this paragraph), in order to prevent theft, loss or any other accident, fixing the edge of a computer with a wire or any other necessary physical measures have been taken.
	(d) Technical safety management measures
	   (i) that appropriate measures have been taken for facilities for handling Satellite Remote Sensing Data so as to prevent unauthorized access (meaning unauthorized access as provided in Article 2, paragraph (4) of the Act on Prohibition of Unauthorized Computer Access (Act No. 128 of 1999).
	   (ii) that measures have been taken to restrict a portable memory device from being connected with a computer or its peripheral equipment.
	   (iii) that operations of computers and terminals relating to the handling of Satellite Remote Sensing Data have been recorded.
	   (iv) that for transfer or telecommunication transmission of Satellite Remote Sensing Data, encryption or any other necessary measures for the appropriate protection of Satellite Remote Sensing Data have been taken.
	   (v) that for processing of Satellite Remote Sensing Data, necessary measures have been taken to ensure that such processing is implemented in an appropriate manner.

(ii) Standard data	(a) Organizational safety management measures
	   Same as (a) for the paragraph of raw data.
	(b) Human safety management measures
	   Same as (a) for the paragraph of raw data.

| | (c) Physical safety management measure |
| | Same as (a) for the paragraph of raw data. |

(2) If a Satellite Remote Sensing Instruments User and Satellite Remote Sensing Data Holder manages the business of handling of Satellite Remote Sensing Data, in whole or part, by the use of external storage service through telecommunication lines, it must expressly provide for the following matters in a contract with the business providing the service (hereinafter referred to as a "service provider" in this paragraph) relating to the use of the service.

(i) that the measures equivalent to those provided in the preceding paragraph are to be taken; and

(ii) that Satellite Remote Sensing Data is not to be stored on a computer located in any of the following countries or regions:

(a) the regions specified in Appended Table 3-2 or 4 of the Export Order; or

(b) the countries or regions determined by a resolution of the United Nations General Assembly or Security Council as being responsible for the occurrence of situations threatening the peace and security of the international community;

(iii) that, upon the cancellation or expiration of the contract, deletion or return of Satellite Remote Sensing Data or any other necessary measures are to be taken; and

(iv) that, if the service provider entrusts all or part of its business to a third party, a contract for the business entrustment provides for a condition that the entrusted party must comply with the matters specified in the preceding three items and that the entrusted party must take any other measures to perform the business in an appropriate and accurate manner.

(3) The measures provided in the preceding two paragraphs do not apply to Satellite Remote Sensing Data to be provided for necessary purposes in view of public interest or for the urgent necessity for rescuing human lives, disaster relief or any other emergencies, as provided in Article 18, paragraph (3) of the Act.

■Article 7, paragraph (1) of the Enforcement Regulation

SRS Instruments Users and SRS Data Holders must take necessary and appropriate measures to prevent divulgence, loss or damage of SRS Data it handles and to otherwise manage the safety of SRS Data.

More specifically, it is necessary to implement, in an appropriate manner, "organizational safety management measures" including establishment of organizational structure, "human safety management measures" including supervision and education of persons engaged in handling of SRS Data, "physical safety management measures" including management of sections for handling SRS Data and prevention of theft of computers and electronic media, and "technical safety

ANNEX 5

management measures" including prevention of unauthorized access from outside and access control.

The following examples of concrete methodologies for the safety management measures are not intended to be limited to only the following.

2.2.3.1. Organizational safety management measures

Organizational safety management measures mean the following safety management measures at the organizational level.
① Establishment of a basic policy for safety management of SRS Data.
② Clearly specifying the authorities and responsibilities as well as the duties of a person in charge of handling SRS Data.
③ Establishment of an organization for business handling in case of divulgence, destruction or damage of SRS Data.
④ Establishment and operation of regulations on safety management measures for SRS Data, as well as assessment and improvement of the operation thereof.

[Applicable persons]
- SRS Instruments User
- SRS Data Holder (Raw Data/Standard Data)

> (i) that a basic policy for safety management of the Satellite Remote Sensing Data is established.

In order to commit to safety management of SRS Data at the organizational level, it is important to provide for a basic policy so as to ensure that persons in charge of handling SRS Data understand such policy.

For this purpose, when drafting a basic policy, SRS Instruments Users and SRS Data Holders are required to make clear the business of handling SRS Data and its position within the organization, present a basic approach for safety management of SRS Data, and ensure compliance with laws, regulations, rules, etc. on SRS Data.

[Examples of methods]
For example, the following items can be provided in the basic policy:
① Compliance with applicable laws, regulations and rules
② Item relating to safety management measures

> (ii) that the responsibilities and authorities as well as businesses of person in charge of handling Satellite Remote Sensing Data are made clear.

Safety management of SRS Data is achieved by making clear the scope of persons in charge of handling SRS Data, and by such persons understanding and fulfilling their respective authorities and responsibilities. In particular, it is important that persona who are determined to be responsible for the safety management of SRS Data understand the importance of the safety management of SRS Data.

To this end, it is necessary to make clear the scope of persons engaged in the handling of SRS Data and their respective authorities and responsibilities by way of a list, etc., and to establish organizations and structures as may be necessary.

"Persons in charge of handling SRS Data" mean officers of SRS Instruments Users and SRS Data Holders engaged in the business of handling of SRS Data (directors, executive officers, auditors, etc.) or their employees having authorities and responsibilities relating to such business, and persons engaged in the business of handling SRS Data under direct or indirect supervision of these persons (e.g. employees of such entity (full-time employees, contract workers, non-regular workers, part-time workers, etc.), and employees of other entities doing business pursuant to an entrustment contract).

[Examples of methods]
① Appointing a person responsible for the business of handling SRS Data and clearly specifying such person's responsibilities.
② Clearly specifying a person in charge of handling SRS Data and such person's roles.
③ Clearly specifying the scope of SRS Data to be handled by the person in charge of handling SRS Data.
④ Clearly specifying allocation of roles and responsibilities of each section, in the case where SRS Data is handled by two or more sections.

> (iii) that an organization for handling business in case of divulgence, loss or damage of Satellite Remote Sensing Data is established.

In the case where the occurrence of divulgence, destruction or damage of SRS Data or indication thereof is discovered, the situation must be immediately identified at the

ANNEX 5

organizational level, and measures including prevention of collateral damage or recurrence of similar incidents must be taken.

For such purpose, SRS Instruments Users and SRS Data Holders are required to establish internal structures necessary for ensuring that such measures are implemented in an appropriate and swift manner.

[Examples of methods]
① Establishment of a reporting structure from a person handling SRS Data to a supervisor, in the case where a fact of violation of applicable laws and regulations or the indication thereof is discovered.
② Investigation into factual relationships and identification of cause.
③ Clearly specifying the person in charge of response to accidents and the supervisor.
④ Creation of a single point of contact for reporting in the case of occurrence of incidents including divulgence.
⑤ Discussion and determination of measures to prevent recurrence.
⑥ Report on factual situations and measures to prevent recurrence, etc.

> (iv) that regulations on safety management measures have been established and implemented, and operation of such regulations is being assessed and improved.

For safety management of SRS Data, it is important to ensure the efficacy thereof.

For such purpose, SRS Instruments Users and SRS Data Holders are required to provide for regulations setting forth the details of the following "human safety management measures," "physical safety management measures" and "technical safety management measures" in addition to these "organizational safety management measures," ensure that persons engaged in handling of SRS Data understand these, and identify and analyze the status of implementation of such regulations and take improvement measures as may be necessary.

[Examples of methods]
For example, the following items can be provided in the regulations:
① Organizational safety management measures
   a) Establishment of a basic policy for safety management of SRS Data
   b) Establishment of organizational structure
   c) Operation in accordance with Handling Regulations, etc.

        d) Establishment of structure to respond to divulgence incident, etc.
        e) Review of safety management measures
    ② Human safety management measures
        a) Supervision of persons in charge of handling SRS Data
        b) Education of persons in charge of handling SRS Data
    ③ Physical safety management measure
        a) Management of facilities for handling SRS Data
        b) Restriction of access to facilities for handling SRS Data
        c) Prevention of theft of computers, electronic devices, etc.
    ④ Technical safety management measures
        a) Prevention of unauthorized access from outside
        b) Control of access, and identification and authorization of persons with access
        c) Recording of operations
        d) Prevention of divulgence of information
        e) Management of appropriate processing of SRS Data
<Attachments>
    ① Organization chart
    ② List of persons in charge of handling SRS Data

2.2.3.2. Human safety management measures

Human safety management measures mean the following safety management measures at human level.
    ① Confirmation that persons in charge of handling SRS Data do not fall under any disqualifying conditions.
    ② Prohibition of use of confidential information relating to SRS Data other than for the prescribed purpose by any person handling SRS Data.
    ③ Circulation of internal regulations and implementation of education and training for persons in charge of handling SRS Data.

[Applicable persons]
SRS Data User, SRS Data Holder (Raw Data/Standard Data)

> (i) that confirmation is made that the person in charge of handling Satellite Remote Sensing Data does not fall under any of Article 5, items (i) to (iv) and Article 21, paragraph (3), item (i)(a) to (d) of the Act

For SRS Instruments Users and SRS Data Holders, and officers and employees having authorities and responsibilities relating to the business of handling SRS Data, a confirmation is to be made as to whether these persons fall under any disqualifying conditions in the process of examination for license and certification; however, persons other than the above who handle SRS Data are not examined.

Therefore, SRS Instruments Users and SRS Data Holders are required to confirm within their organizations that persons engaged in handling of SRS Data do not fall under any disqualifying conditions.

[Examples of methods]
Employer, etc. is to confirm that persons engaged in the handling of SRS Data within the organization do not fall under any disqualifying conditions by way of a commitment letter or confirmation letter.

> (ii) that a person in charge of handling Satellite Remote Sensing Data has taken has taken measures to ensure that information on Satellite Remote Sensing Data handled by such person in the course of business and any other special confidential information (meaning unpublished information which such person may learn in the course of business) will not be used for any purpose other than for the ensuring of appropriate operation of such business or any other purpose found to be necessary.

SRS Data involves the risk of harming the Ensuring of the Peace of the International Community, etc. and therefore requires appropriate safety management. As such, SRS Instruments Users and SRS Data Holders need to take measures to ensure that confidential information pertaining to the SRS Data they handle in the course of business will not be used for any other purpose.

[Examples of methods]
① Conclusion of non-disclosure contract upon recruiting employees or execution of service contract
Preferably, a non-disclosure provision under an employment contract or service contract should remain effective for a certain period even after the termination of the contract.
② Establishment of regulations on the measures in case of violation of a non-disclosure contract

> (iii) that necessary education and training are provided to a person in charge of handling Satellite Remote Sensing Data.

Even if regulations on safety management of SRS Data are established in an appropriate manner, it is impossible to achieve appropriate management if the contents of such regulations are not understood and complied with by persons engaged in the handling of SRS Data.
Therefore, SRS Instruments Users and SRS Data Holders are required to provide education and training for persons engaged in handling of SRS Data to gain deeper understanding as to the appropriate handling of SRS Data.

[Examples of methods]
① Regular training on points in attention relating to applicable laws and regulations, rules, etc.
② Implementation of necessary and appropriate education and training for persons in charge of handling SRS Data.

2.2.3.3. Physical safety management measures

Physical safety management measures mean the following safety management measures at physical level.
① Clearly designating facilities for handling SRS Data.
② Control of entry into and exit from facilities and bringing of equipment into facilities
③ Prevention of theft, etc. by physical protection of equipment and device

[Applicable persons]
SRS Data User, SRS Data Holder (Raw Data)

> (i) that facilities for handling Satellite Remote Sensing Data are clearly distinguished.

If SRS Data is obtained by any person intending to use it for an inappropriate purpose, due to leaking and other reasons, Ensuring of the Peace of the International Community, etc. may be harmed.

Therefore, SRS Instruments Users and SRS Data Holders are required to control entry into and exit from facilities, restriction of bringing equipment into or taking it out from facilities, and, in order to ensure that these measures are implemented in an appropriate manner, clearly designate facilities for handling of SRS Data.

[Examples of methods]
① Establishment of wall, door with lock, partition, etc.
② Clearly designating facilities for handling of SRS Data in regulations, etc.
③ Determine the scope of facilities taking into account the actual status of operation, and only the scope which is capable of being managed in an appropriate manner should be designated.
④ In the case of handling SRS Data at a portable facility (e.g. vehicle mounted earth station), specify information enabling clear identification of the equipment, such as the name of such equipment

> (ii) that measures have been taken to restrict entry into and bringing any device into facilities for handling Satellite Remote Sensing Data.

If facilities for handling SRS Data and server system for storing such data and terminals are located in the environment enabling contact by unspecified, multiple persons, there is a risk of impersonation by a person with improper intent or physical destruction of the system, as well as divulgence of information by unauthorized transfer of data from server rooms and terminals.

Therefore, SRS Instruments Users and SRS Data Holders are required to take measures for control of entry into and exit from facilities handling SRS Data and bringing device into such facilities, so as to secure safety of SRS Data they handle.

The term "device" means a portable information communication/storage device, and any other devices with function of storing electronic or magnetic records.

[Examples of methods]
① Creation of control system by such means as IC cards and number keys
② Record of entry/exit in case of control by means of lock (creation of entry/exit management list)
③ Installing locker, etc. for storing devices outside the facilities for handling data, and restricting bringing devices into the facilities.

> (iii)that for a computer and portable memory device (meaning a portable media or device capable of being inserted into or connected to a computer or its peripheral equipment to store information; hereinafter the same applies in this paragraph), in order to prevent theft, loss or any other accident, fixing the edge of a computer with a wire or any other necessary physical measures have been taken.

Even if entry into and exit from facilities is controlled or bringing of equipment into facilities is restricted, it is meaningless if a computer handling SRS Data or portable memory device storing SRS Data located in such facilities is physically taken out from facilities.

Therefore, SRS Instruments Users and SRS Data Holders are required to take physical measures to prevent such computers and portable memory device from being taken out from facilities.

[Examples of methods]
① Keeping portable memory device storing SRS Data in a lockable cabinet, library, etc., and manage preparation and quantities thereof.
② In the case where the information system for handling of SRS Data is operated only with equipment, fixing such equipment by anti-theft wire lock.
③ Prohibiting leaving on desks, etc. media storing SRS Data and portable computers
④ Ensure that workers activate passwords and screen savers when leaving their desks

2.2.3.4. Technical safety management measures

Technical safety management measures mean the following technical management measures at the technical level.
① Appropriate measures for prevention of unauthorized access
② Measures relating to restriction of connection of portable memory devices to computers, etc. for handling SRS Data
③ Recording of operations of computers and terminal devices
④ Encryption of SRS Data at the time of transfer and transmission and any other necessary protective measures
⑤ Management of appropriate processing of SRS Data

ANNEX 5

[Applicable persons]
SRS Data User, SRS Data Holder (Raw Data/Standard Data)

> (i) that appropriate measures have been taken for facilities for handling Satellite Remote Sensing Data so as to prevent unauthorized access (meaning unauthorized access as provided in Article 2, paragraph (4) of the Act on Prohibition of Unauthorized Computer Access (Act No. 128 of 1999).

Possibility of divulgence, etc. of SRS Data increases if any persons other than those handling SRS Data have access to SRS Data or if there is deficiency in the security measures of the information system.
Therefore, SRS Instruments Users and SRS Data Holders are required to restrict access to computers for handling SRS Data, and to appropriately manage information which identifies persons engaged in handling of SRS Data. In addition, measures must be also taken against possible attacks concerning vulnerabilities of computer security.

[Examples of methods]
① Limiting the persons with authority to give access to SRS Data and the authority to be granted to such persons
② Verifying the effectiveness of the access control function introduced into the information system for handling SRS Data (for example, checking vulnerabilities of OS and web applications)
③ Identification by a user ID, password, one-time password, IC card, etc. (note that user ID, etc. is to be provided so as to identify individual persons handling data).
   * In order to prevent unauthorized access, when creating passwords, take measures such as prohibiting passwords identical with user IDs, setting the effective period of passwords, restricting reuse of the identical or similar passwords, setting a minimum number of characters of passwords, and suspending IDs in the case of fault in log-in attempts exceeding more than a certain number.
④ Setting firewalls, etc. at the connecting point of the information system and outside network.
⑤ Installing anti-virus software and verifying of effectiveness and stability of such software (e.g. confirming updating of pattern files and hotfix)
⑥ Applying hotfix for security measures (namely, security patch) to operating

systems (OS), middleware (e.g. DBMS), applications, etc. to computers and servers.
⑦ Preventing installation of software not authorized by the organization.

> (ii) that measures have been taken to restrict a portable memory device from being connected with a computer or its peripheral equipment.

For measures to prevent SRS Data from being taken out from facilities without authorization, it is important to restrict access to computers for handling SRS Data; however, unauthorized acquisition of data is also possible by way of connecting with external storage devices such as external hardware, CD-Rs and USB flash drives.
Therefore, SRS Instruments Users and SRS Data Holders are required to take measures to restrict connection of portable memory devices to computers for handling SRS Data and their peripheral equipment.

[Examples of methods]
① Restriction and management of connection of external memory devices, such as CD-Rs and USB flash drives.
② Restriction and management of connection of equipment with a storage function, such as smartphones and PCs.

> (iii) that operations of computers and terminals relating to the handling of Satellite Remote Sensing Data have been recorded.

Recording of operation means recording of the history of operation and user access of the system and any other necessary information.
Such recording is serve as important information for detecting security incidents such as unauthorized access and unauthorized operation by a third party with malicious intent (including the indication thereof). Grasping such information is important as such information is useful for identification of the cause when any incidents such as information divulgence occur.
Therefore, SRS Instruments Users and SRS Data Holders are required to keep records of operations of information systems in an appropriate manner, and ensure that such records are stored in the environment enabling appropriate preservation without the risk of being tampered with.

ANNEX 5

[Examples of methods]
① Preservation of status of use of the information system for handling SRS Data (e.g. log-in history, access logs) and regular-basis monitoring
② Monitoring of access to SRS Data (including the details of operations)
③ Measures to prevent alteration and unauthorized deletion of collected logs
④ Monitoring access to the information system for handling SRS Data from outside by the use of an intrusion detection system, intrusion protection system, etc.

(iv) that for transfer or telecommunication transmission of Satellite Remote Sensing Data, encryption or any other necessary measures for the appropriate protection of Satellite Remote Sensing Data have been taken.

It is expected that SRS Data are provided by means of: (i) portable memory devices and (ii) telecommunication lines. In either case, data is transferred outside information systems, etc. for which protection measures are implemented.

Therefore, in order to prevent divulgence, etc. of SRS Data, SRS Instruments Users and SRS Data Holders are required to take measures against physical theft and loss including the lock of containers, in the case of transfer by means of portable memory devices, or to secure secrecy by encryption or other means, in the case of transmission through telecommunication lines. Even if data is obtained by a third party, measures must be taken such that SRS data to be transferred or transmitted may not be viewed.

[Examples of methods]
① Use a lockable container upon the transfer, and store the SRS Data on electronic memory device after encryption, coding and password protection.
② In the case of transmission through telecommunication, encryption of the telecommunication route by SSL (TLS), IPSec, etc. or by using a dedicated line or VPN line.
③ For emails summarizing details of data transmission, transmit an email message encrypted by S/MIME, etc., or attach to an email an encrypted document summarizing details.

(v) that for processing of Satellite Remote Sensing Data, necessary measures have been taken to ensure that such processing is implemented in an appropriate manner.

SRS Data are categorized according to the degree of their processing. In the case where the processing is incomplete, electronic or magnetic records which would otherwise have been processed into non-SRS Data might be considered as SRS Data. Provision of such data may result in harm to the Ensuring of the Peace of the International Community, etc.

Therefore, SRS Instruments Users and SRS Data Holders are required to establish systems to ensure that appropriate processing are applied according to the relevant categories.

[Examples of methods]
① Clearly specifying the responsible person in charge of processing SRS Data.
② Clearly specifying the person engaged in the business of processing SRS Data.
③ Clearly specifying the procedures for processing SRS Data.
④ Implement the processing according to the prescribed procedures (in the case of automatic processing using a system, take measures such as ensuring that unauthorized persons cannot make modification to the setting)
⑤ Implementing processing at the designated facilities.
⑥ Clearly specifying the purpose of processing SRS Data.

*For research and development for which procedures are not established, ensure that processing is appropriately managed by a supervisor.

■Article 7, paragraph (2) of the Enforcement Regulation
In the case of managing business of handling SRS Data, in whole or in part, by the use of third party storage service provided through telecommunication lines, it is required to take measures equivalent to the safety management measures referred to in Article 7, paragraph (1) under a service contract with a provider of such service, ensure that SRS Data is not stored on a computer located in country with potential risks, and expressly provide for the above measures in the case of subcontracting such business to another party.

The details of the measures are as provided in the Enforcement Regulation.

■Article 7, paragraph (3) of the Enforcement Regulation
For SRS Data provided for the public interest or for the urgent necessity for responding to the need to rescue human lives or other emergencies, requiring the same level of safety management measures as general SRS Data would increase the burden of a holder of SRS Data, leading to a negative impact on service for the public interest and swift disaster response. As such, these safety management measures are not required

ANNEX 5

for data possessed within the scope necessary to achieve the above objectives.

2.2.4. Preparation and Management of Logs

Article 23 of the Act (Log)

(1) A person who obtained a certification under Article 21, paragraph (1) of the Act must keep a log and record the matters specified by Cabinet Office Order in relation to the status of handling of Satellite Remote Sensing Data, pursuant to the provisions of Cabinet Office Order.

(2) The log under the preceding paragraph must be preserved pursuant to the provisions of Cabinet Office Order.

Article 30 of the Enforcement Regulation (Matters to be Stated in Logs)

(1) The matters specified by Cabinet Office Order, as referred to in Article 23, paragraph (1) of the Act, are as set forth in the following items:
  (i) Identification Codes of Satellite Remote Sensing Data for receiving or providing Satellite Remote Sensing Data;
  (ii) Categories of Satellite Remote Sensing Data;
  (iii) date and time of receiving or providing the Satellite Remote Sensing Data;
  (iv) the name of the recipient or provider, and, the certificate number of the recipient or provider if the recipient or provider has obtained a certificate under Article 21, paragraph (4) of the Act; and
  (v) status of processing or deleting the Satellite Remote Sensing Data.

(2) When a person who obtained a certification under Article 21, paragraph (1) of the Act creates electronic or magnetic records for the log under Article 23, paragraph (1) of the Act, the person must do so by a method of recording the created electronic or magnetic records on a file stored in a computer used by the person who obtained certification under Article 21, paragraph (1) of the Act, or by a method of preparing the records using Magnetic Disks, etc.

(3) A person who obtained a certification under Article 21, paragraph (1) of the Act must record in the log the matters set forth in the items of paragraph (1) without delay for, each instance of receiving or providing the Satellite Remote Sensing Data, or processing or deleting of Satellite Remote Sensing Data, and must keep the record for five years from the entry of the information.

A person who obtained a certification under Article 21, paragraph (1) of the Act is required to prepare a log to record the status of handling of SRS Data and preserve such log.

More concretely, in relation to the following matters, if there occurs any event which requires entry of information into the log, the relevant person is required to make such entry without delay and to preserve the log for the designated period.

■Article 30, paragraph (1), items (i) to (iv) of the Enforcement Regulation
Provision of SRS Data to other parties must be made in accordance with the category of such data. In order to verify that the SRS Data is provided in an appropriate manner, it is necessary that the category of data and the name of recipients, in addition to the Identification Codes of the SRS Data and date and time of provision, are recorded in the log.

■Article 30, paragraph (1), item (v) of the Enforcement Regulation
In order to identify the status of SRS Data, it is necessary that the status of the processing and deletion, as well as provision to third parties, is recorded in the log.
More concretely, the relevant person is required to enter into the log the Identification Codes of the SRS Data to be processed and the details of such processing, in the case of processing; or the Identification Codes of the SRS Data to be deleted and the date and time of method of the deletion, in case of deletion.

■Article 30, paragraph (2) of the Enforcement Regulation
Information on handling of SRS Data is generally created and managed on a computer. As such, creation of such information by way of electronic or magnetic records is acceptable.

■Article 30, paragraph (3) of the Enforcement Regulation
In order to verify that the SRS Instruments User is handling the SRS Data in an appropriate manner, it is necessary that each of the activities specified in the Enforcement Regulation are recorded in the log without delay.

Annex 5

3. Review of Guidelines

Details of measures relating to the use of SRS Instruments and treatment of SRS Data are subject to change due to technological development and international situations. These Guidelines will be reviewed as appropriate, according to the change in various circumstances.

ANNEX 6

Application Manuals for Act on Ensuring Appropriate Handling of Satellite Remote Sensing Data

November 14, 2017

National Space Policy Secretariat, Cabinet Office

Table of Contents

1. Introduction 2
2. Applications 3
3. Application for license to use SRS Instruments 5
    3.1. Information items to be entered in application form and guides for preparation 5
    3.2. Attachments 13
4. Application for certification for handling SRS Data 19
    4.1. Information items to be entered in application form and guides for preparation 19
    4.2. Attachments 23
5. Example of completed application form 27

[Explanatory Notes]

Unless otherwise provided, the terms used in these Manuals have the meanings as defined in the Act and Enforcement Regulation. The abbreviations as used in these Manuals have the following meanings:

- Act: Act on Ensuring Appropriate Handling of Satellite Remote Sensing Data (Act No. 77 of 2016)
- Enforcement Regulation: Regulation for Enforcement of the Act for Ensuring Appropriate Handling of Satellite Remote Sensing Data (Cabinet Office Order No. 41 of 2017)
- SRS Instruments: Satellite Remote Sensing Instruments
- SRS Data: Satellite Remote Sensing Data

Annex 6

1. Introduction

A person who intends to obtain a license for use of the SRS Instruments or a certification for handling of SRS Data is required to make an application to the Prime Minister pursuant to the Act and Enforcement Regulation.

These Manuals provide for explanation on matters necessary for the application.

2. Applications

[Application for license to use SRS Instruments]

Article 4 of the Act (License)

(1) A person who intends to conduct the Use of Satellite Remote Sensing Instruments by use of a Ground Radio Station for Command and Control located in Japan (excluding Specified User Organization) must obtain a license from the Prime Minister per Satellite Remote Sensing Instruments.

(2) A person who intends to obtain a license under the preceding paragraph must submit a written application to the Prime Minister, pursuant to the provisions of Cabinet Office Order, specifying the following information, attaching the documents specified by Cabinet Office Order.

Article 4 of the Enforcement Regulation (Application for License)

(1) A person who intends to obtain a license under Article 4, paragraph (1) of the Act must submit a written application in Form 1 to the Prime Minister.

(2) The following documents must be attached to the written application under the preceding paragraph.

[Application for certification for handling SRS Data]

Article 21 of the Act (Certification)

(1) A person handling Satellite Remote Sensing Data (excluding Specified Data Handling Organization) may, upon application, obtain a certification from the Prime Minister that states that person is found to be capable of properly handling Satellite Remote Sensing Data, according to the categories of Satellite Remote Sensing Data specified by Cabinet Office Order having regard to circumstances such as Distinguishing Accuracy of Target, the scope and degree of information changed as a result of processing of Electromagnetic Data of Detected Information, or the time elapsed since the relevant Electromagnetic Data of Detected Information was recorded.

(2) A person who intends to obtain a certification referred to in the preceding paragraph must submit a written application to the Prime Minister, pursuant to the provisions of Cabinet Office Order, specifying the following information, attaching the documents evidencing that the certification standards set forth in each item of the following paragraph are met and other documents specified by Cabinet Office Order.

Article 23 of the Enforcement Regulation (application for certification)

(1) A person who intends to obtain a certification under Article 21, paragraph (1) of the Act must submit a written application in Form 13 to the Prime Minister.

> (2) The following documents must be attached to the written application under the preceding paragraph.

> [Languages to be Used for Application, etc.]
> Article 33 of the Enforcement Regulation (Terms, etc. of documents)
> (1) Written applications, written notifications and documents under Article 21, paragraph (1) as provided in this Cabinet Office Order must be prepared in Japanese; provided, however, that address, name and contact information may be written in a foreign language.
> (2) Documents to be attached to written applications, written notifications and documents under Article 21, paragraph (1) as provided in this Cabinet Office Order must be prepared in Japanese or English; provided, however, that the Japanese translations must be submitted for any documents written in English.
> (3) Due to special circumstances, if a person is unable to submit the documents under the preceding paragraph in a language provided in that paragraph, notwithstanding the provisions of that paragraph, the person may submit the document together with its Japanese translations.

A person who intends to obtain a license for use of the SRS Instruments or a certification for handling of SRS Data is required to submit to the Prime Minister written applications in Form 1 and Form 13, respectively, attaching necessary documents.

Information items to be stated in an application form and guidelines for preparation thereof is shown in 3.1 and 4.1, and the necessary attached documents are shown in 3.2 and 4.2. For a detailed example, please see 5.

An application form must be written in Japanese; however, an address, name and contact information may be written in a foreign language. Documents to be attached to the application form must be written in Japanese or English. For English documents, Japanese translations must be attached. If, due to any special circumstance, the applicant cannot submit attached documents in Japanese or English, documents in another language, together with the Japanese translation thereof, may be submitted.

3. Application for license to use SRS Instruments

3.1. Information items to be entered in application form and guides for preparation

A person who intends to obtain a license for the use of the SRS Instruments is required to submit a written application containing the following matters to the Prime Minister.

> ① Address, name and contact information
> ② Name, type, configuration and performance of the SRS Instruments
> ③ Orbit of the Earth Orbiting Satellite installed with the SRS Instruments
> ④ Place, configuration and performance of Ground Radio Station for Command and Control, as well as the method of management thereof.
> ⑤ Place, configuration and performance of Receiving Station, as well as the method of management thereof.
> ⑥ Method of management of SRS Data.
> ⑦ In the case where the applicant is an individual person, the name or address of the Successor.
> ⑧ Name and address of an officer or employee engaged in the business pertaining to the Use of SRS Instruments
> ⑨ If a person other than the applicant is to manage the Ground Radio Station for Command and Control, the name and address of the manager.
> ⑩ If a person other than the applicant is to manage the Receiving Station, the name and address of the manager.
> ⑪ Purpose and method of use of SRS Data
> ⑫ Name of investors, investment ratio and nationalities
> ⑬ Major customers

The following is the guide for preparation for each of the information items.

> ① Address, name and contact information

In the case where a person intending to use SRS Instruments is an individual person, state the name and address as stated in the residence certificate. If such person is a foreign national, state the name and address as stated in a document issued by the foreign government approved by the government of Japan or an equivalent document.
In the case of a corporation, state the corporation name and address as stated in the

ANNEX 6

certificate of registered information.

For contact information, state the address, name, corporation name, section in charge, person in charge, etc. to enable receiving of mail.

② Name, type, configuration and performance of the SRS Instruments

➢ "Name"

Distinguishing Accuracy of Target and an orbit of Earth Orbiting Satellite installed with SRS Instruments, as well as Operational Radio Station for the operation, may vary with respective SRS Instruments. Therefore, a license is needed for respective SRS Instruments.

In the case where two or more SRS Instruments with different types, configurations and performances are boarded on the same Earth Orbiting Satellite, as an individual license is required, state the respective names of the Systems.

Example: Case of on board three different SRS Instruments on a single Earth Orbiting Satellite named "CAOSAT": CAOSAT-1α, -1β, -1γ

In the case of integral operation of two or more Earth Orbiting Satellites installed with SRS Instruments of the same types, configurations and performances, state the model, series, etc. so that each of SRS Instruments may be identified.

※ Case of integrated operation of four Earth Orbiting Satellites named "CAOSAT," each on board the same SRS Instruments: CAOSAT-1、-2、-3、-4

➢ "Type"

Choose one of the four categories provided in Article 2 of the Enforcement Regulation.

For detectable scope of sensors, see the following:
- Optical sensor: 0.03 μm to 8 μm
  (Ultraviolet: 0.03 μm to 0.4 μm, Visible light: 0.4 μm to 0.7 μm,
  Near-infrared: 0.7 μm to 1.3 μm, Mid-wavelength: 1.3 μm to 8 μm)
- Thermal infrared sensor: 8 μm to 14 μm
- SAR sensor: not less than 1 mm

If the detectable range covers both optical sensor and thermal infrared sensor, choose both the optical sensor and thermal infrared sensor.

For a hyperspectral sensor, applicability of the Act is judged if not only Distinguishing Accuracy of Target for the sensor does not exceed 10 meters, but also detectable wavelength bands of the sensor exceed 49.

> "Structure"

State attitude control method, whether the SRS Instruments is equipped with orbit control function (thruster), launch weight, power generation, design life, communication method (frequencies for both uplink and downlink) and manufacturer.

> "Performance"

State the matters relating to imaging performance, including Distinguishing Accuracy of Target (surface resolution), so as to identify that the performance is subject to the regulation by the Act.

In case of an optical sensor for both panchromatic and multi-spectral imagers, state the resolutions of both sensors. For a hyperspectral sensor, state the resolution and detectable wavelength bands (number of bands). In case of a SAR sensor with two or more imaging modes, state resolutions for each of the imaging modes.

Also specify the swath, pointing angle, onboard memory capacity, accuracy of position, etc.

③ Orbit of the Earth Orbiting Satellite installed with the SRS Instruments

In order to confirm that two or more satellite systems do not exist in the same orbit, specify the semi-major axis, eccentricity, inclination, right ascension of the ascending node, argument of perigee and time of perigee passage.

④ Place, configuration and performance of Ground Radio Station for Command and Control, as well as the method of management thereof.

In the case where the applicant uses SRS Instruments using Ground Radio Station for Command and Control managed by itself, fill in this section.

If the Ground Radio Station for Command and Control managed by any person other than the applicant is to be also used, specify the name and address of the manager in ⑨ in addition to the statement of this section.

If the application for a license of the radio station is pending, or if the applicant intends to file such application in the future, specify to that effect in this section.

- "Place"

    If an Operational Radio Station is located in Japan, such installation is subject to the application of the Act. As it is necessary to clearly identify the location, specify the address of the location of such installation.

    Generally, a Ground Radio Station for Command and Control is constituted by a telecommunication facility (control system) and transmission station (antenna), which may be located in different places. In such case, specify their respective locations.

    Specify all communication facilities and transmission stations if they are located in different places.

- "Structure"

    Specify configuration of Ground Radio Station for Command and Control

    Configuration of Ground Radio Station for Command and Control are not the same as they vary with satellite systems. For example, the following configuration is assumed.

    If the station, etc. is to be also used as a receiving station, specify to that effect.

    ① Communication facility (control system): creation of commands, monitoring of telemetry, etc.

    ② Transmission station (antenna): modulation and demodulation as well as transmission of telemetry and commands

- "Performance"

    State that the Ground Radio Station for Command and Control has a function to create conversion codes and to monitor the status of orbit as required under the Act.

- "Method of Management"

    State the manager, managing system, management/update of conversion codes, etc., method of implementing measures against unauthorized use as set forth in the items of paragraph (1) of Article 10 of the Enforcement Regulation, and method of update of these, etc.

    If the applicant has regulations, etc. providing for the method of such management, the applicant may state the name of such documents.

> ⑤ Place, configuration and performance of Receiving Station, as well as the method of management thereof.

If the applicant uses the Receiving Station managed by the applicant itself, specify the "location," "configuration" and "performance", as well as the "method of management thereof" in this section.

If the Receiving Station managed by any person other than the applicant is to be also used, specify the "location" of such Receiving Station managed by such person other than the applicant, in addition to the abovementioned information pertaining to the Receiving Station managed by the applicant itself. Such statement shall be made in a way that enables distinction between the Receiving Stations managed by the applicant itself and by the person other than the applicant.

In ⑩, state the name and address of the manager.

If the application for a license of the radio station is pending, or if the applicant intends to file such application in the future, specify to that effect in this section.

➢ "Place"

Generally, a Receiving Station is constituted by a receiving processing station (receiving processing system) and receiving station (antenna), which may be located in different places. In such case, specify their respective locations.

Specify all receiving processing stations (receiving processing system) and receiving stations (antenna) if they are located in different places.

➢ "Structure"

State the configuration of the Receiving Station. Configuration of Receiving Station are not the same as they vary with satellite systems. For example, the following configuration are possible.

① Receiving processing station (Receiving processing system): processing, etc. of received data

② Receiving station (antenna): receiving and demodulation of Electromagnetic Records of Detected Information (imaging data)

If the facility is to be also used as a transmission station, specify to that effect.

➢ "Performance"

State the performance of the Receiving Station and state that it has a restoration function by corresponding data conversion codes as required by the Act.

ANNEX 6

> ➤ "Method of Management"
> State the manager, managing system, method of implementing measures as set forth in the items of Article 11 of the Enforcement Regulation, and method of update of these, etc.
> If the applicant has regulations, etc. providing for the methods of such management, the applicant may state the name of such documents.

⑥ Method of management of SRS Data

State the names of documents such as regulations, etc. on safety management measures of SRS Data under Article 7 of the Enforcement Regulation.
If the applicant uses the service provided in paragraph (2) of such Article to manage the SRS Data, state to that effect.

⑦ In the case where the applicant is an individual person, the name or address of the Successor.

In the case where the Successor is an individual person, state the name and address as stated in the residence certificate, or, in the case of a corporation, state the corporation name and address as stated in the certificate of registered information.

⑧ Name and address of an officer or employee in charge of business of using the SRS Instruments.

State the names and addresses of officers or employees (persons with authority and responsibilities) as stated in their residence certificates.
An "employee" means an employee of the applicant who has authority and responsibilities for the business relating to the use of SRS Instruments (See Article 5 of the Enforcement Regulation).

⑨ If a person other than the applicant is to manage the Ground Radio Station for Command and Control, the name and address of the manager.

For example, a case where a ground transmission/receiving station (including foreign ground stations) managed by any person other than the applicant is used for transmission to the SRS Instruments (uplink), and an Ground Radio Station for Command and Control by a person other than the applicant is to be used.

State the names of respective managers, if two or more Operational Radio Stations, etc. are to be used.

The following cases do not fall under the case where the Ground Radio Station for Command and Control is managed by a person other than the applicant.

- Use of lines between the systems which are configuration of the Ground Radio Station for Command and Control (i.e. between the control station and transmission/receiving station)
- If the applicant uses third party storage service (e.g. cloud service) through telecommunication lines, as a part of the system comprising the Ground Radio Station for Command and Control (in such case, state to that effect in ④ "Method of Management.")

⑩ If the Receiving Station is to be managed by any person other than the applicant, the name and address of such manager.

For example, a case where a ground receiving station (including foreign ground receiving stations) managed by any person other than the applicant is used for transmission from the SRS Instruments to the ground (downlink), and a Receiving Station managed by a person other than the applicant is to be used.

State the names of respective managers, if two or more Receiving Stations are to be used. Also likewise state the relevant information if the location of the Receiving Station is different from the address of the manager of the Receiving Station.

The following cases do not fall under the case where the Receiving Station is managed by a person other than the applicant.

- Use of lines between the systems which are configuration of the Receiving Stations (i.e. between the receiving processing station and receiving station)
- If the applicant uses third party storage service (e.g. cloud service) through telecommunication lines, as a part of the system comprising the Receiving Station (in such case, state to that effect in ⑤ "Method of Management.")

⑪ Purpose and method of use of SRS Data

When using the SRS Instruments, handling of SRS Data obtained from such SRS Instruments is anticipated.

ANNEX 6

The purpose of use of SRS Data includes, for example, business and academic activities, and method of use thereof includes, for example, provision to third parties and processing. State the subject field, purpose of use, details, whether data is intended to be provided to third parties or to be processed. If there are two or more fields of use, fill in information for all information items, respectively.

---
⑫ Name of investor, investment ratio and nationalities
⑬ Major customers and suppliers

In order to ascertain the effect from third parties in terms of capital configuration of the applicant, it is important to check the following matters relating to the applicant.

- "Name of Investors"
  State the major investor of the applicant.
  For example, in the case of a listed company, fill in the information according to the status of major shareholders as set forth in the latest annual securities report, etc.

- "Investment Ratio"
  State the investment ratio of major investors referred to in the preceding section.

- "Nationalities"
  State the nationalities of major investors referred to in the preceding section.

- "Major Customers and Suppliers"
  State the major purchasers or suppliers.

3.2. Attachments

A person who intends to obtain a license for use of the SRS Instruments is required to also submit the following documents in relation to the information items in the application form as referred to above.

---
① Documents pertaining to the applicant
② A document setting forth the type, configuration and performance of SRS Instruments;
③ A document pertaining to the Ground Radio Station for Command and Control
  ③-1 a document containing the location, configuration and performance of the Ground Radio Station for Command and Control, as well as the management method thereof;
  ③-2 if a person other than the applicant is to manage the Ground Radio Station for Command and Control, the following documents pertaining to the manager:
④ Documents pertaining to the Receiving Station
  ④-1 a document containing the location, configuration and performance of the Receiving Station, as well as the management method thereof;
  ④-2 if a person other than the applicant manages the Receiving Station, a copy of the certification for such manager;
⑤ Documents on safety management measures
⑥ Any other document which the Prime Minister determines necessary.

The following is the guide of the documents for each of the information items.

Please notify detailed contents of the change to Cabinet Office in advance, if a person who obtained license intends to change any of the matters regarding the application for the license and its attachments.

---
① Documents pertaining to the applicant

The applicant is ineligible to obtain a license for SRS Instruments if such person, as well as any of its officers, employees and Successors, falls under any of the disqualifying conditions provided in Article 5 of the Act.

Therefore, in order to confirm that none of these persons fall under the disqualifying

ANNEX 6

conditions, the applicant is required to submit a document according to the applicant's category.

○ in the case where the applicant is an individual, the following documents:
1) a copy of a residence certificate or a document in lieu thereof;
   Limited to a certificate containing the registered domicile, or, in case of a foreign national, his/her nationality, etc. provided in Article 30-45 of the Residential Basic Book Act (Act No. 81 of 1967);
2) a document to pledge that such applicant does not fall under any of the items (i) to (iv) of Article 5 of the Act;
3) the following documents relating to employees and Successor:
   a) a copy of a residence certificate or a document in lieu thereof;
   b) in the case of an employee, a document to pledge that such employee does not fall under any of the items (i) to (iv) of Article 5 of the Act, or, in the case of a Successor, a document to pledge that such Successor does not fall under any of the items (i) to (vi) of Article 5 of the Act.

○ In the case where the applicant is a corporation, the following documents:
1) Its articles of incorporation and certificate of registered information, or a document equivalent thereto;
   In the case of a foreign corporation, a document issued by the foreign government approved by the government of Japan or competent international organization or an equivalent document, e.g., certificate of registered information or registered seal certificate of such corporation, a document issued from competent authorities or any other equivalent documents, containing the name of such corporation and its head office or principal office.
2) a document to pledge that such applicant does not fall under any of the items (i) to (iv) of Article 5 of the Act;
3) the following documents pertaining to officers and employees as set forth in Article 5, item (v) of the Act
   a) a copy of a residence certificate or a document in lieu thereof.
   b) a document to pledge that the relevant person does not fall under any of the items (i) to (iv) of Article 5 of the Act;

---

② A document setting forth the type, configuration and performance of the SRS Instruments;

As documents specifying the types, configurations and performances of the SRS Instruments, submit a document describing the details of "types," "configurations" and "functions" in ② of page 5.

For example, submit a copy of a specification, design, system block diagram, test report, etc. containing such information.

An applicant is not required to submit the entire copy of the document, and only the portion enabling confirmation of the relevant part, such as important specification list, should be submitted.

> ③ Documents pertaining to the Ground Radio Station for Command and Control
> ③-1 a document containing the location, configuration and performance of the Ground Radio Station for Command and Control, as well as the management method thereof;

Submit a document describing the details of "types," "configurations," "functions" as well as "method of management" in ④ of page 5.

For Ground Radio Station for Command and Control regulated by the Radio Act, submit the following documents:

- A copy of the license certificate of such radio station (or, a copy of the written application for license of such radio station, in the case where the application is pending).
- A copy of the management regulations, etc. (only the portion relating to the Ground Radio Station for Command and Control, if such regulations are integrated with regulations on other device such as SRS Instruments)
- A document specifying appropriate management of information relating to measures against unauthorized use as set forth in the items of paragraph (1) of Article 10 of the Enforcement Regulation

For a Ground Radio Station for Command and Control not regulated by the Radio Act, the following documents need to be submitted in lieu of "a copy of the license certificate of such radio station."

- A copy of the specifications, design drawings, performance test results, etc. relating to the device (only the portion enabling confirmation of the relevant part, such as the specification list).

Annex 6

> ③ Documents pertaining to the Ground Radio Station for Command and Control
> ③-2 if a person other than the applicant is to manage the Ground Radio Station for Command and Control, the following documents pertaining to the manager:

1) in the case where the manager is an individual, the following documents:
   a) a copy of a residence certificate or a document in lieu thereof;
   b) a document to pledge that such manager does not fall under any of the items (i) to (iv) of Article 5 of the Act;
2) in the case where the manager is a corporation, the following documents:
   a) its articles of incorporation and certificate of registered information, or a document in lieu thereof;
   b) a document to pledge that such manager does not fall under any of the items (i) to (iii) of Article 5 of the Act;

If the manager has obtained a certification for handling SRS Data, and if the Receiving Station so certified is to be also used as Ground Radio Station for Command and Control, a copy of such certificate may be submitted in lieu of the abovementioned document.

Further, there may be a case where an application for license to use SRS Instruments and an application for certification for handling SRS Data are submitted at the same time. In such case, by submitting a copy of written the application for certification for handling by the relevant Receiving Station, an application for license for use of the SRS Instruments may be examined at the same time.

> ④ Documents pertaining to the Receiving Station
> ④-1 a document containing the location, configuration and performance of the Receiving Station, as well as the management method thereof;

For example, as documents pertaining to the Receiving Station, submit a system block diagram, specification list, management regulations, etc.

If the receiving station for the Receiving Station is to be also used as a transmission station for a Ground Radio Station for Command and Control, and if such station is regulated by the Radio Act, submit the following documents:

- A copy of the license certificate of a radio station pertaining to such Receiving Station (or, a copy of the written application for license of such radio station, in the case where the application is pending).
- A copy of the management regulations, etc. (only the portion relating to the Receiving Station, if such regulations are integrated with regulations on other devices)
- A document specifying appropriate management of information relating to measures against unauthorized use as set forth in the items of paragraph (1) of Article 10 of the Enforcement Regulation

For a receiving station not regulated by the Radio Act, the following documents need to be submitted in lieu of "a copy of the license certificate of the radio station pertaining to the station."

- A copy of the specifications, design drawings, test reports, etc. relating to the Receiving Station (only the portion enabling confirmation of the relevant part, such as the specifications list).

---

④ Documents pertaining to the Receiving Station
    ④-2 if a person other than the applicant manages the Receiving Station, a copy of the certificate for such manager;

If the Receiving Station is to be managed by a person other than the applicant, such manager is required to obtain a certification for handling SRS Data. Thus, submit a copy of such certification.

Further, there may be a case where an application for a license to use the SRS Instruments and an application for certification for handling SRS Data pertaining to the management of the Receiving Station may be submitted at the same time. In such case, by submitting a copy of the written application for certification for handling by the relevant Receiving Station, an application for license for use of the SRS Instruments may be examined at the same time.

---

⑤ Documents relating to safety management measures

Submit regulations for safety management measures as provided in Article 7 of the Enforcement Regulation or documents to be able to confirm detailed contents for the measures (including a name list of employees and a floor plan of the facility to handle

ANNEX 6

SRS Data).

If the regulations already established by the applicant would be satisfactory, submit the copy of the relevant portion.

⑥ any other document which the Prime Minister determines necessary.

1) In the case where a third party storage service through telecommunication lines is used

    In the case where a third party storage service through telecommunication lines is used pursuant to the provisions of Article 7, paragraph (2) of the Enforcement Regulation, the applicant is expected to submit a copy of the contract, etc. enabling confirmation that the applicant has established safety management measures with such service provider (or, if the conclusion of such contract is planned, a document enabling confirmation of the contents thereof).

2) If it is considered necessary in light of circumstances of the applicant

    An applicant is expected to submit a document relating to the status of the equity structure.

Further, additional documents may be requested in the course of examination.

4. Application for certification for handling SRS Data

4.1. Information items to be entered in application form and guides for preparation

A person who intends to obtain a certification for handling the SRS Instruments is required to submit a written application containing the following matters to the Prime Minister.

> ① Address, name and contact information
> ② Categories of SRS Data
> ③ Purpose and method of use of SRS Data
> ④ Method of management of SRS Data
> ⑤ If SRS Data is received at the Receiving Station, the place thereof
> ⑥ Names and addresses of officers and employees in charge of business of handling SRS Data
> ⑦ If a person other than the applicant is to manage the Receiving Station, the name and address of the manager
> ⑧ Name of investors, investment ratio and nationalities
> ⑨ Major customers and suppliers

The following is the guide for preparation for each of the information items.

> ① Address, name and contact information

In the case where a person intending to obtain a certification of handling SRS Data is an individual person, state the name and address as stated in the residence certificate. If such person is a foreign national, state the name and address as stated in a document issued by the foreign government approved by the government of Japan or an equivalent document.

In the case of a corporation, state the corporation name and address as stated in the certificate of registered information.

For contact information, state the address, name, corporation name, section in charge, person in charge, etc. to enable receiving of mail.

> ② Categories of SRS Data

ANNEX 6

Please choose the category of SRS Data to be handled, as in the example.
If the applicant intends to handle two or more categories of SRS Data, tick each of the items.
For the details of categories of SRS Data, see Article 22 of the Enforcement Regulation.

---
③ Purpose and method of use of SRS Data

The purpose of use of SRS Data include, for example, business and academic activities, and method of use thereof include, for example, provision to third parties and processing. State the subject field, purpose of use, details, whether data is intended to be provided to third parties or to be processed. If there are two or more fields of use, fill in the information for all information items, respectively.

---
④ Method of management of SRS Data

State the names of documents such as regulations, etc. on safety management methods of SRS Data under Article 7 of the Enforcement Regulation or any other regulations providing for the method of management of SRS Data.
If the applicant uses the service provided in paragraph (2) of such Article to manage the SRS Data, state to that effect.

---
⑤ If SRS Data is received at the Receiving Station, the place thereof.

If the applicant uses the Receiving Station managed by itself, specify its location in this section.
If the Receiving Station managed by any person other than the applicant is to also be used, specify the "location" of such Receiving Station managed by such person other than the applicant, in addition to the abovementioned information pertaining to the Receiving Station managed by the applicant itself. Such statement shall be made in a way that enables distinction between Receiving Stations managed by the applicant itself and by the person other than the applicant.
In ⑦, state the name and address of the manager.
State all addresses if the locations are different.

> "Place"
  Generally, a Receiving Station is constituted by a receiving processing station

(receiving processing system) and receiving station (antenna), which may be located in different places. In such case, specify their respective locations.

Specify all receiving processing stations (receiving processing systems) and receiving stations (antennas) if they are located in different places.

⎯⎯⎯
⑥ Names and addresses of officers in charge of business of handling SRS Data
⎯⎯⎯

Names and addresses of officers or employees (persons with authority and responsibilities) as stated in their residence certificates.

An "employee" means an employee of the applicant who has authority and responsibilities for the business relating to the handling of SRS Data (See Article 24 of the Enforcement Regulation).

⎯⎯⎯
⑦ If the Receiving Station is to be managed by any person other than the applicant, the name and address of such person in charge of management.
⎯⎯⎯

For example, a case where a ground receiving station (including foreign ground receiving stations) managed by any person other than the applicant is used for transmission from the SRS Instruments to the ground (downlink), and a Receiving Station managed by a person other than the applicant is to be used.

State the names of respective managers, if two or more Receiving Stations are to be used. Also likewise state the relevant information if the location of the Receiving Station is different from the address of the manager of the Receiving Station.

The following cases do not fall under the case where the Receiving Station is managed by a person other than the applicant.

- Use of lines between the systems which are configuration of the Receiving Stations (i.e. between the receiving processing station and receiving station)
- If the applicant uses third party storage service (e.g. cloud service) through telecommunication lines, as a part of the system comprising the Receiving Station (in such case, state to that effect in ④ "Method of Management.")

⎯⎯⎯
⑧ Name of investor, investment ratio and nationalities
⑨ Major customers and suppliers
⎯⎯⎯

In order to ascertain the effect from third parties in terms of capital structure of the applicant, state the following matters relating to the applicant.

ANNEX 6

- "Name of Investors"
  State the major investors of the applicant.
  For example, in the case of a listed company, fill in the information according to the status of major shareholders as set forth in the latest annual securities report, etc.

- "Investment Ratio"
  State the investment ratio of major investors referred to in the preceding section.

- "Nationalities"
  State the nationalities of major investors referred to in the preceding section.

- "Major Customers"
  State the major purchasers or suppliers.

4.2. Attachments

A person who intends to obtain a certification for handling of SRS Data is required to also submit the following documents in relation to the information items in the application form as referred to above.

---
① Documents pertaining to the applicant
② Documents on safety management measures
③ The following documents pertaining to the Receiving Station
  ③-1 a document containing the location, configuration and performance of the Receiving Station, as well as the management method thereof;
  ③-2 if a person other than the applicant manages the Receiving Station, a copy of the license or certificate for such manager;
④ any other document which the Prime Minister determines necessary.

The following is the guide of the documents for each of the information items.

Please notify detailed contents of the change to Cabinet Office in advance, if a person who obtained certification intends to change any of the matters regarding the application for the certification and its attachments.

---
① Documents pertaining to the applicant

The applicant is ineligible to obtain a certification for handling SRS Data if such person, as well as any of its officers, employees and Successors, falls under any of the disqualifying conditions provided in Article 21, paragraph (3), item (i) of the Act.

Therefore, in order to confirm that none of these persons fall under the disqualifying conditions, the applicant is required to submit a document according to the applicant's category.

○ in the case where the applicant is an individual, the following documents:
1) a copy of a residence certificate or a document in lieu thereof; limited to a certificate containing the registered domicile, or, in case of a foreign national, his/her nationality, etc. provided in Article 30-45 of the Residential Basic Book Act (Act No. 81 of 1967);
2) a document to pledge that the applicant does not fall under any of Article 21, paragraph (3), item (i)(a) to (d);
3) The following documents pertaining to employees

ANNEX 6

   a) a copy of a residence certificate or a document in lieu thereof;
   b) a document to pledge that the employee does not fall under any of Article 21, paragraph (3), item (i)(a) to (d);

○ in the case where the applicant is a corporation, the following documents:
  1) its articles of incorporation and certificate of registered information, or a document in lieu thereof;
     In the case of a foreign corporation, a document issued by the foreign government approved by the government of Japan or competent international organization or an equivalent document, e.g., certificate of registered information or registered seal certificate of such corporation, a document issued from competent authorities or any other equivalent documents, containing the name of such corporation and its head office or principal office.
  2) a document to pledge that the applicant does not fall under any of Article 21, paragraph (3), item (i)(a) to (d);
  3) The following documents pertaining to officers and employees
     a) a copy of a residence certificate or a document in lieu thereof;
     b) a document to pledge that the relevant person does not fall under any of Article 21, paragraph (3), item (i)(a) to (d);

② Documents relating to safety management measures

Submit regulations for safety management measures as provided in Article 7 of the Enforcement Regulation or documents to be able to confirm detailed contents for the measures (including a name list of employees and a floor plan of the facility to handle SRS Data).
If the regulations already established by the applicant would be satisfactory, submit the copy of the relevant portion.

③ the following documents pertaining to the Receiving Station:
   ③-1 a document containing the location, configuration and performance of the Receiving Station, as well as the management method thereof;

For example, as documents pertaining to the Receiving Station, submit a system block diagram, specifications list, management regulations, etc.
If the receiving station for the Receiving Station is to be also used as a transmission

station, and if such Receiving Station is regulated by the Radio Act, submit the following documents:
- A copy of the license certificate of a radio station pertaining to such Receiving Station (or, a copy of the written application for a license of such radio station, in the case where the application is pending).
- A copy of the management regulations, etc. (only the portion relating to the Receiving Station, if such regulations are integrated with regulations on other devices)
- A document specifying appropriate management of information relating to measures against unauthorized use as set forth in the items of paragraph (1) of Article 10 of the Enforcement Regulation

For a receiving station not regulated by the Radio Act, the following documents need to be submitted in lieu of "a copy of the license certificate of the radio station pertaining to the station."
- A copy of the specifications, design drawings, test reports, etc. relating to the Receiving Station (only the portion enabling confirmation of the relevant part, such as the specifications list).

> ③ the following documents pertaining to the Receiving Station:
> ③-2 if a person other than the applicant manages the Receiving Station, a copy of the license or certificate for such person in charge of management;

If the Receiving Station is to be managed by a person other than the applicant, it is necessary to confirm that such management will be performed in an appropriate manner. Thus, submit a copy of such license certificate or certificate pertaining to such manager.

Further, there may be a case where an application for a certification for handling SRS Data and an application for certification for handling SRS Data by such Receiving Station by any person other than the applicant may be submitted at the same time. In such case, by submitting a copy of written the application for license, etc. for such Receiving Station, an examination relating to the handling of SRS Data may be conducted at the same time.

> ④ any other document which the Prime Minister determines necessary.

1) In the case where a third party storage service through telecommunication lines is used

   In the case where a third party storage service through telecommunication lines is used pursuant to the provisions of Article 7, paragraph (2) of the Enforcement Regulation, submit a copy of the contract, etc. enabling confirmation that the applicant has established safety management measures with such service provider (or, if the conclusion of such contract is planned, a document enabling confirmation of the contents thereof).

2) If it is considered necessary in light of circumstances of the applicant

   An applicant is expected to submit document relating to the status of the equity configuration.

Further, additional documents may be requested in the course of examination.

5. Example of completed application form

The following pages are the examples of forms for applications for license to use the SRS Instruments and for certification for handling SRS Data.

ANNEX 6

Example (Use of System)

Form 1 (Re: Article 4)

Application for License

November 16, 2017

To: Prime Minister

                (Postal Code) 100-0013

                Address: ○○○, Kasumigaseki, Chiyoda-ku, Tokyo

                Name

                (Corporation name, in case of a corporation)

                _____ Co., Ltd. (Company seal)

                Contact address: *-*-* Otemachi, Chiyoda-ku,

                      Tokyo 100-81**

                Taro Naikaku, General Affairs Section, General Affairs Division

                                    _____ Co., Ltd.

                      Tel: 03-6205-**** Extension: 9999

                      Email address: naikaku-taro@xxx.co.jp

The applicant hereby submits an application for license for use of the Satellite Remote Sensing System pursuant to the provisions of Article 4, paragraph (2) of the Act on Ensuring Appropriate Handling of Satellite Remote Sensing Data.

1. Matters relating to the use of the Satellite Remote Sensing System

Name, type, configuration and performance of the Satellite Remote Sensing System	Name: CAO-OP1
	Type: ■ Optical sensor □ SAR sensor
	      □ Hyperspectral sensor
	      □ Thermal infrared sensor
	Structure: attitude control, 3-axis control method, with a thruster
	Launch weight: 150kg
	Power generation: 400W
	Design life: 3 years
	Communication mode: S-band (uplink), X-band (downlink)
	Manufacturer: Tokyo Satellite Manufacturing Co.,

	Ltd. Performance: surface resolution (nadir) 0.5m (panchromatic image) 2m (multi-spectral image) Band: Four bands Swath: 30km Onboard memory capacity: 120GB Position accuracy: 10m CE90
Orbit of the Earth Orbiting Satellite installed with the Satellite Remote Sensing System	Semi-major axis: ○○km Eccentricity: ○○ Inclination: ○○° Right ascension of the ascending node: ○○° Argument of perigee: ○○° Time of perigee passage: ○○
Place, configuration and performance of Ground Radio Station for Command and Control, as well as the method of management thereof	Place: ① ○○, Chiyoda-ku, Tokyo ② ○○, ○○-cho, Hokkaido ③ ○○ (Country), ×× (State), △△ Structure: ① Radio facility (satellite control system) ② Transmission station (antenna, modulation facility, etc.) ③ Transmission station (antenna, modulation facility, etc.) Performance: The system has a function to generate conversion codes. The system has a function to control orbit. Method of management: ① CAO-OP1 Management Regulations (C01-0011) ② CAO Transmission Station Management Regulations (C01-0001) (Application for a license of radio station is currently pending.) ③ XXX Transmission Station Management Regulations (X01-0001)
Place, configuration and performance of the Receiving Station,	Places: ① ○○, Yakumo-cho, Hokkaido ② ○○, Nagoya-shi, Aichi ③ ○○ (Country), □□ (State), ◇◇ (Manager: ○○○○

Annex 6

as well as the method of management thereof	Ltd.) Structure: ① receiving antenna ② receiving antenna Performance: ① and ②: Receive X band. Has a function to restore corresponding data. Method of management: ① and ②: According to CAO Transmission Station Management Regulations (C01-0001)
Method of management of Satellite Remote Sensing Data	CAO Satellite Remote Sensing Data Management Regulations (C01-0101) Use of service under Article 7, paragraph (2) of the Enforcement Regulation
In the case where the applicant is an individual person, the name and address of Successor	Name: Address:
Name and address of officer or employee in charge of business relating to use of the Satellite Remote Sensing System	Name: ○○○○ Address: ○○, Chiyoda-ku, Tokyo
In the case where the Ground Radio Station for Command and Control is managed by any person other than the applicant, the name and address of such manager	Name: ○○○○ Ltd. Address: ○○ (Country), □□ (State), △△
In the case where the Receiving Station is managed by any person other than the applicant, the name and address of such	Name: ○○○○ Ltd. Address: ○○ (Country), □□ (State), △△

manager Address:	
Purpose and method of use of Satellite Remote Sensing Data	Purpose: Business activity (geospatial information) Method: - Provision of data (sale of data) - Provision of value-added product and information (for agricultural industry businesses, provision of information relating to disaster prevention)

2. Matters pertaining to applicant

Name of investors, investment ratio and nationalities	Name: ① ○○ Heavy Industries Co., Ltd. ② ○○ Electronics Co., Ltd. ③ ○○ Aerospace Ltd. Investment ratio: ①40%, ②40%, ③20% Nationalities: ① Japan ② Japan ③ Canada
Major customers	○○ Ministry, ○○ Trading Company, ○○ Shipping Company

Note: 1. The size of the paper must be Japan Industrial Standards (JIS) A4.
    2. An applicant may affix its signature instead of affixing its name and seal. In such case, the applicant must affix the signature in person.
    3. Attach documents set forth in the items of paragraph (2) of Article 4 of the Regulation for Enforcement of the Act for Ensuring Appropriate Handling of Satellite Remote Sensing Data

ANNEX 6

Example (Handling of data)

Form 13 (Article 23)

Application for Certification

November 16, 2017

To: Prime Minister

                          (Postal Code) 100-****
                          Address: *-*-*, Shibaura, Minato-ku, Tokyo
                          Name: ○○○○
                          (Corporation name, in case of a corporation)
                          _____ Co., Ltd. (Company seal)
                          Contact Address: *-*-*, Shibaura, Minato-ku,
                          Tokyo
                                    □□□□ Co., Ltd., General Affairs Group, Hifumi Chiyoda
                                    Tel: 03-6205-**** Extension: 9999
                                    Email: hifumi@xxx.com

The applicant hereby submits an application for certification for handling Satellite Remote Sensing Data pursuant to the provisions of Article 21, paragraph (2) of the Act on Ensuring Appropriate Handling of Satellite Remote Sensing Data.

1. Matters relating to Satellite Remote Sensing Data to be handled

Categories of Satellite Remote Sensing Data	(i) (Optical sensor/raw data) (v) (Optical sensor/standard data)
Purpose and Method of Use of Satellite Remote Sensing Data	Purpose: Business activity (provision of receiving station to system users, geospatial information) Method: - Provision of data (① provision of data received from SRS Instruments to system users without decryption; ② provision of data purchased from system users after processing) - Provision of value-added product and information (for civil engineering industry businesses, provision of information relating to disaster prevention)

Method of management of Satellite Remote Sensing Data	□□□□ Satellite Image Management Regulations (C-0119) Use of service under Article 7, paragraph (2) of the Enforcement Regulation
If Receiving Satellite Remote Sensing Data is received at the Receiving Station, the place thereof	Address: ① ○○, Chitose, Hokkaido ② ○○, Kamiamakusa, Kumamoto
Names and addresses of officers and employees in charge of business of handling Satellite Remote Sensing Data	Name: _____ Address: ○○○○, Sapporo, Hokkaido
In the case where the Receiving Station is managed by any person other than the applicant, the name and address of such manager	Name: Address: ○○○-2, Chiyoda-ku, Tokyo

2. Matters pertaining to applicant

Name of investors, investment ratio and nationalities	Name: □□□□ Holdings Investment ratio: 100% Nationality: Japan
Major customers	□□ Civil engineering, □□ Construction, □□ Ocean Development

Note: 1. The size of the paper must be Japan Industrial Standards (JIS) A4.
    2. An applicant may affix its signature instead of affixing its name and seal. In such case, the applicant must affix the signature in person.
    3. Attach the document set forth in the items of Article 23, paragraph (2) of the Regulation for Enforcement of the Act on Ensuring Appropriate Handling of Satellite Remote Sensing Data.

Annex 7 – Basic Space Law (Law No. 43 of 2008)

Table of Contents

Chapter I General Provisions (Articles 1 to 12)

Chapter II Basic Measures (Articles 13 to 23)

Chapter III Basic Space Plan (Article 24)

Chapter IV Strategic Headquarters for Space Development (Articles 25 to 34)

Chapter V Preparing Legal Frameworks for Space Activities (Article 35)

Supplementary Provisions

Chapter I General Provisions

Article 1 (Purpose)
The purpose of this Law is, in view of the fact that the development and use of space (hereinafter referred to as "Space Development and Use") have become increasingly important as a result of changes in circumstances in and outside the country, to implement policy measures with respect to Space Development and Use in a comprehensive and systematic manner by establishing basic principles and matters for their realization, clarifying responsibilities, etc. of the national government, providing for the formulation of the Basic Space Plan, and establishing the Strategic Headquarters for Space Development, etc., in order to extend the role of Space Development and Use in our country based on the fundamental principle of pacifism of the Constitution of Japan and in harmony with the environment, and thereby to contribute to the improvement of the lives of citizens and the development of the economy and society, as well as to the improvement of world peace and the welfare of humanity.

Article 2 (Peaceful Use of Space)
Space Development and Use shall be implemented in accordance with treaties and international agreements concerning Space Development and Use, such as the Treaty

on Principles Governing the Activities of States in the Exploration and Use of Outer Space, including the Moon and Other Celestial Bodies, and in accordance with the fundamental principle of pacifism of the Constitution of Japan.

Article 3 (Improvement of the Lives of the Citizens, etc.)
Space Development and Use shall be implemented to contribute to improving the lives of citizens, developing a society where one can live safely and securely, mitigating disasters, poverty and other various threats to the survival and lives of humanity, ensuring international peace and security, and enhancing our country's national security.

Article 4 (Promotion of Industries)
Space Development and Use shall be implemented to strengthen technical competence and international competitiveness of the space industry and other industries of our country, by promoting Space Development and Use proactively and systematically and the smooth commercialization of the results of research and development concerning Space Development and Use, and thereby to contribute to promotion of industries of our country.

Article 5 (Development of Human Society)
Space Development and Use shall be implemented, in view of the fact that the accumulated knowledge of space is an intellectual asset of humanity, to contribute to the realization of humanity's dreams for space and the development of human society, by promoting state-of-the-art Space Development and Use as well as developing space science, etc.

Article 6 (International Cooperation, etc.)
Space Development and Use shall be implemented to enable our country to play a proactive role and advance the interests of our country in international society by promoting international cooperation, diplomacy, etc. regarding Space Development and Use.

Article 7 (Consideration for the Environment)
Space Development and Use shall be implemented in consideration of the effect on the environment caused by Space Development and Use.

Article 8 (Responsibilities of the National Government)
The national government shall be responsible for formulating and implementing comprehensive policy measures concerning Space Development and Use in accordance

with the basic principles provided in Articles 2 to 7 (hereinafter referred to as the "Basic Principles").

Article 9 (Endeavor of Local Governmental Organizations)
In accordance with the Basic Principles, local governments shall endeavor to formulate and implement policy measures concerning Space Development and Use on their own initiative, taking into consideration the division of roles between the national government and themselves and making use of characteristics of their respective areas.

Article 10 (Strengthening Coordination)
The national government shall take policy measures necessary for strengthening coordination among the national government, universities, private business operators, etc. in view of the fact that their cooperation based on mutual coordination encourages the effective promotion of Space Development and Use.

Article 11 (Legislative Measures, etc.)
The government shall implement legislative, financial, tax or financial services measures, or other measures as necessary to implement the policy measures concerning Space Development and Use.

Article 12 (Streamlining Administrative Organizations, etc.)
The national government shall endeavor to streamline administrative organizations and improve administrative management in implementing policy measures concerning Space Development and Use.

Chapter II Basic Measures

Article 13 (Use of Artificial Satellites for the Improvement of the Lives of Citizens, etc.)
The national government shall promote the establishment of stable networks for information and telecommunications using artificial satellites, information systems concerning observation, information systems concerning positioning, etc., in order to contribute to improving the lives of citizens, developing a society where one can live safely and securely, and mitigating disasters, poverty and other various threats to the survival and lives of humanity.

Article 14 (Ensuring International Peace and Security and the National Security of Our Country)

The national government shall implement policy measures as necessary for the promotion of Space Development and Use to contribute to ensuring international peace and security and to the national security of our country.

Article 15 (Independent Launching, etc. of Artificial Satellite, etc.)
In view of the importance for the national government to have the capability to independently develop, launch, track and operate artificial satellites, etc., the national government shall promote the research and development of equipment (including components), technologies, etc., and install the equipment, facilities, etc., to make radio frequencies available for Space Development and Use by our country, as well as implement other necessary policy measures.

Article 16 (Promotion of Space Development and Use by Private Business Operators)
In view of the important role that the private sector plays in Space Development and Use, the national government shall, when engaging in its own Space Development and Use activities, not only give consideration to procuring goods and services in a systematic manner utilizing the skills of the private sector, but also to improving launch sites (facilities for launching rockets), laboratories, other equipment and facilities, etc., promoting the transfer of the results of research and development concerning Space Development and Use to the private sector, promoting the commercialization of the results of research and development concerning Space Development and Use in the private sector, and implementing tax and financial services measures to facilitate the private sector's investment in activities concerning Space Development and Use, as well as other necessary measures.

Article 17 (Maintenance and Improvement of Credibility)
In view of the importance of maintaining and improving the technologies concerning Space Development and Use, the national government shall promote basic research and development of fundamental technologies concerning Space Development and Use and implement other necessary policy measures.

Article 18 (Promotion of State-of-the-art Space Development and Use)
The national government shall implement policy measures necessary for promoting academic research, etc. concerning Space Development and Use, including for space exploration, etc. and space science.

Article 19 (Promotion of International Cooperation)
The national government shall promote international coordination, international, technological cooperation and other international cooperation and implement measures

necessary for deepening foreign countries' understanding of Space Development and Use of our country, in order to enable our country to play a proactive role and promote our country's interests in international society.

Article 20 (Preservation of the Environment)
(1) The national government shall implement policy measures necessary to promote Space Development and Use in harmony with the environment.
(2) The national government shall endeavor to ensure international coordination to preserve the environment of space.

Article 21 (Securing Human Resources)
In order to promote Space Development and Use, the national government shall implement policy measures necessary to securing and train human resources for Space Development and Use and improve their capabilities, while coordinating and cooperating with universities, private business operators, etc.

Article 22 (Promotion of Education and Learning)
The national government shall take measures to promote education and learning, public relations activities, and other measures necessary for Space Development and Use, so that citizens across the country deepen their understanding of and interest in Space Development and Use.

Article 23 (Management of Information Regarding Space Development and Use)
The national government shall, in view of the characteristics of Space Development and Use, implement policy measures necessary for the appropriate management of information regarding Space Development and Use.

Chapter III Basic Space Plan

Article 24
(1) The government shall prepare a basic plan on Space Development and Use (hereinafter referred to as the "Space Basic Plan") in order to implement policy measures concerning Space Development and Use in a comprehensive and systematic manner.
(2) The Space Basic Plan shall provide for the following matters:
    (i) basic policies concerning the promotion of Space Development and Use,
    (ii) policy measures which the national government shall implement in a comprehensive and systematic manner, and

(iii) in addition to the matters provided for in the preceding two items, matters necessary for the national government's implementation of policy measures concerning Space Development and Use.
(3) Regarding the policy measures stipulated in the Basic Space Plan, concrete goals and schedules for achieving them shall, in principle, be provided for.
(4) The Prime Minister shall seek a Cabinet decision regarding the draft of Basic Space Plan prepared by Strategic Headquarters for Space Development.
(5) If there is the Cabinet decision as provided for in the preceding item, the Prime Minister shall publicize the Basic Space Plan without delay by the Internet or other appropriate means.
(6) The Prime Minister shall check the progress of the goals provided for in paragraph (3) and publicize this by the Internet or other appropriate means in a timely manner.
(7) Taking into consideration the progress of Space Development and Use, the effect of the policy measures taken for Space Development and Use, etc., the government shall review the Basic Space Plan when appropriate and revise it as necessary. In such case, the provisions of the paragraphs 4 and 5 shall apply *mutatis mutandis*.
(8) In every fiscal year, the government shall endeavor to take measures necessary for implementing the Basic Space Plan smoothly, including appropriating national funds within the limit permitted by the national budget, in order to secure funds to cover the expenses necessary for implementing the Basic Space Plan.

Chapter IV Strategic Headquarters for Space Development

Article 25 (Establishment)
The Strategic Headquarters for Space Development (hereinafter referred to as the "Headquarters") shall be established in the Cabinet in order to promote the implementation of Space Development and Use in a comprehensive and systematic manner.

Article 26 (Affairs under its Jurisdiction)
The Headquarters shall administer the following affairs:
(i) matters concerning the drafting of the Basic Space Plan and promoting its implementation; and
(ii) in addition to what is listed in the preceding item, matters concerning research and deliberation on important projects of the policy measures relating to Space Development and Use, and the promotion of their implementation and comprehensive coordination.

Article 27 (Organization)
The Headquarters shall consist of the Director-General of the Strategic Headquarters for Space Development, the Deputy Director-General of the Strategic Headquarters for Space Development, and the Members of the Strategic Headquarters for Space Development.

Article 28 (Director-General of the Strategic Headquarters for Space Development)
(1) The head of the Headquarters shall be the Director-General of the Strategic Headquarters for Space Development (hereinafter referred to as the " Director-General "), the post of which shall be assumed by the Prime Minister.
(2) The Director-General shall exercise control and supervision over affairs of the Headquarters and the personnel under its jurisdiction.

Article 29 (Deputy Director-General of the Strategic Headquarters for Space Development)
(1) The Headquarters shall establish Deputy Director-General of Strategic Headquarters for Space Development (hereinafter referred to as the "Deputy Director-General "), the post of which shall be assumed by the Chief Cabinet Secretary and the State Minister in charge of Space Development (the State Minister who performs the duties of assisting the Prime Minister with regard to Space Development and Use by his/her order).
(2) The Deputy Director-General shall assist the duties of the Director-General.

Article 30 (Members of the Strategic Headquarters for Space Development)
(1) The Headquarters shall establish Members of Strategic Headquarters for Space Development (hereinafter referred to as the "Members").
(2) The Members shall be assumed by all the State Ministers other than the Director-General and the Deputy Director-General.

Article 31 (Submission of Materials and Other Cooperation)
(1) The Headquarters may, if it finds necessary, request the heads of relevant administrative organs, local public entities, and Incorporated Administrative Agency(ies) (Incorporated Administrative Agency provided for by Article 2, paragraph (1) of Act on General Rules for Incorporated Administrative Agencies (Law number: Act No. 103 of 1999)) and the representatives of government-affiliated corporations (corporations which have been established directly by a law or by special incorporation in accordance with a special law and to which Article 4, paragraph (4), item (ix) of Act for Establishment of the Ministry of Internal Affairs and Communications (Act No. 91 of 1999) applies) to submit materials, express opinions, provide explanation, and offer any other necessary cooperation.

(2) The Headquarters may request persons other than those provided for in the preceding paragraph to provide necessary cooperation, if it finds this to be necessary.

Article 32 (Affairs)
Affairs regarding the Headquarters shall be handled by the Prime Minister.

Article 33 (Competent Minister)
With regard to matters relating to the Headquarters, the Competent Minister referred to in the Cabinet Act (Act No. 5 of 1947) shall be the Prime Minister.

Article 34 (Delegation to Cabinet Orders)
In addition to the matters provided for in this Law, necessary matters regarding the Headquarters shall be provided for by a Cabinet Order.

Chapter V Enacting Legislation for Space Activities

Article 35
(1) The government shall enact legislation for matters, etc. necessary to implement regulations concerning space activities as well as treaties and other international agreements concerning Space Development and Use in a comprehensive, systematic and prompt manner.
(2) The legislation in the preceding paragraph shall be prepared to contribute to promoting the interests of our country in international society and promoting Space Development and Use in the private sector.

Supplementary Provisions

Article 1 (Effective Date)
This Law shall come into effect at a date determined by a Cabinet Order within three months from the date of promulgation.

Article 2 (Enacting Legislation for the Cabinet Process for Affairs Regarding the Headquarters)
The government shall enact legislation and implement other measures to address the Cabinet process for affairs regarding the Headquarters approximately one year after the effective date.

Article 3 (Review Regarding Japan Aerospace Exploration Agency, Incorporated Administrative Agencies, etc.)

Regarding the Japan Aerospace Exploration Agency and other organizations concerning Space Development and Use, the government shall review and revise their purposes, functions, scopes of operation, how they should be structured, administrative organizations having jurisdiction over them, etc., approximately one year after the effective date.

Article 4 (Review Regarding Administrative Organizations to Implement Measures Concerning Space Development and Use in a Comprehensive and Integrated manner)
The government shall review administrative organizations to implement measures concerning Space Development and Use in a comprehensive and integral manner and, based on the result, implement necessary measures.

Supplementary Provisions (Act No. 66 of September 11, 2015) excerpt

Article 1 (Effective Date)
This Law shall come into effect on April 1, 2016; provided, however, that the provisions listed in the following items shall come into effect on the dates prescribed respectively as to those items.
(1) Article 7 of the Supplementary Provisions: the date of promulgation

Article 7 (Delegation to Cabinet Orders)
In addition to Article 2 to the preceding article, transitional measures necessary for implementing this Law shall be determined by a Cabinet Order.

Annex 8 – Summary of Space Industry Vision 2030 Creating Space Use in the Fourth Industrial Revolution

May 29, 2017
Committee on National Space Policy

1   Domestic and Foreign Conditions Concerning Space Industry

Significant waves of change are occurring in the development and use of outer space. A paradigm shift has begun as technological innovations in the space field are combining with changes in other fields in what is known as the Fourth Industrial Revolution and are pushing innovation further. In addition, lower costs for spacecraft and rockets are attracting larger numbers of people to use outer space, and the private sector is contributing to additional innovations, which are both signs that the changes are accelerating.

The space equipment industry in Japan is not necessarily internationally competitive compared with leading foreign companies, and new industry participants are not sufficiently activating the market. As the pace of change in the space industry has accelerated, two pieces of space legislation were enacted in November 2016 ("Act on Launching of Spacecraft, etc. and Control of Spacecraft" and "Act on Ensuring Appropriate Handling of Satellite Remote Sensing Data"). Henceforth, growing existing businesses and creating new players in the industry are key to promoting the space industry.

2   Direction of the Space Industry (Towards a Virtuous Cycle of Growth in Japan's Space Industry)

Regarding the space use industry, the quality and quantity of information obtained from satellite data, such as high-frequency and high-resolution satellite remote sensing information and high-precision positioning services using Quasi-Zenith satellites, have improved. Analysis of big data from these satellites as well as from Earth through means including artificial intelligence is expected to provide solutions to problems in various fields. Thus, satellite data is driving the Fourth Industrial Revolution, which helps strengthen the competitiveness of many industries.

As to the space equipment industry, new opportunities for growth are occurring, including in small satellites and rockets. At the same time, Japan will also boost its international competitiveness with large satellites and rockets by developing technologies that meet market needs, capture growing demand overseas, and expand the scale of our businesses, thereby achieving a virtuous cycle of growth that will stimulate additional investment.

Furthermore, in order to expand so-called "New Space," including orbital services and space resources development, Japan will pursue new business opportunities by creating and maintaining a favorable business environment, such as providing "risk money" and institutional development.

As one goal, the 2015 Basic Plan on Space Policy has stated that the scale of Japan's space equipment industry in the public and private sectors combined will reach a total of JPY 5 trillion in 10 years. In addition, measures will be taken to double the market size (currently JPY 1.2 trillion) of the space industry as a whole, including the space use industry, by the early 2030s.

3   Space Use Industry

There are several challenges to promoting solution businesses using satellite data: securing continuity of satellite data, achieving higher frequency of observation data, and solving difficulties in accessing necessary information from among diverse satellite data. In Japan, there are not enough service providers to analyze satellite data and provide value-added solutions, which limits the introduction of satellite data-based solutions. In the United States and Europe, various governmental organizations purchase data collected by private satellites, stimulating related businesses. In the future, Japan also needs to develop and expand the end-user market, including for public use.

In view of these challenges, it is necessary to develop and maintain environments to provide continuous satellite data by establishing mechanisms to identify the needs for satellite data use and to reflect such needs in satellite projects. In order to improve accessibility to satellite data, it also is necessary to develop catalogs of satellite data and to make governmental satellite data available to people free of charge. In doing so, data use hubs (data centers) should be established to contribute to regional economic revitalization. Furthermore, potential users, including IT business operators in non-space industries and the national and local governments, should work together to promote model projects that use satellite data. At the same time, the government and the public sector should actively use satellite data to promote efforts to invigorate and industrialize satellite data use.

4 Space Equipment Industry

The space equipment industry is the foundation of Japan's self-sustaining space activity capabilities. However, Japan's space equipment industry is inferior to that of the United States and Europe in terms of international competitiveness. Therefore, Japan keenly needs to achieve high technological development capacity, as well as better quality, performance, and lower cost. While most technological development and support are concentrated in JAXA and universities, other governmental organizations also are expected to offer support.

As for space-related parts and components, Japan relies heavily on foreign countries, which may reduce its international competitiveness in satellite manufacture and cause a vicious cycle of losing opportunities to obtain private and foreign demand. There have been a series of cases where companies cannot remain viable and withdraw from the business, jeopardizing the supply base.

As for large-sized rockets, Japan occupies a leading position globally, including in terms of launch success rate. However, issues remain concerning cost and additional launch opportunities.

Regarding equipment development, government procurement does not provide developers with sufficient profits proportionate to the nature of the business, including development risks, and therefore investment capacity for technological development cannot be secured.

In regard to small-sized satellites and rockets, capital, technology, and speed are required, and it is important to provide financial and other assistance, as well as to cooperate with major companies. It also is a major issue that currently there is no launch site for private small-sized rockets.

In light of these challenges, and in order to enhance international competitiveness, it is important to continue R&D and experimentation in the government's satellite development which meet market needs. In addition, it is necessary to steadily develop and commercialize the new mainstay rocket (H3) with the aim of halving launching costs, and to continue assisting the development of inexpensive small-sized rockets using commercially manufactured parts. As for parts and components, in accordance with the road map based on the "Parts and Components Technology Strategy" prepared in March 2016, measures are required such as promoting R&D and mutual certification with overseas space organizations. With respect to procurement systems, it is important to be innovative in securing business operators' capacity to invest in areas such as technology development, including by measures such as further consideration the introduction of fixed contracts and establishing appropriate and reasonable expense ratios, and promoting procurement which will lead to venture companies to enter into the industry and create innovation.

5   Overseas Expansion

In order to expand Japan's space industry, it is essential to access foreign markets, which are growing especially in emerging countries. In August 2015, the government established a taskforce to expand space systems overseas so both the public and private sectors could better develop commercial space markets. However, because space-related projects tend to take a long time, issues arise such as loss of personal connections with key counterparts overseas, and long-term, sustained promotion strategies are required. In addition, because the need for satellites and ancillary services varies depending on the developmental stage and environment in partner counties, different strategies should be adopted with the partners. Moreover, international collaboration in the space industry needs strengthening in some cases.

Faced with these challenges, it is necessary to establish a continuous support coordination function in order to resolve various problems. In addition, it is important to develop and strengthen packages meeting the needs of partner countries so that such packages include equipment, satellite service businesses, and personnel education. With regard to Quasi-Zenith satellites, which are expected to comprise four satellites in fiscal 2018, discussions are needed to realize new businesses and services in areas including not only Japan, but also in the Asia-Pacific region. Furthermore, in order to strengthen international collaboration, besides partnering with space organizations such NASA, DLR, and CNES, it is necessary to cooperate further with APRSAF and ERIA.

6   Improving the Environment for New Space Businesses

In Europe and the United States, paradigm change is led and driven by venture companies, which are involved not only with satellite data solution services and small-sized satellites and rockets, but also taking on the challenges of projects such as orbital services and space resources. In Japan as well, it is important to expand the base of new entrants, including venture companies, and to actively support their project development and growth.

In order to achieve the above under current conditions where "risk money" available for investment in space businesses is overwhelmingly lacking, it is necessary to create an environment so that the supply of "risk money" for the space sector such as private venture capital will increase, as well as to encourage the participation of governmental financial institutions, including the Development Bank of Japan (DBJ). In addition to bolstering efforts of the Space New Economy Creation Network (S-NET), including by fortifying its connection to policies of relevant ministries and agencies, support should be provided for the discovery and commercialization of new ideas, and venture companies should be promoted by strengthening space development and use prizes and business idea contests through the creation of S-Booster.

It also is necessary to comprehensively reinforce the human basis in order to maintain and strengthen the space industry infrastructure. Japan's space industry is suffering from a vicious cycle in which mobility of human resources in and around the industry is low, it is difficult to develop new businesses and the industry does not scale up, and the inflow of human resources into the industry declines as a result. In addition, the industry needs persons capable of "matching needs and seeds."

In order to resolve these challenges, human resources mobility should be heightened by strengthening networking through activities such as S-NET.

Recently, more and more private business operators are focusing on space businesses in new areas, including businesses in deep space such as developing asteroid resources and orbital businesses, such as monitoring and removing debris, mainly in Europe and the United States. Legislation in light of these commercial opportunities is occurring overseas: in the United Kingdom and the Netherlands, laws for orbital compensation have been enacted; and in the United States and Luxembourg, laws for developing resources in outer space.

Based on this trend abroad, Japan also will consider necessary measures with a view to new businesses.

7   Final Comments

Implementing a three-pronged approach to national security, civilian use, and scientific technology will increase the breadth of the Japanese space industry and lead to a space industry in Japan comparable to that of the world.

Active moves in line with this vision are expected, seeing that the space industry's current state is undergoing transformation and this is an opportunity for the development of Japan's space industry.

Annex 9 – Changes to the Basic Space Plan

June 30, 2020 Debate
    Decision
    Definite

In accordance with the provisions of Article 4 of the Basic Space Law (Law No. 43 of 2008), pursuant to Article 24, Clause 7, all of the Basic Space Plan (decided by the Cabinet on April 1, 2016) will be changed as attached.

Table of Contents
Preamble
1. Current Situation Surrounding Our Space Policy
    1. Increasing Importance of Outer Space in Security
    2. Increasing Dependence on Society's Space Systems
    3. Increasing Risks that Hinder the Sustainable and Stable Use of Outer Space
    4. Activation of Space Activities in Other Countries
    5. Strengthening Private Space Activities and the Rise of New Business Models
    6. Expansion of Space Activities
    7. Rapid Evolution of Science and Technology
2. Goals of Our Space Policy
    1. Contributing to Diverse National Interests
        1. Ensuring Space Security
        2. Contributing to Disaster Management, Infrastructure Maintenance, and Resolving Global Issues
        3. Creation of New Knowledge through Space Science and Exploration
        4. Realizing Economic Growth and Innovation Driven by Space
    2. Strengthening the Comprehensive Foundations Supporting Japan's Space Activities, including the Industrial, Scientific and Technological Foundations
3. Basic Stances in Promoting Our Space Policy
    1. Goal Oriented Space Policy Based on Needs for Security, Industrial Use, etc.
    2. Space Policy that Provides Predictability for Investment and Maximizes the Vitality of the Private Sector
    3. Space Policy that Effectively and Efficiently Utilizes Resources such as Human Resources, Funds, and Intellectual Property
    4. Space Policy in Strategic Collaboration with Allies and Friends
4. Specific Approaches to Our Space Policy
    1. Ensuring Space Security

1. Basic Concept
2. Major Efforts
    i. Quasi-Zenith Satellite System
    ii. X-Band Defense Satellite Communications Network
    iii. Information-Gathering Satellites
    iv. Small Rapid Response Satellite Systems
    v. Uses of various Satellites including Commercial Satellites
    vi. Early Warning Functions, etc.
    vii. Understanding Maritime Conditions
    viii. Space Situational Awareness
    ix. Enhanced Mission Assurance of the Space System as a Whole
    x. Formulation of International Rules in Strategic Partnership with Allies and Friends
2. Contributing to Disaster Management, Infrastructure Maintenance, and Resolving Global Issues
    1. Basic Concept
    2. Major Efforts
        i. Meteorological Satellites
        ii. Greenhouse Gas Observation Satellites
        iii. Earth Observation Satellites and Sensors
        iv. Quasi-Zenith Satellite System
        v. Information-Gathering Satellites
        vi. Utilization of Satellite Data for Disaster Management and Infrastructure Maintenance
        vii. Resource Exploration Sensors (Hyperspectral Sensor)
3. Creation of New Knowledge through Space Science and Exploration
    1. Basic Concept
    2. Major Efforts
        i. Space Science and Exploration
        ii. Participation in International Space Exploration
        iii. Low Earth Orbit Activities, including ISS
4. Realizing Economic Growth and Innovation Driven by Space
    1. Basic Concept
    2. Major Efforts
        i. Expanded Use of Satellite Data (Satellite Remote Sensing and Geo-positioning)
        ii. Open & Free Government Satellite Data
        iii. Government Satellite Data Platform
        iv. Construction of Systems to Provide Space Situational Awareness to the Private Sector

- v. Expanded Procurement from the Private Sector, including Venture Companies, in Government Projects
- vi. Strengthen JAXA Initiatives for Business Creation and Open Innovation (Use of Investment Functions, etc.)
- vii. Promote the Entry of Companies from Other Industries and Venture Companies into the Space Industry
- viii. Improve the Institutional Environment
- ix. Launch Sites/Space Ports
- x. Overseas Market Development
- xi. Promoting Private Sector Participation in Lunar Exploration Activities
- xii. Promotion of Economic Activities in Low Earth Orbit, including ISS

5. Enhancing the Comprehensive Foundations Supporting Space Activities, including the Industrial, Scientific and Technical Foundations
   1. Basic Concept
   2. Major Efforts
      - i. Develop and Operate Mainstay Rockets (H-IIA/B rockets, H3 rockets, Epsilon rockets)
      - ii. Research and development on Future Space Transport Systems
      - iii. Establishment of a Framework (Satellite Development and Demonstration Platform) to Strategically Promote Satellite Development and Demonstrations
      - iv. Development of Innovative and Fundamental Satellite-related Technologies
      - v. Review of Manned Space Activities
      - vi. Space Debris Countermeasures
      - vii. Research and Development on Space Photovoltaic Power Generation
      - viii. Space Environment Monitoring (Space Weather)
      - ix. Strengthening the Human Resources Base that Supports Space Activities
      - x. Development of the Environment for Intellectual Property Activities in the Space Sector
      - xi. Strengthening the Space Industry Supply Chain
      - xii. Promotion of International Rulemaking
      - xiii. Strengthening International Space Cooperation
      - xiv. Strengthening Research, Analysis, and Strategic Planning functions
      - xv. Promoting Public Understanding

Preamble

Today, the role of space systems is becoming more important than ever in the context of our national security, economy, and society. This trend is likely to further develop in the years to come. Under these circumstances, we are entering an era in which space activities are driven by public-private collaborations instead of conventional government-led initiatives. As a result, wide-ranging fields of industries are being invigorated through space utilization. Moreover, progress in space exploration expands the sphere of human activities beyond the earth's orbit to reach the lunar surface, and further beyond into deep space. **Space becomes more important as a frontier of science and technology, and as a driving force for economic growth. It has also potential to be a major driver for the economic growth of Japan.**

On the other hand, space security is a matter of urgency, as an increasing number of countries, including the United States, **regard outer space as a "Warfighting Domain" or an "Operational Domain"**, while there is suggestion of the increasing threats in outer space, including the development of various means of disruption, not limited to destruction of satellites by missile strikes or other ways as has been discussed. In addition, a growing number of small and micro-satellite constellations is growing as a "game changer" in the space industry. Japan's space equipment industry is, however, lagging behind these trends. While space-related technologies are undergoing a rapid evolution, **the reinforcement of the industrial, scientific and technological basis is an immediate challenge to maintain the autonomy of space activities that Japan has built up since the end of the Second World War**.

The Government of Japan, recognizing such huge potential of outer space and the severe situation that it is facing, hereby decides **a basic plan on space policy for coming ten years with the view of the next two decades, and will secure sufficient budgetary allotments and other necessary resources, and effectively and efficiently utilize these resources to strengthen its space policy through the whole-of-government- approach**, while promoting public-private collaborations.

For the time being, Japan faces one of the worst socio-economic challenges ever in the postwar era, posed by the impact of Coronavirus disease 2019 (COVID-19). While the response to this crisis is the highest priority in the current situation, it is necessary to build a resilient social structure foreseeing the future especially in such time. **Space systems providing positioning, and timing information, images, communication and other functions, are expected to further contribute to the society and economy, since they constitute the basis for a digitalized and remotely-controlled society, while ensuring safety.**

Annex 9 – Changes to the Basic Space Plan

On the other hand, the basis of our space industry, including local and start-up business has been considerably damaged by the severe situation surrounding our economy and society. Therefore the Government of Japan needs to continuously support its maintenance and growth **with a view to ensuring sustainable development of the space systems**.

Moreover, particularly in such a difficult time, it is important to pursue initiatives with a long-term perspective which gives hopes to people. Through exploring the intellectual horizon, Japan needs to continue making its best endeavor on the frontiers of outer space.

1. Current Situation Surrounding Our Space Policy
   1. Increasing Importance of Outer Space in Security

      The importance of outer space in security has grown significantly. Without the use of space systems, contemporary security ceases to exist, and the United States, Europe, Russia, China, and other countries have deployed diverse satellites in outer space for security purposes, enabling advanced military operations. In the United States, space has been positioned as a "Warfighting Domain," and in December 2019 a Space Force was established as an independent military branch alongside the Army, Navy, Air Force, and Marine Corps. In September 2019, a Space Command was established in France, and the North Atlantic Treaty Organization (NATO) declared space to be an "Operational Domain" in December 2019.

      In the *National Defense Program Guidelines for FY 2019 and beyond* (approved by the Cabinet in December 2018), Japan has recognized that the rapid expansion of the use of outer space in security is fundamentally changing the state's security policy, which has been emphasizing responses in the traditional physical domains of land, sea, and air. Based on this recognition, the acquisition of superiority in outer space is vitally important. In order to ensure superiority of outer space use at all times, from peacetime to emergencies, including continuous monitoring of outer space conditions and functional assurance (mission assurance), a new Space Squadron has been formed in the JASDF. As the importance of outer space in terms of security is expected to continue to increase, it is necessary to continue to expand and strengthen the functions of intelligence-gathering satellites, to make effective use of various satellites, and to establish a system for surveillance of outer space in light of the National Security Strategy (approved by the Cabinet in December 2013). In promoting the use of space development, we will also pay attention to contributing to national security from a mid-to long-term perspective.

      United States deterrence is essential for maintaining peace and stability in the Indo-Pacific region. The United States space system is indispensable for the

maintenance and exercise of United States deterrent power and operations, and the SDF's activities to protect Japan also depend heavily on this system. As part of efforts to strengthen the Japan-U.S. Alliance, it is necessary to further strengthen Japan-U.S. space security cooperation by sharing roles with the United States in the maintenance of space systems, focusing on areas such as geo-positioning, space situational awareness, and maritime status monitoring.

2. Increasing Dependence on Society's Space Systems
Space-system geo-positioning, communications, and broadcasting have already taken root in everyday life and are one of the important foundations of our economic and social activities. In the event of a disaster, they play an important role as a means of understanding the state of the disaster and communicating in case of an emergency, and their importance as infrastructure that supports society is expected to increase further in the future.

In the forthcoming Society 5.0, the acquisition and distribution of diverse data will be crucial, and space systems, in conjunction with ground systems, provide three-dimensional geo-positioning data, which is a key component of big data, and remote sensing data, which captures various conditions on the ground. In order to realize the advanced fusion of cyberspace and physical space, it is important to smoothly distribute data in all areas where people and goods interact. As people move globally and become more active in outer space and the oceans, it is expected that an advanced and secure information and telecommunications network will be achieved that seamlessly connects the oceans, air, and space. While 5G has been put into practical use for terrestrial radio communications and efforts have begun toward a future Beyond 5G (B5G), the importance of space systems, including communications satellites, is steadily growing in order to expand such future advanced radio networks into the oceans, sky, and space.

The importance of space systems is expected to increase in times of disaster. In recent years, disasters have become more frequent and severe, as evidenced by the enormous damage caused by Typhoons No. 15 (Boso Peninsula Typhoon) and No. 19 (East Japan Typhoon). Many lives have been lost and the functioning of critical infrastructures has been hindered, which has had a great impact on Japan's economy and society. Under such circumstances, space systems' potential is large, where functions continue without being influenced by the conditions on the ground, and wide-area observations and communications are possible. Actually, the use of satellite data has begun, for example, through SIP4D (Basic Disaster Prevention Information Distribution Network) for disasters. However, further expansion of the use of space data in disaster prevention, disaster mitigation, and infrastructure maintenance is expected in the future.

In addition, the wide-ranging functions of space systems are expected to be utilized in resolving global issues. For example, data on greenhouse gas emissions obtained by our GOSAT are used to combat climate change. As a means of contributing to the resolution of global issues such as energy, climate change, the environment, food, public health, and large-scale natural disasters, as well as the achievement of the United Nations' SDGs, it is vital to actively utilize Japan's excellent space systems as a means to strengthen our diplomatic capabilities.

3. Increasing Risks that Hinder the Sustainable and Stable Use of Outer Space
While outer space is expected to become increasingly important for security, the economy and society in the future, there are concerns that congestion in outer space will rise because of an increase in the number of satellites due to the lower costs of launching satellites and the emergence of a vast number of small and micro-satellite constellations, as well as an increase in space debris that may significantly impair the functions of satellites in the event of collisions. It is also pointed out that, in some countries, the development of low-cost technologies to interfere in space, such as using laser beams to interfere with a satellite's functions, is progressing.
As Japan's security, economy and society become increasingly dependent on outer space systems, addressing the risks that hinder the sustainable and stable use of outer space has become an extremely important and urgent issue. Japan needs to play an active role in the face of calls for active efforts by other countries for confidence-building and international rulemaking.

4. Activation of Space Activities in Other Countries
The world of space activities has changed from the former "United States-Soviet bipolar structure" to a "multipolar structure." China has developed for global use the Hokuto Satellite Positioning System with the goal of establishing itself as a powerful space nation. In addition to this effort, in January 2019 China became the first country to succeed in landing on the far side of the Moon. In 2019 China also had the largest number of satellite launches of any country. Meanwhile, to maintain its superiority, the United States is promoting policies such as the creation of the Space Force and international space exploration. India, which has its own satellite geo-positioning system, is boldly endeavoring to land on the Moon and would be the fourth country in the world to do so if it succeeds.
Efforts in other countries have also become more vigorous. For example, in September 2019 the United Arab Emirates (UAE) sent astronauts to the International Space Station (ISS), and the UAE's Mars exploration program is progressing steadily. The countries of Asia and Africa are also actively promoting the development and use of outer space through the use of small satellites, and Japan is actively supporting such moves through technical

cooperation and other means. The number of countries that have established government agencies responsible for space development and use (space agencies) is steadily increasing. Against the backdrop of these developments, the space-related market is expected to expand in the future. The scale of the global market will continue to grow, and some say it will expand by several times in 20 years.

As the space activities of other countries are becoming more active, it is necessary for Japan to make use of this opportunity to further strengthen its position as an advanced space nation. Competition in the field of science and technology is expected to intensify in the future, and in order to achieve significant results in science and technology, it is necessary for Japan to strengthen its own efforts and to take a strategic approach to collaborate with allies and friends that share common values.

5. Strengthening of Private Space Activities and the Rise of New Business Models

In recent years, private companies, including venture companies, have been active in space activities and have had a major impact on the existing space industry and the space policies of each country. With the participation of substantial capital such that from as the United States, the price of rocket launch services is becoming lower, and new business models for communications satellites and observation satellites are emerging through small and micro-satellite constellations. In response to this trend, there has also been diversification of the means of launching satellites, such as small-sized rockets. The development of low orbit satellite constellations has brought innovation opportunities to related industries, including unprecedented cost reductions, expanded practical opportunities for new technologies, and shortened technology renewal periods, through mass production and frequent launches of satellites. Future space policies need to be considered based on such a major game changer, including from a security and other perspectives.

The space equipment industry, which has been an important industrial foundation supporting the space activities of our country, has begun to lag behind Europe and the United States in responding to this drastic change in the environment, including in terms of technology. Traditionally, demand for the space equipment industry in our country has been insufficient to maintain the supply chain, including the parts business, and maintaining and strengthening demand has been a major issue. Accordingly, in the Basic Space Plan projects have been clarified on a national level, and efforts have been made to strengthen the space equipment industry by enhancing the "predictability" of investment. However, as technological innovations are rapidly progressing worldwide, our vision for the future is not sufficiently delineated, and development of advanced technology is stagnating. If the competitiveness gap is allowed to continue to expand, it is feared that there will be a serious impact on the space equipment industry, which supports the independence of space activities that Japan has cultivated since the war.

On the other hand, private companies such as venture companies that take on new challenges in the space business are also active in our country. Fundraising of JPY 10-20 billion has continued for three consecutive years, the investment side is highly motivated, and entry from other industries is also advancing. However, in order for venture companies to grow, it is necessary to secure funds continuously, and rules are lacking for new business areas such as space debris removal services. Many challenges still remain before these venture companies, etc. will be able to support the space industry, including limited entry opportunities to enter and the need to strengthen technological capabilities. Further improvement of the business environment will be indispensable in the future.

In order to maintain and strengthen the industrial foundation that supports the autonomy of our space activities, it is necessary to promote collaboration between the existing space equipment industry and new entrants such as venture companies, while working together with industry, academia, and government to address these issues by expanding domestic demand, incorporating external demand, and promoting research, development, and practical demonstrations.

6. Expansion of Space Activities

Until now, the use of outer space has consisted mainly of geo-positioning, communications and broadcasting, and observation, operated by satellites with each of these functions being launched on rockets. However, technological innovation is in the background and new space activities are expected to appear in the future. For example, sub-orbital flight, which departs from the ground and reaches an altitude of approximately 100 km, is being developed as a means of satellite orbit deployment services and manned transportation. Studies are also underway on orbital services such as removal of space debris and repair of defective satellites. Satellite data being used is also expected to add more value to products through advances in data analytics utilizing AI and other technologies.

In October 2019, the Government of Japan decided to participate in the International Space Exploration (Artemis Project) proposed by the United States, which aims at sustainable activities on the Moon and with a view to Mars missions. This plan is quite different in character from conventional space science and exploration in that it aims to carry out continuous activities on the Moon. In light of the fact that territory on the Moon and Mars will become the stage of human activities, it is necessary to consider this plan from a wide range of perspectives, including future economic activities, diplomacy and security.

7. Rapid Evolution of Science and Technology

   In recent years, technological advances that lead to innovative changes in space activities have been rapidly advancing. As mentioned above, small and micro-satellites and satellite constellations are leading to various innovations.

   The flow of digitization is also having a major impact on space systems. In addition to the digitization of devices installed in satellite systems, the development of software for a variety of functions is progressing, and this will enable flexible functional changes after launch. Process innovation through digitization is also advancing in the development, manufacture, and operation of satellites, improving design and manufacturing flexibility, ensuring further reliability, reducing costs, and shortening development periods. In addition, advanced science and technologies such as space optical communications, quantum cryptographic communications, AI, and robotics are rapidly evolving, and if we are unable to move forward, our space industry and scientific and technological infrastructure will be undermined.

   In Europe and North America, these advanced science and technologies have been developed in a manner that meets security needs, and subsequently they have been utilized for the benefit of the space industry and related industries and to enhance industrial competitiveness. In this way, a mechanism for effective collaboration between the public and private sectors has been established. On the other hand, such a mechanism has not been firmly established in Japan, as it has not long provided an environment where the use of space for security could be actively implemented, and related research and development have been insufficient. As a result, it cannot be said that these advanced science and technologies have been adequately addressed. Therefore, it is urgent to boost R&D for security and strengthen the mechanism for the flow of appropriate technologies to the private sector. In Japan, some advanced technologies developed in space science and exploration have been diverted to the private sector, but this process needs to be bolstered.

2. Goals of Our Space Policy

   It is necessary to continuously strengthen our capabilities to independently develop and operate space transport systems, such as satellites, which play an essential role in securing Japan's security, disaster management, and enhancing infrastructure maintenance through geo-positioning, communications, and intelligence gathering, and which will play a fundamental role in future societies, as well as the core rockets necessary for launching such satellites.

   In the future, it will be important for Japan to strengthen its ability to address the capabilities necessary for the independent exploration and development of the Moon and Mars. In order to effectively utilize space systems, it is also essential to be able to define and analyze their purposes appropriately.

In order to contribute to the following diverse national interests, the government aims to become an independent space-using power that realizes a virtuous cycle of fortifying and expanding the foundations of the use of space by strengthening the industrial, scientific and technological bases that support the autonomy of space activities and expanding the use of space, while strategically cooperating with allies and other countries

1. Contributing to Diverse National Interests
    1. Ensuring Space Security
       As mentioned in 1., the importance of outer space in Japan's security is increasing, and society's dependence on outer space systems is rising. At the same time, the risks that hinder the sustainable and stable use of outer space are becoming more serious, and ensuring outer space security is an urgent issue. Based on the *National Defense Program Guidelines for FY 2019 and beyond*, the following objectives will be achieved while ensuring security.
       a. Ensure the sustainable and stable use of outer space by enhancing space situational awareness and mission assurance, as well as by playing a greater role in international rulemaking.
       b. Further improve various abilities such as information collection, communications, and geo-positioning using space, and strengthen the capacity for ensuring superiority in space use at all times from peacetime to armed contingencies, including the ability to assure functionality and block an adversary's command and control telecommunications.
       c. As part of efforts to strengthen the Japan-United States alliance, Japan will comprehensively bolster Japan-United States space cooperation in the area of security, including the sharing of roles with the United States in maintaining space systems. Japan will also pursue wide-ranging collaboration and cooperation in the area of space with friends and additional countries other than the United States. In particular, Japan will strengthen its efforts in the Indo-Pacific region with a view to contributing to the maintenance and promotion of a free and open Indo-Pacific.
    2. Contributing to Disaster Management, Infrastructure Maintenance, and Resolving Global Issues
       We will strengthen our advanced space systems, such as geo-positioning, communications and broadcasting, and observation.
       a. Disaster management and infrastructure maintenance will be promoted by contributing to the response to large-scale disasters and accidents such as earthquakes, tsunamis, volcanic eruptions, typhoons, tornadoes, and

torrential rain, as well as the maintenance and management of aging infrastructures.

    b. In cooperation with the international community, Japan will demonstrate its leadership and contribute to the resolution of global issues that are becoming more serious, such as the global energy situation, climate change, the environment, food, public health, and large-scale natural disasters, etc., thereby contributing to the achievement of SDGs.

3. Creation of New Knowledge through Space Science and Exploration
With regard to our space science and exploration, which is highly praised internationally for disseminating outstanding research results both domestically and internationally and has greatly contributed to ensuring Japan's place in the international community, efforts will be strengthened to lead international cooperation and global results will be achieved that will lead to the creation of new knowledge.

4. Realizing Economic Growth and Innovation Driven by Space
In order to respond to the increasing dependence of the economy and society on space systems and to further expand the scope of human activities, we will further strengthen and increase the use of space systems, which are important infrastructures, and will make maximum use of these as a driving force in our country's economic growth and innovation.

2. Strengthening the Comprehensive Foundations Supporting Japan's Space Activities, including the Industrial, Scientific and Technological Foundations
As mentioned in 1., recognizing that maintaining the independence of our space activities faces great challenges, the government will strengthen the comprehensive foundations that support our space activities, including the industrial, scientific, and technological foundations, in order to achieve the items listed in (1). In particular, we will actively promote collaboration where international cooperation is required, and comprehensively aim for self-reliance where independence is necessary. We will revamp our space industry ecosystem while actively incorporating new entrants.

In particular, under a strategy based on thorough investigation and analysis, we will: (1) upgrade space transportation systems as the basis of space activities, (2) confidently develop and demonstrate satellites with advanced technologies, such as space optical communications, quantum cryptography communications, AI and simulation, micro-satellite systems, and satellite constellations, (3) develop space use environments such as collection and analysis of geospatial information including satellite data, (4) develop human resources responsible for the development of technologies and mobilize human resources, including exchange with non-space fields, and (5) promote international rulemaking and international space cooperation.

At the same time, while strengthening the role of JAXA, which is positioned as a core implementation organization that supports technological development/utilization of the government as a whole, efforts will be made to strengthen the foundations of industry and science/technology through cooperation and sharing of roles among JAXA, industry, academia, and government organizations.

3. Basic Stances in Promoting Our Space Policy
   1. Goal-Oriented Space Policy Based on Needs for Security, Industrial Use, etc.
      a. Users' need for space systems in security, disaster management, infrastructure maintenance, industrial use, etc. will be carefully considered, based on a wide range of stakeholders and with users participating on their own initiative. In addition, the government will make thorough efforts to clarify in strategic outcomes that necessary resources will be devoted under an appropriate division of roles between the public and private sectors and relevant ministries, that the space systems to be developed in the future will function effectively within the entire network of users, and that plans for sharing this are prepared. In doing so, close coordination will be pursued with related measures, including growth strategies, promotion of Society 5.0, science/technology policies, and maritime policies.
      b. Based on a thorough survey and analysis of global trends in space systems, the SDF will confidently take on challenges in the development of innovative space systems that anticipate future needs. At the same time, the SDF will take strategic measures, such as timely verification of technologies and their effectiveness, in order to ensure that development results are delivered while accurately responding to changes in needs and technological trends.

   2. Space Policy that Provides Investment Predictability and Maximizes the Vitality of the Private Sector
      a. As far as possible, a Process Chart will announce specific measures (projects, etc.) of the government that will be implemented over a 10-year period, with a view to the next 20 years. While providing predictability for private-sector investment, the government will promote its space policy after securing the necessary financial resources. After the Process Chart has been decided by the Strategic Headquarters for Space Development based on the Basic Space Plan, the Committee on National Space Policy will examine the status of progress each year, and make flexible reviews and additions in response to changes in circumstances, etc., which will be revised by the Strategic Headquarters for Space Development. Through the revision of Process Chart, the Basic Space Plan will continually evolve.

b. The vitality of the private sector will be maximized in order to accomplish all the necessary space activities effectively and efficiently. For this reason, the basic policy of the government in its specific national measures is to procure as much as possible from the private sector that which the private sector can supply. In order to enhance predictability for private-sector investment, the government will make a concrete announcement in the Process Chart as soon as possible.

3. Space Policy that Effectively and Efficiently Utilizes Resources such as Human Resources, Funds, and Intellectual Property
    a. In light of the importance of outer space to security and the increasing dependence of the economy and society on outer space systems, Japan will strengthen its efforts in the space sector as a whole. At the same time, Japan will maximize the results of its space policy by making effective and efficient use of limited resources, such as human resources, funds, intellectual property, and satellite data, as much as possible. For this reason, clear performance targets will be set for each policy item for the next 10 years, and thorough ex-ante and ex-post evaluations will be carried out to ensure comprehensive efficiency, streamlining, and refinement of existing projects, while maximizing policy effects.
    b. In addition to strengthening the development of advanced technologies for space security and space science and exploration, the government will actively promote the space industry and convert advanced technologies that have been developed for effective use in other industries.
    c. The government will promote human resource exchanges with non-space sectors, whether public or private, and foster and utilize core human resources in the space sector. In addition, fund flows between different industries will be stimulated, and activities in the space sector will be further revitalized.

4. Space Policy in Strategic Collaboration with Allies and Friends
    a. In cooperation with our allies, such as the United States and Europe, Japan will actively work on international rulemaking. At the same time, Japan will promote the sharing of roles and international cooperation with its allies and other countries that make use of Japan's strengths in the space sector. Japan will take the initiative in realizing economic prosperity through expanded use of space, as well as in ensuring peace and stability through capacity building and problem solving. In particular, Japan will strengthen its efforts in the Indo-Pacific region with a view to contributing to the maintenance and promotion of a free and open Indo-Pacific.

b. In order effectively and efficiently to promote the development of technologies, etc. necessary for the development and utilization of space, appropriate themes will be strategically coordinated with allies, friendly nations, etc., including the development of an environment for standardization, etc., while making use of Japan's strengths.

The National Security Strategy will be considered when promoting the use of space, and precautions will be taken to contribute to national security from a medium- to long-term perspective.

4. Specific Approaches to Our Space Policy
   1. Ensuring Space Security
      1. Basic Concept
      Considering the increasing importance of the use of space systems for geo-positioning, communications, information gathering, and understanding maritime conditions, efforts will be made to develop these space systems and to further improve their capabilities.
      In addition, in view of the growing risk to the sustained and stable use of space, the government will develop systems necessary to understand the status of space and other issues in collaboration with allies, friends, and other countries. In addition, it will work to strengthen capabilities for mission assurance of the entire space system, as well as be involved in international rulemaking.
      The Artemis Plan is also important from a diplomatic and security perspective, as it aims for sustainable activities on the Moon.
      2. Major Efforts
         i. Quasi-Zenith Satellite System
         In order to independently ensure the ability to recognize and mark positions and synchronize time that is essential for maintaining and strengthening Japan's security capabilities, the government will steadily advance the development of three additional satellites, which are necessary to establish a system of seven satellites capable of continuous positioning of the Quasi-Zenith Satellite System, to start operation by FY 2023, and begin the development of successors necessary to maintain and improve continuous positioning capabilities. In addition, based on technological trends overseas and the needs of domestic and foreign countries, strategic and continuous advances will be made in positioning technology, such as improving accuracy and reliability and enhancing resilience. In addition, Japan-United States cooperation will be promoted to improve space situational awareness

capabilities by incorporating United States sensors. (Cabinet Office, Ministry of Foreign Affairs, Ministry of Education, Culture, Sports, Science and Technology, Ministry of Defense)

ii. X-Band Defense Satellite Communications Network

We will steadily develop the X-band satellite communications network and aim to launch a third satellite by FY 2022. Through the development of this satellite communications network, the SDF will strengthen its command and control and information and communications capabilities, and it will work to further bolster its resilience. In addition, in light of the trends in space communications system technology and the status of examining the strength of the mission assurance of the entire space system, enhancement of the satellite communications network will continue to be evaluated and necessary measures will be taken. (Ministry of Defense)

iii. Information-Gathering Satellites

Currently there are four optical and radar satellites (core satellites), including time-axis diversified satellites and data relay satellites. We will steadily increase the number of satellites with the addition of relay satellites in order to improve immediacy and readiness through the establishment of a 10-satellite system. At the same time, we will expand and strengthen the functionality and improve the quality of information through research and development of advanced technologies, etc. Functional guarantees will also be strengthened by making use of empirical research on short-term-launch small satellites and efforts related to understanding space conditions. While paying attention to ensuring necessary functionality in development, costs will be reduced through the creation of a competitive environment and lump-sum procurement of similar satellites in bulk. (Cabinet Secretariat)

iv. Small Rapid Response Satellite Systems

With regard to small rapid response satellites, attention will be given to the operational needs and concepts, etc. of space systems as a whole, considering the strengthening of mission assurance and the progress of private business, and taking necessary measures. (Cabinet Secretariat, Cabinet Office, Ministry of Defense)

v. Uses of Various Satellites including Commercial Satellites

Information collection through the frequent use of small satellite constellations is advanced. Based on the status of the deliberations by the Cabinet Office and each Ministry, the concrete operational circumstances of small satellites, including the use of commercial satellites in Japan (satellite remote sensing data, etc.), and the needs of small satellites, will be examined and necessary measures will be taken. In the defense field, including quasi-zenith satellites with a view to enhancing of C4ISR,* redundancy will be ensured by receiving multiple positioning satellite signals and using commercial satellites, including micro-satellites (Ministry of Defense)

* An abbreviation for Command, Control, Communication, Computer, Intelligence, Surveillance, Reconnaissance, and a generic term for "command, control, communications, computers, intelligence, surveillance, and reconnaissance" functions.

vi. Early Warning Functions, etc.
Technological trends related to missile detection and tracking functions, such as early warning and small satellite constellations, will be examined in cooperation with the United States, necessary measures will be taken, and research will be conducted on advanced infrared sensors, such as highly sensitive, broadband infrared detection devices. (Cabinet Office, Ministry of Defense)

vii. Understanding Maritime Conditions
With the cooperation of relevant ministries and agencies, the government will comprehensively examine maritime conditions using space technologies, such as various satellites and other space technologies owned by Japan, including in conjunction with aircraft, ships, ground infrastructure and in collaboration with the United States, and necessary measures will be taken.

In coordination with the Basic Maritime Plan and the Process Chart, capabilities (time, spatial resolution, etc.) required in the operation of systems and initiatives related to the collection and acquisition of maritime data will be strengthened with a view to using various governmental satellites and small private satellites (optical satellites and SAR satellites), etc. (Cabinet Secretariat, Cabinet Office, Ministry of Foreign Affairs, Ministry of Land, Infrastructure and Transport, Ministry of Defense, etc.)

viii. Space Situational Awareness
The Ministry of Defense will establish a system for space situational awareness and improve its capabilities through the operation of an integrated government system, including ground radar and artificial satellites (space-based optical telescopes). At the same time, the Ministry will advance discussion on how to strengthen international cooperation with the United States and other friendly countries and to contribute to ensuring the sustainable and stable use of outer

space. (Cabinet Office, Ministry of Foreign Affairs, Ministry of Education, Culture, Sports, Science and Technology, Ministry of Defense)

ix. Enhanced Mission Assurance for the Space System as a Whole

Measures will be considered to comprehensively and continuously maintain and strengthen the mission assurance of all space systems (including civilian use) operated by Japan and its allies, and necessary measures will be taken. (Cabinet Secretariat, Cabinet Office, MPHPT, METI, MLIT, Meteorological Department, MPHPT and MOD)

x. Formulation International Rules in Strategic Partnership with Allies and Friends

In order to ensure Japan's space security and the sustainable and stable use of outer space, Japan will play a greater role in effective rulemaking from a comprehensive perspective, including regarding measures against space debris, while strategically cooperating with its allies, friends, and other nations. Japan will also call on other nations to take responsible action in outer space. In addition, the government will discuss the importance of enhancing communication among the concerned countries and implementing transparency and confidence-building measures (TCBM) in outer space in order to avoid risks caused by misunderstanding and miscalculation. (Cabinet Office, Ministry of Foreign Affairs, Ministry of Defense, etc.)

2. Contributing to Disaster Management, Infrastructure Maintenance, and Resolving Global Issues

   1. Basic Concept

   Japan will steadily develop and utilize space systems for geo-positioning, communications and broadcasting, meteorology, environmental observation, and Earth observation based on the needs of the international community. In addition, Japan will improve disaster prevention and post-disaster response capabilities. In cooperation with the international community, Japan will contribute to the resolution of global issues and the achievement of SDGs by actively providing data. In addition to steadily promoting the utilization of satellites currently in development and use, the development of new satellites and the advancement and miniaturization of sensor technologies will be promoted under an appropriate division of roles among relevant ministries and agencies, taking into account Japan's technological advantages, requests from the academic and user communities, international cooperation, and diplomacy.

2. Major Efforts
   i. Meteorological Satellites
   Himawari-8 is currently in operation, and operation of Himawari-9 is planned to begin by the end of fiscal year 2022. In order to ensure seamless meteorological satellite observation systems aimed at ensuring people's safety and security, by monitoring and forecasting typhoons and torrential rains, safe navigation of aircraft and ships, and monitoring the global environment and volcanoes, production of successor satellites will commence by fiscal year 2023 in order to start operation of these successors at the end of fiscal year 2029. These successor satellites will incorporate the latest technologies, such as high-density observation, and will upgrade disaster prevention weather information to reduce the damage from natural disasters. (MLIT)

   ii. Greenhouse Gas Observation Satellites
   In order to contribute to understanding the greenhouse gas emission reduction policies of countries and their status in achieving the goals of the Paris Agreement, Units 1 and 2 (GOSAT) will be operated appropriately and necessary measures will be taken to ensure that these satellites do not become space debris. In addition, in order to improve the capability to identify GHG emission sources and the accuracy of emission estimation, and to maintain the present observation system for greenhouse gases worldwide, Unit 3 is equipped with a high-performance microwave radiometer (AMSR3) designed to observe changes in water cycles, and launch is targeted for fiscal 2023 (GOSAT-GW). In addition, the government will continue to study the concept of future greenhouse gas observation missions, strengthen cooperation with relevant countries, international organizations, and the private sector, and contribute to the Paris Agreement. (Ministry of Education, Culture, Sports, Science and Technology and Ministry of the Environment)

   iii. Earth Observation Satellites and Sensors
   The advanced Earth observation satellite ALOS-3, which leverages Japan's technological strengths, will be available in 2020. The advanced Earth observation satellite ALOS-4 will be steadily developed with a launch target of fiscal year 2021. (Ministry of Education, Culture, Sports, Science and Technology)
   In order to maintain the satellites seamlessly, based on the design lives and development times of optical and radar satellites, ALOS-3's successor satellite will be developed by fiscal year 2022 and will be in

operation by fiscal year 2026. In addition, we will begin developing successor satellites for ALOS-4 by the end of fiscal year 2023 and will begin operating them by the end of fiscal year 2027. (Ministry of Education, Culture, Sports, Science and Technology, etc.)

With regard to the ideal successor satellite, the basic policy will be to identify several possible options for satellite systems from the viewpoint of strengthening security, industrial development, maintenance and upgrading of scientific and technological foundations, etc., based on the satellite development and demonstration platforms described in 4. (5)(2)(iii), and fully align these with utilization needs and technological trends (technologies that are superior, unique technologies, and technologies that Japan should maintain and advance). At the same time, the basic policy will be to identify several possible options for satellite systems from the viewpoint of ensuring flexibility that reflects changes through the start of development and the early results of the ALOS-3 and ALOS-4 operations, while incorporating perspectives in terms of international cooperation, development costs, user burdens, etc. (Ministry of Education, Culture, Sports, Science and Technology, etc.)

In addition, a satellite for observing greenhouse gases and hydrological cycles will be launched in fiscal year 2023 (GOSAT-GW). With regard to technologies such as precipitation radar and microwave radiometry, in which our country is strong in terms of continuously securing global observation data, etc., we will continuously upgrade this fundamental satellite technology, which is essential for rulemaking and policymaking to resolve global issues and for critical business judgments such as ESG investment decisions that contribute to the achievement of SDGs, etc. (Ministry of Education, Culture, Sports, Science and Technology)

The government will strengthen the analytical environment of the Data Integration and Analysis System (DIAS) to accumulate, integrate and analyze global environmental data such as Earth observation information, and will improve its sophistication (development of basic technologies for integrating and analyzing big data, etc.), as well as utilize the framework of GEO (Intergovernmental Meeting on Global Observations: Group on Earth Observations) to contribute to solutions for global issues such as disaster prevention and mitigation, infrastructure maintenance, and climate change. (Ministry of Education, Culture, Sports, Science and Technology, etc.)

iv. Quasi-Zenith Satellite System
Even in the event of disrupted ground communications due to a large-scale disaster, the government will steadily develop and operate the "Disaster and Crisis Management Notification Services" to distribute disaster information via quasi-zenith satellites, and the "Satellite Safety Confirmation Service" for collecting disaster victim information from evacuation centers via quasi-zenith satellites, and promote their use by disaster prevention and disaster response organizations, etc. The Satellite Safety Confirmation Service will be distributed to around 20 prefectures by fiscal year 2021 and will be coordinated with the Second Phase of the Strategic Innovation Creation Program (SIP: Strategic Innovation Promotion Program). Practical demonstration projects using the quasi-zenith satellite system will contribute to disaster management and infrastructure maintenance. (Cabinet Secretariat, Cabinet Office, Ministry of Internal Affairs and Communications, Ministry of Land, Infrastructure and Transport, etc.)

v. Information-Gathering Satellites
In the event of a large-scale disaster, the relevant government ministries and agencies will share information and appropriately use image data in order to quickly understand the situation of the disaster and contribute to the prompt rescue and evacuation of affected people. (Cabinet Secretariat)

vi. Utilization of Satellite Data for Disaster Management and Infrastructure Maintenance
Empirical research will be conducted on new models that contribute to disaster management and infrastructure maintenance using satellite geo-positioning data and remote sensing data. In the second term of SIP, "Strengthening of National Resilience (Disaster Prevention and Mitigation)," satellite data will be utilized for disaster countermeasures and infrastructure maintenance by, for example, developing a system by fiscal year 2022 that uses various types of remote sensing data from domestic and overseas sources to quickly understand the damage situation and share data products through SIP 4D, etc. in collaboration with users in the field of disaster prevention and mitigation, and by implementing social applications of such systems. (Cabinet Office, etc.)

vii. Resource Exploration Sensors (Hyperspectral Sensors)
Hyperspectral sensors were launched at the end of 2019 and mounted on the ISS. Regular operation of HISUI will start as soon as possible, and the data that is obtained will be evaluated and verified. Data

acquired by these sensors will be provided to users through the Governmental Satellite Data Platform (Tellus), etc. in a form that contributes to Japan's energy security. In addition, utilizing the characteristics of hyperspectral sensors, the government will work not only to explore resources, but also to promote R&D and use of analytical methods in a wide range of fields, including the environment, agriculture, forestry, and disaster prevention. (METI)

3. Creation of New Knowledge through Space Science and Exploration
   1. Basic Concept

   Space science and exploration combine the wisdom of mankind to create intellectual assets and expand the scope of activities in outer space. As the amount of data increases dramatically in the future, new developments in fields such as planetary science are expected. We will further develop our space science and exploration by taking the lead in international missions, including exploring the origins of space and life, and create global results that will lead to the creation of new knowledge. In so doing, Japan will contribute to enhancing its role by promoting international cooperation. At the same time, Japan will further advance its space science and technologies and strengthen its efforts for derivations (spin-offs) to ground technologies. In order to promote innovative technological development and develop human resources, the government will also emphasize the creation of an environment in which employees can confidently take on challenges.

   In view of the fact that the Artemis Plan is different in character from conventional space science and exploration in terms of aiming for sustainable activities on the Moon, the government will address the Artemis Plan from the perspective of economic activities, diplomacy, security, and additional areas other than space science and exploration. We will promote participation that ensures the independence of our country. In so doing, Japan will participate effectively and efficiently based on clarification of strategies, such as the fields in which it will cooperate and what it will gain, and Japan will also consider measures to obtain the active participation of private companies, universities and research institutes.

   The ISS program will be used to further improve operational efficiency and to acquire and strengthen the capabilities required for the aforementioned lunar and Martian exploration. In addition, taking into account trends in the United States and Europe, which are aiming to extend the ISS's operational period, the government will consider future scenarios for

activities in low earth orbit, including the condition of the ISS after 2025, and take the required measures.
2. Major Efforts
   i. Space Science and Exploration
      With regard to space science and exploration as an academic discipline, in order to generate global achievements and intellectual knowledge, Japan will continue to make long-term efforts based on its past accomplishments and technological capabilities, which have earned it a high reputation worldwide, including for the deep-space exploration technologies developed through the "Hayabusa" and "Hayabusa 2" missions. Japan will further advance its role in this respect, and in addition will actively promote the spin-off of these technologies to ground-based technical applications. (Ministry of Education, Culture, Sports, Science and Technology)
      For this reason, based on bottom-up proposals from researchers, JAXA's space science and exploration road map will be used as a reference, and the government will continue to secure and promote funding in the future. In the next 10 years, the government will aim to launch three spacecraft based on the strategic mid-term plan and five spacecraft every two years based on the published short-term plan. The government will also secure opportunities to generate results in smaller-scale missions by actively participating in overseas-led missions based on the overseas joint plan strategy and the short-term plan. (Ministry of Education, Culture, Sports, Science and Technology)
      For missions that require strategic and long-term efforts, including exploration of the solar system and unmanned exploration, efficient and effective activities will be developed with a bottom-up approach and a comprehensive program.
      In programming, with a view to applying key technologies in various future projects, Japan will promote research and development of common and innovative technologies that should be acquired ahead of the world (frontloading of technologies). In addition, for deep space exploration (especially beyond Jupiter), basic research will be promoted to develop innovations in the power and propulsion systems of deep space exploratory probes. With regard to sample returns, which is one of our country's strengths, we will strive to establish a system that enables appropriate follow-up, such as rapid sample analysis.
      It should be noted that astronomical observations from outer space are expected to become even more important in the future due to the

congestion of outer space and other factors. (Ministry of Education, Culture, Sports, Science and Technology)

In addition, JAXA will continue to conduct academic research on space science and exploration, promote personnel exchanges between JAXA and universities, etc. and develop human resources from a long-term perspective through the integrated implementation of research, education, and projects for graduate students. By providing opportunities for students, young researchers, and other diverse human resources to participate in space science and exploration projects, the government will promote initiatives that contribute to the development of space science and the space industry, such as human resource development, mobility of human resources, collaboration with other fields, and exchanges with private companies. (Ministry of Education, Culture, Sports, Science and Technology)

ii. Participation in International Space Exploration

Since the Moon is the nearest celestial body to Earth, it has advantages in terms of transport and communications and is the first celestial body to become an area of human activity outside Earth. In particular, it is an important place to acquire and demonstrate technologies for future solar system exploration, such as landing and return technology to and from celestial bodies as well as planetary surface exploration robot technology. With this in mind, Japan will take full advantage of the opportunity to participate in the Artemis Project, which aims to achieve sustainable lunar exploration and secure an opportunity for Japanese astronauts to play active roles. The government will take the initiative that is strategically and efficiently meaningful for Japan. (Cabinet Office, Ministry of Education, Culture, Sports, Science and Technology, etc.)

Specifically, by leveraging our experience in the ISS program, we will participate in areas where Japan is strong (manned stay technology and supply), and will endeavor to demonstrate technologies for the construction, operation, and use of Gateway, a manned lunar base, and the utilization of Gateway to acquire the capabilities required for deep space exploration. In so doing, the government will use technologies from universities and other institutions that been developed during the design of micro-satellites for low Earth orbit, etc., and, in cooperation with private businesses, promote the development and improvement of basic technologies necessary for sustainable exploration activities beyond the Moon, in order to strengthen and expand the foundations supporting international space exploration. (Ministry of Education, Culture, Sports, Science and Technology)

Since the presence or absence of water resources on the Moon and the difficulty of extracting any such resources greatly affect the manner of participation in this plan, data about water resources will be obtained independently by mobile exploration so that Japan can make a pinpoint landing in a region of the Moon where water resources are expected to exist and study future lunar exploration on its own initiative. (Ministry of Education, Culture, Sports, Science and Technology)

Based on the level of water resources and other factors, we will consider measures to be taken for the next 20 years of space science and exploration, including the utilization of Gateway, and promote them with the participation of a wide range of scientific fields. Topics of consideration include positioning on the Moon, communications, remote sensing, multipoint exploration by micro-probes, three-dimensional exploration, sample returns, data science, and astronomical observation. In addition, with the participation of private companies including from the non-space industry, we will work on the construction of a system that is essential for lunar activities, such as Gateway and transportation on the Moon. We will transmit pioneering results to the world, including demonstrations of fundamental technologies. (Cabinet Office, Ministry of Education, Culture, Sports, Science and Technology, etc.)

iii. Low Earth Orbit Activities, including ISS

The ISS activities that have contributed to the enhancement of our international presence, will continue to be used to create knowledge through the use of the space environment, while improving cost effectiveness. In addition, ISS will be utilized as a platform to demonstrate the technologies required for international space exploration, and scientific research and technological development in ISS will be seamlessly and efficiently linked to lunar exploration and future low earth orbit activities through international cooperation. With regard to activities after 2025 in low Earth orbit, including ISS, the government will continue to study and take necessary measures, while paying close attention to the status of deliberations by each country. (Ministry of Education, Culture, Sports, Science and Technology)

4. Realizing Economic Growth and Innovation Driven by Space
   1. Basic Concept
      Against the backdrop of the arrival of Society 5.0, an increase in emerging space nations, and the expansion of space activities, Japan will, in addition to strengthening space systems and expanding their use, promote necessary measures to realize economic growth and innovation in Japan with space as a driving force. Specifically, the government will take measures such as: promoting the use of satellite data in various fields, including agriculture, disaster prevention, transportation and logistics, and advanced use of geospatial information; strengthening coordination among databases that contribute to this data; actively using contributions and procurement by research institutions and incorporation of these into different space sector industries; promoting the mutual use of terrestrial and lunar technologies; developing necessary institutional environments; and strengthening systems to develop overseas markets while paying attention to national security. In so doing, the aim is to create foundations that support our space activities and expand the use of space with regional revitalization measures that contribute to the renewal of regional economies.

      The Artemis Plan is also important from the viewpoint of economic activity, etc., in terms of achieving sustainable endeavors on the Moon.

      In the Basic Plan for Space, the government and private sectors have set a total target of ¥5 trillion for the space equipment industry over a 10-year period. While striving to achieve this goal, we aim by the early 2030s to double the size of the space industry (currently about ¥1.2 trillion), including Japan's space use industry, by maximizing the expansion of the global space equipment industry and new space activities in Japan's economic growth and promoting the development of space system-based industries. The expansion of space use will accelerate the spread of autonomous driving and smart agriculture and contribute to realizing a wide range of economic results.
   2. Major Efforts
      i. Expanded Use of Satellite Data (Satellite Remote Sensing and Geo-positioning)
         In order to take the initiative in the appropriate use of satellites to improve the efficiency and sophistication of national and local government operations, the Task Force on the Use of Satellite Remote Sensing Data (tentative name) composed of relevant ministries and agencies will be established to share the actual status and issues in using data and will promote measures for use

of satellite remote sensing data in administrative services.
The relevant ministries and agencies will examine the possibility of using satellite remote sensing data for their respective services and, in principle, will use remote sensing data where appropriate, and will clarify the specifications required for satellite remote sensing data according to the field of use. In addition, we will boost demonstration projects to accelerate the use of satellite remote sensing data. Specifically, efforts will be made to set practical themes through collaboration among relevant ministries and agencies, strengthen cooperation with local governments, create a model for horizontal deployment through participation by multiple local governments, verify cost effectiveness, and train human resources where the data is used. (Cabinet Office, MPHPT, MAFF, METI, MLIT, MPIIPT, Ministry of the Environment, and MOD)
Geospatial information, including satellite remote sensing and geo-positioning data, is key to supporting the Fourth Industrial Revolution. The government will promote business in a wide range of fields, including disaster prevention and mitigation, transportation and logistics, the natural environment, regional revitalization, and overseas expansion, including representative projects in the Basic Plan for Promoting the Use of Geospatial Information. The government will also promote the active use of the G-Space Information Center and strive to realize a "Geospatial Information Advanced Utilization Society (G-Space Society)." In particular, in the area of disaster prevention and mitigation, the government will promote the construction of an integrated G-space disaster prevention/mitigation system in which relevant ministries and agencies cooperate with each other in efforts to use technologies and geospatial information to contribute to disaster prevention and mitigation. (Cabinet Secretariat, Cabinet Office, MPHPT, MAFF, METI, METI, MLIT, etc.)
Regarding the Quasi-Zenith Satellite System, the "Task Force for Promoting the Utilization of Quasi-Zenith Satellite Systems," which consists of relevant ministries and agencies and private companies, will be used to share issues and promotional measures concerning the use of geo-positioning data between the public and private sectors, including for autonomous driving, to demonstrate the use of such data in society and economic activities across various fields, such as agriculture, transportation and logistics,

and construction, and to further accelerate utilization through the creation of advanced usage models. In addition, based on technological trends overseas and domestic needs and the needs of foreign countries, strategic advances will be made continuously in geo-positioning technology, such as improving its accuracy and reliability and strengthening its resilience. (Cabinet Secretariat, Cabinet Office, MPHPT, MAFF, METI, METI, MLIT, etc.)

We will consider autonomous driving technology that utilizes satellite geo-positioning techniques, such as the quasi-zenith satellite "Michibiki", etc., assessment of crops and farmland using satellite images, and cooperation between the government satellite data platform "Tellus" and WAGRI, an agricultural data collaboration platform, and will accelerate the implementation of productive smart agriculture by expanding private services. (Ministry of Agriculture, Forestry and Fisheries, etc.)

ii. Open & Free Governmental Satellite Data

Government satellite data includes much public data that can be widely used and is expected to contribute to addressing global issues such as large-scale natural disasters, energy, and climate change. With regard to this public government satellite data, we will establish an "Open & Free" system that provides data free of charge in a form that is easy to process and analyze, with the exception of data with security concerns, to promote the use of satellite data in various fields and for the convenience of satellite data users. In so doing, care will be taken not to interfere with satellite data sales businesses of private companies, etc.

Regarding future government satellites, the relevant government agencies will plan their satellite projects from the development stage so as to provide public data with the requisite processing. For government satellites that have already begun development or are in operation, the relevant government agencies will provide processed public data as much as possible. (Cabinet Office, Ministry of Education, Culture, Sports, Science and Technology, Ministry of Economy, Trade and Industry, Ministry of Land, Infrastructure, Transport and Tourism, Ministry of the Environment, etc.)

iii. Government Satellite Data Platform

In order to enable the stable and permanent provision of satellite data, the government-based satellite data platform "Tellus" will continue to expand various data, including satellite data, from fiscal year 2020 onward, while making maximum use of the vitality of the private sector. We will promote functional improvements, such as cooperation with other platforms and expansion of analytical tools.

We will promote the utilization of satellite data (anchor tenancy) through the active use of "Tellus" by governmental and public organizations, promote the international sharing of satellite data through cooperation with foreign satellite data platforms, and encourage private companies to create new businesses using satellite data. (METI, etc.)

iv. Construction of Systems to Provide Space Situational Awareness to the Private Sector

In cooperation with the Ministry of Defense and other government programs for space situational awareness, and using the excellent capabilities of the private sector, the government establish a plan in early fiscal year 2020 to construct systems that integrate and analyze observation data of outer space objects and appropriately provide the trajectory information of these objects to private business operators, etc. Under such a system, after commencement of the operation of the Ministry of Defense's space situational awareness system, the Ministry of Defense will promote development of related systems after examining the optimal architecture and operational modalities, while considering the appropriate division of roles between the public and private sectors. (Cabinet Office, Ministry of Education, Culture, Sports, Science and Technology, Ministry of Economy, Trade and Industry, Ministry of Defense, etc.)

v. Expanded Procurement from the Private Sector, including Venture Companies, in Government Projects

Government projects, including those of national research institutes such as JAXA, etc., will procure from the private sector that which the private sector can provide, and this will be announced in the Process Chart, etc. as soon as possible. In addition, we will actively review procurement by and contracts of governmental agencies by using the new SBIR system, introducing flexible contract forms such as milestone payments, and accelerating the disclosure of the required specifications of technologies and services. The government will promote independent efforts by the private sector by expanding procurement from private sector entities, such as venture companies, and taking into account the development trends of small rockets and various satellites in the private sector. (Cabinet Office, Ministry of Education, Culture, Sports, Science and Technology, and METI)

vi. Strengthen JAXA Initiatives for Business Creation and Open Innovation (Use of Investment Functions, etc.)

We will promote the use of investment functions, etc. to strengthen efforts to create businesses that utilize JAXA's R&D results and stimulate open innovation. In addition, in order to enhance a common technological infrastructure and promote the entry of new private companies into the space sector, the functions of JAXA as a platform for technology will be strengthened and the effective use of intellectual property will be promoted. Research and development will be promoted not only in the space sector, but also through collaboration and joint research with researchers and engineers from universities and private companies in other fields. Through these efforts, Japan will continue actively to contribute to the promotion of the space industry. (Cabinet Office, MPHPT, MPHPT, MEXT, and METI)

vii. Promote the Entry of Companies from Other Industries and Venture Companies into the Space Industry

We will support the commercialization of space businesses through, among other things, facilitating matches of venture companies that have business ideas with investors interested in the space sector (S-Matching), promoting a space use business idea contest that strengthens ties with other Asian countries (S-Booster), supporting financing, such as collaboration with foreign investors, further expanding the supply of "risk money" by both the public and private sectors, and collaborating with private sector initiatives aimed at promoting the space industry. (Cabinet Office, METI)

In order to promote the entry of companies from other industries and venture companies into the space sector, accelerate commercialization, and strengthen competitiveness, the government will diversify and expand opportunities for research and development and the demonstration of government technologies useful for businesses in space and on the ground. (MEXT, METI)

From the viewpoint of promoting the advancement and use of information and communications technology (ICT) that will support the future development of the space industry, we will promote the formation of a community in which enterprises from different industries and venture companies, etc. interested in the space use industry and space systems participate, as well as encourage new entrants into the space industry and collaboration among related parties. (Ministry of Internal Affairs and Communications)

Looking ahead to the expansion of manned space activities over the next decade, the government will promote the development of industries that support living in outer space by strengthening industry-academia-government collaboration to develop foods that dramatically enhance quality of life in enclosed spaces and a complete resource-recycling food supply system, promoting joint research, and creating fora for cooperative endeavors, etc. (Ministry of Agriculture, Forestry and Fisheries)

viii. Improve the Institutional Environment

We will accelerate the development of the institutional environment necessary for new space businesses, which are expected to grow in the future. With regard to sub-orbital flights, which are being considered for small satellite launches and space travel, we will accelerate fostering an environment conducive to future business development, centering on the Public-Private Sector Council, and taking into account the status of efforts and trends internationally of domestic and overseas private businesses aiming for commercialization in the first half of the 2020. In addition, based on the trends of international discussions regarding the exploration and development of outer space resources including the Moon by private business operators, orbital activities, and space traffic management (STM), the government will establish a review system by the relevant ministries and agencies as soon as possible, study the development of requisite systems, and take necessary measures. The current system will also be operated appropriately to contribute to strengthening the space industry infrastructure. (Cabinet Office, MOFA, MEXT, METI, and MLIT, etc.)

ix. Launch Sites/Space Ports

With regard to the development of space ports for the realization of new transport businesses, such as the development of launch sites and sub-orbital flights, in anticipation of the future expansion of launch demand by private businesses and local governments, necessary measures will be considered and taken, including to ensure the functioning of space systems, regional revitalization, and support for private small rocket operators. (Cabinet Office, Ministry of Education, Culture, Sports, Science and Technology, Ministry of Economy, Trade and Industry, Ministry of Land, Infrastructure, Transport and Tourism, and Ministry of Defense, etc.)

x. Overseas Market Development

In addition to exporting space equipment, we will strengthen the joint efforts of the public and private sectors to expand overseas solution businesses using space and application of satellite positioning technologies such as the quasi-zenith satellite "Michibiki." In particular, with regard to emerging space nations in Asia and elsewhere, we will promote comprehensive efforts by enhancing the offering with capacity building for development of legal regimes and human resources and support to solve various problems in order to contribute to the creation of a sustainable space industry that meets the needs of partner countries. We also will focus on efforts to secure and train human resources who can take the lead in such activities. In expanding Japan's space systems overseas, we will fully consider the government's security policy under the National Security Strategy, the Development Cooperation Charter, and SDGs. (Cabinet Secretariat, Cabinet Office, Ministry of Internal Affairs and Communications, Ministry of Foreign Affairs, Ministry of Education, Culture, Sports, Science and Technology, Ministry of Agriculture, Forestry and Fisheries, Ministry of Economy, Trade and Industry, Ministry of Land, Infrastructure, Transport and Tourism, Ministry of the Environment, Ministry of Defense, etc.)

xi. Promoting Private Sector Participation in Lunar Exploration Activities

In order to obtain the active involvement of various private companies in future lunar exploration activities, the government will build a community in which private companies interested in developing lunar projects can exchange information and encourage their willingness to participate. In addition, we will collaborate with private companies to promote development of core technologies for lunar exploration activities aimed at benefiting Japan's private sector. (Ministry of Education, Culture, Sports, Science and Technology, etc.)

xii. Promotion of Economic Activities in Low Earth Orbit, including the ISS

We will promote the creation of new business services in low Earth orbit, including the ISS. In order to boost the participation of private businesses in the ISS, we will conduct concrete discussions regarding measures to increase the options and roles of private companies as users and for procurement and consignment activities, and support systems needed to expand demand. In addition, the government will conduct concrete discussions about supporting the continued implementation and expansion of Japanese economic activities in low Earth orbit, including the possibility of international cooperation that takes advantage of Japan's strengths, and will conduct research and development of required basic technologies and systems. (Ministry of Education, Culture, Sports, Science and Technology)

5. Enhancing the Comprehensive Foundations Supporting Space Activities, including the Industrial, Scientific and Technological Foundations
   1. Basic Concept

      The government and private sector will continue to strengthen Japan's capability to independently develop and operate satellites, which play an indispensable role in ensuring Japan's security, enhancing disaster countermeasures and national infrastructure, and play a fundamental role in society in geo-positioning, communications, and intelligence-gathering, as well as strengthening the space transport systems necessary for satellite launches based on ongoing research and analysis of domestic and overseas technologies, markets, and policies. Mainstay rockets will be prioritized for launching government satellites.

      In addition, with the aim of expanding the scope of the space sector, Japan will strengthen its comprehensive foundations to support the country's outer space activities. For example, Japan will enhance human resources development and ongoing education to support all types of outer space activities. In addition, Japan will improve the environment for intellectual property activities in the space sector, and, in cooperation with its allies and friends, promote international rulemaking and international space collaboration.

   2. Major Efforts
      i. Develop and Operate Mainstay Rockets (H-IIA/B rockets, H3 rockets, Epsilon rockets)

         In order to ensure the independence of Japan's space activities, we will continue to develop and upgrade our mainstay rockets, including the H-IIA/B rockets, the Epsilon rocket, a solid-fuel rocket that is crucial as a strategic technology, and complete development of the new mainstay H3 rocket. At the same time, the H3 and Epsilon rockets will be transferred to the private sector while arranging a division of roles between the public and private sectors to steadily take over mainstay rocket technology. For this reason, the government will continue preferentially to use mainstay rockets when launching government satellites and take necessary measures to operate mainstay rockets effectively and efficiently in light of trends in the competitive rocket launch market environment. In addition, necessary measures, such as measures against wear and tear, will continue to be implemented in order to appropriately maintain and manage mainstay rocket launch sites and test equipment. (Cabinet Secretariat, Cabinet Office, the Ministry of Education, Culture, Sports, Science and Technology, the Ministry of Economy, Trade and Industry, the Ministry of Land, Infrastructure and Transport, the Ministry of the Environment, and the Ministry of Defense, etc.)

ii. Research and Development on Future Space Transport Systems

In order to continuously ensure the independence of our space transport systems and to strengthen competitiveness in the future market, we will examine innovative future space transport system technologies (reuse technologies, innovative materials technologies, propulsion technologies (LNG, air breathing) and reliability improvement technologies that contribute to manned transport) aimed at drastically reducing costs, while taking into account business models of domestic private-sector launchers and the significance in foreign countries' space transport policies and security, and we will formulate a concrete roadmap specifying the time for realizing this as clearly as possible. In addition, under an ongoing promotion effort in which a wide range of actors from industry, academia, and government, including users, participate, we will conduct challenging R&D while appropriately monitoring progress, including focusing on issues from the perspective of business viability, cost, advantages of our country, and future developmental potential. In order to drastically reduce costs, we will need to expand transportation demand significantly, as expected in P2P (Point to Point: Point-to-Point Transportation). In addition, attention will be paid to the importance of manned transport in the future. (Ministry of Education, Culture, Sports, Science and Technology, etc.)

The "Long-Term Vision for Space Transport Systems (April 2014 Committee on Space Policy)" will be reviewed based on domestic and international trends concerning future space transport systems. (Cabinet Office, Ministry of Education, Culture, Sports, Science and Technology, etc.)

iii. Establishment of a Framework (Satellite Development and Demonstration Platform) for the Strategic Promotion of Satellite Development and Demonstrations

In order to ensure the independence, and strengthen the competitiveness, of satellites, such as geo-positioning, communications, and earth observation satellites, a new framework will be established consisting of key industry-academia-government actors, including satellite users, with continuous research, analysis, and strategic planning functions and strong leadership in domestic and overseas technologies, markets, and policies. In this system, we will promote projects systematically by rallying relevant organizations, setting up themes for the development of innovative and ambitious satellite technologies that anticipate future user needs and look ahead to the exit. In so doing, (1) demonstrations will be repeated confidently and in a timely manner, and (2) the program will be revised in a timely and appropriate manner as necessary to achieve targets based on the demonstration results and overseas trends, etc. Proposed projects will be promoted by appropriately allocating the necessary resources to the participating ministries and agencies, universities and research institutes, and private companies. including venture companies, and effective cooperation under this system. (Cabinet Office, MPHPT, MAFF, METI, MLIT, the Ministry of the Environment, and the Ministry of Defense, etc.)

iv. Development of Innovative and Fundamental Satellite-related Technologies

The government will promote the development of innovative satellite-related technologies, including the following initiatives, by utilizing the framework in item iii above.

With the aim of realizing practical quantum cryptographic communications between satellites and the ground, we will promote the development of the basic technology by fiscal year

2022 and research and development for the realization of global quantum cryptographic communications satellite networks. (Ministry of Internal Affairs and Communications)

The directions for satellite demonstrations (securing launch opportunities, satellite size, etc.) will be coordinated by the relevant ministries and agencies, etc., efforts will be promoted to conduct early satellite demonstrations under an appropriate division of roles, and future uses in the field of security will be considered. (Cabinet Office, Ministry of Internal Affairs and Communications, Ministry of Education, Culture, Sports, Science and Technology, Ministry of Defense, etc.)

Field demonstrations will be conducted on 10Gbps-class space optical communications technology, which is the world's highest level between geostationary satellites and the ground. In addition, the government will work toward establishing basic technologies such as ultra-long-distance, large-capacity space optical communications in preparation for miniaturization technologies that will also be used in satellite constellations and for the future development of Japan's activities in deep space. (Ministry of Internal Affairs and Communications, Ministry of Education, Culture, Sports, Science and Technology, etc.)

In order to acquire satellite communications technologies that can quickly and flexibly respond to the demand for rapidly increasing communications in the event of a disaster, etc., we will continue to develop and test various satellite buses and mission equipment for the technology test satellite (Unit 9), and to conduct demonstration experiments for the realization of next-generation high-throughput satellites. In addition, we aim to acquire the fundamental satellite technologies required to further strengthen international competitiveness and respond to diversifying needs for the use of space, industry, academia, and the government will work together to study and develop the technical themes for the next-generation technology test satellite (Unit 10) based on trends in advanced technologies (artificial intelligence, IoT, light/quantum, flexibility, digitization, etc.) and technologies in which Japan has strengths. (Cabinet Office, MPHPT, MPHPT, MEXT, and METI)

In order to realize an advanced information-and-communications network that seamlessly connects space to the ground, ocean, and sky, R&D and demonstrations of fundamental technologies will be conducted on technologies related to satellite constellations, technologies related to satellite communications and radio networks (heterogeneous networks) with different ground transmission characteristics such as 5G/B5G, and communication technologies in the terahertz frequency band, under the appropriate division of roles among private companies and related ministries and agencies. (Cabinet Office, Ministry of Internal Affairs and Communications, Ministry of Education, Culture, Sports, Science and Technology, Ministry of Defense, etc.)

In order to adopt new development methods that lead to rapid progress, shorter development periods, energy savings, and cost reductions in both public and private satellites, we will build a new satellite technology innovation program to flexibly develop and demonstrate small and micro-satellites, in addition to conventional technology test satellites. At the same time, fundamental technologies such as AI, robotics, power storage, and semiconductor technologies that support the program will be steadily developed. (Ministry of Education, Culture, Sports, Science and Technology, etc.)

In order to realize a terahertz sensor capable of observing and analyzing the composition of the Earth's atmosphere in more detail, we will promote R&D on fundamental technologies, such as developing a terahertz wave propagation model and sensing technology, and studies will be conducted by the relevant ministries and agencies on the use of these technologies. (Ministry of Internal Affairs and Communications, etc.)

In order to respond to the acceleration of technological progress and provide quick and effective demonstration opportunities for space parts and components using leading technologies in the private sector, from fiscal year 2020 we will support the orbital demonstration of space parts and components using micro-satellites, with a view to the possibility of using domestic private small-sized rockets, etc. (METI)

We will continue to provide opportunities for universities and research institutes to demonstrate new fundamental technologies using micro-satellites, etc., and to demonstrate technologies that lead to new businesses. (Ministry of Education, Culture, Sports, Science and Technology)

v. Review of Manned Space Activities

Based on our past achievements, the government will examine the future modalities of manned space activities in low Earth orbit, the

Moon, Mars, and other areas, including the ISS, to maintain and improve our international standing, strengthen our diplomatic capability, contribute to the expansion of future fields of human activity, and apply this to technological development on Earth. (Cabinet Office; Ministry of Education, Culture, Sports, Science and Technology)

vi. Space Debris Countermeasures

In view of the fact that the space debris issue is an extremely important and urgent one, and in order to ensure the sustainable and stable use of outer space, the relevant ministries and agencies will work together to promote the following measures with a sense of speed based on the Ministerial Meeting of the Task Force on Space Debris (May 2019).

In addition to steadily developing space debris observation and space debris removal technologies, including space debris removal satellites, as well as developing technologies to reduce debris from orbit launch rockets and reduce the debris created by satellites themselves in orbital services, etc., efforts will be made to prevent the generation of new debris, including monitoring the space environment and research on simulation models of changes in atmospheric density that may affect satellite and debris orbits, in cooperation with universities and private businesses, and including the consistent implementation of existing guidelines. (Ministry of Internal Affairs and Communications, Ministry of Education, Culture, Sports, Science and Technology, and Ministry of Defense)

In order to reduce space debris and address the challenges associated with space debris removal, the government will take the lead in creating international rules and contribute Japan's advanced efforts to ensure the sustainable and stable use of outer space, including anti-debris measures. (Cabinet Office, Ministry of Foreign Affairs, Ministry of Education, Culture, Sports, Science and Technology, etc.)

In order to encourage voluntary measures against space debris by satellite operators, etc., and in collaboration with Japanese businesses, we will actively participate in and contribute to international discussions on establishing a system (rating scheme) that will evaluate businesses, etc. working to mitigate space debris and contribute to the promotion of Japan's space industry. (METI)

With regard to government satellites, the government will continue to take steps such as measures to dispose of government satellites after their operations are completed, in accordance with the Space Debris Mitigation Guidelines issued by the United Nations Committee on the Peaceful Uses of Outer Space (COPUOS). (Cabinet Secretariat, Cabinet Office, MPHPT, MPHPT, MLIT, METI, MOD, and Ministry of the Environment)

vii. Research and Development on Space Photovoltaic Power Generation

The government will steadily advance efforts, including consideration of the transition to the space experimental phase, based on the Roadmap for Research and Development of Space Photovoltaic Power Generation Systems, with the aim of commercializing space solar power systems, which have the potential to address global issues faced by mankind, such as energy, climate change, and the environment, and which can also be applied to power supply systems, such as for space structures. In so doing, it should be kept in mind that space solar power generation R&D can be expected to derive from ground-based technologies, such as wireless power supply to IoT sensors, drones, robots, etc. (MEXT, METI)

viii. Space Environment Monitoring (Space Weather)

In order to contribute to the stable utilization of space, satellite development and operation, terrestrial communication and broadcasting, satellite positioning, etc., the government will continuously observe and analyze the ionosphere, magnetosphere, and solar activities and continue to distribute space weather forecasts through manned operations 24 hours a day, 365 days a year. In addition, in order to further enhance the ability to respond to changes in the space environment, we will strengthen observation and analysis systems for the ionosphere and solar activity, etc. in cooperation with relevant organizations both domestically and abroad, as well as conduct research on simulation technologies using observation data to improve the accuracy of space weather forecasting systems, etc. (Ministry of Internal Affairs and Communications)

ix. Strengthening the Human Resources Base that Supports Space Activities

With a view to expanding the base of people involved in space, we will implement human resource development in conjunction with school education. In addition, we will work to strengthen the development of next-generation human resources through practical measures in space technology for university students, etc., and build research bases through industry-university collaboration. In order to develop human resources that will lead future space development and use, we will also seek to enhance short-term space projects that enable young people to participate in a central way. (Ministry of Education, Culture, Sports, Science and Technology)

In addition to expertise in the space field, we will promote the discovery and development of human resources that bridge other fields and have advanced knowledge in the humanities and social sciences that will lead the formation of international rules, collaboration with other countries and market expansion, analysis of social and economic ripple effects, and the creation of new industries. In addition, in developing and strengthening human resources we will make full use of existing experimental platforms that involve actual space conditions, such as the ISS. (Ministry of Education, Culture, Sports, Science and Technology)

In order to secure the human resources needed to revitalize the space industry and increase people's mobility, the Space Specialist Platform (S-Expert) will support the matching of space ventures with specialists, as well as take advantage of opportunities such as domestic and overseas seminars, idea contests, demonstration projects, and joint research to bring diverse human resources into the space sector. (Cabinet Office, METI)

x. Development of the Environment for Intellectual Property Activities in the Space Sector

Based on the "Direction of Measures and Support for Intellectual Property in the Space Field" formulated in fiscal year 2019, support will be provided for the establishment of a mechanism to collect and provide information on trends in applications for space-related patents both domestically and internationally, and private companies will be encouraged to formulate intellectual property strategies, including open / closed strategies. (Cabinet Office, METI, etc.)

xi. Strengthening the Space Industry Supply Chain

In order to maintain and strengthen the space industry infrastructure for ensuring the autonomy of our space activities, we will start investigations in fiscal year 2020 to identify important technologies necessary for efficient, low-cost, and short-term development and production of space systems, and will work on providing intensive research and development support and demonstration opportunities for such important technologies. (METI)

xii. Promotion of International Rulemaking

In order to realize the rule of law in outer space and to ensure Japan's outer space security and the sustainable and stable use of outer space, Japan will actively engage in multilateral discussions, etc., concerning the future of outer space activities, and play an even greater role in effective rulemaking. Japan will also ensure its influence in international discussions by holding international conferences on the sustainable and stable use of outer space on an ongoing basis. (Cabinet Office, Ministry of Foreign Affairs, Ministry of Economy, Trade and Industry, Ministry of Defense, etc.)

xiii. Strengthening International Space Cooperation

The government will promote multilayered international cooperation in the field of outer space through bilateral dialogue, etc., while sharing recognition of the importance of outer space in security and the increasing dependence of the economy and society on outer space systems. In particular, Japan and the United States will contribute to the strengthening of the Japan-United States alliance by cooperating comprehensively in all areas, including security, civilian use of space, and space science and exploration. In addition, with friends and other countries that share strategic interests, we will establish multilayered cooperative relationships, including joint development of advanced technologies, mission equipment synergies, and joint use of satellite data, while taking into account the needs and capabilities of partner countries. (Cabinet Secretariat, Cabinet Office, MPHPT, MEXT, METI, and MOD)

By actively utilizing multilateral cooperative frameworks such as the Global Earth Observation System of Systems (GEOSS: Global Earth Observation System of Systems) and the Asia-Pacific Regional Space Agency Forum (APRSAF: Asia-Pacific Regional Space Agency Forum), we will promote broad-based international space cooperation and thereby strengthen Japan's leadership and diplomatic capabilities. (Cabinet Secretariat, Cabinet Office, Ministry of Internal Affairs and Communications, Ministry of Foreign Affairs, Ministry of Education, Culture, Sports, Science and Technology, Ministry of Economy, Trade and Industry,

Ministry of Defense, etc.)

By supporting capacity building to meet the needs of space agencies in emerging space nations, the establishment of space-related legislation and policies in each country, and the implementation of international norms in Japan, etc., we will strive to form an international space policy community and contribute to the improvement of transparency and confidence-building of space activities in the international community. In particular, we will strengthen our regional efforts with a view to contributing to the maintenance and promotion of a free and open Indo-Pacific. (Cabinet Office, Ministry of Foreign Affairs, Ministry of Education, Culture, Sports, Science and Technology, etc.)

We will promote international cooperation that leverages Japan's strengths in the field of space and will contribute to resolving global issues such as energy, climate change, the environment, food, public health, and large-scale natural disasters, as well as to achieving SDGs. Japan will also strengthen its maritime surveillance capabilities through maritime- space cooperation and contribute to ensuring the safety of navigation in Japanese sea lanes. For this purpose, we will comprehensively utilize various assistance measures and implement international cooperation in space policy and space science and technology, combining human resource development and capacity building according to the needs of partner countries with the provision of related equipment and services, etc. In particular, Japan will bolster its efforts in the Indo-Pacific region with a view to contributing to the maintenance and promotion of a free and open Indo-Pacific. (Cabinet Secretariat, Cabinet Office, Ministry of Internal Affairs and Communications, Ministry of Foreign Affairs, Ministry of Education, Culture, Sports, Science and Technology, Ministry of Health, Labour and Welfare, Ministry of Agriculture, Forestry and Fisheries, Ministry of Economy, Trade and Industry, Ministry of Land, Infrastructure, Transport and Tourism, Ministry of the Environment, Ministry of Defense, etc.)

In order to develop technologies, etc. necessary for Japan's use of space effectively and efficiently, the government will use the platform in item iii and identify areas to strategically cooperate with allies and friendly countries while using Japan's strengths, from the viewpoint of comprehensive cooperation on dual-use technology, division of labor by field, mutual certification of technologies, international standardization, functional guarantees, etc. In particular, we will promote strategic cooperation with Europe in the areas of Earth observation, greenhouse gas observation and satellite positioning, including areas of use and application. (Cabinet Secretariat, Cabinet Office, Ministry of Internal Affairs and Communications, Ministry of Foreign Affairs, Ministry of Education, Culture, Sports, Science and Technology, Ministry of Agriculture, Forestry and Fisheries, Ministry of Economy, Trade and Industry, Ministry of Land, Infrastructure, Transport and Tourism, Ministry of the Environment, Ministry of Defense, etc.)

xiv. Strengthening Research, Analysis, and Strategic Planning Functions

In addition to the research and analysis functions to be developed through the satellite development and demonstration platform in item iii, Japan should investigate and analyze the space industry policies and trends of other countries in cooperation with overseas diplomatic missions, etc., and strengthen the planning function for examining Japan's long-term strategies (Cabinet Office, MOFA, MEXT, and METI)

xv. Promoting Public Understanding

In promoting Japan's development and use of space, the government aims to gain the broad understanding and support of the public and appropriately disseminate information about the significance of the use of space as well as the value and importance of the results generated by the use of space, and to foster understanding from the public. It is important for Japanese astronauts to play an active role in space to gain the public's understanding and support for space development and use, and to give the public a wide range of dreams and hopes. As such, Japan will promote efforts that make full use of these values. (Cabinet Office, Ministry of Education, Culture, Sports, Science and Technology, etc.)

About the Authors

Editors: Masataka Ogasawara and Joel Greer

Co-authors: A team of lawyers and other professionals from ZeLo (Foreign Law Joint Enterprise) researched and drafted this book: Joel Greer, Satoshi Nomura, Koichi Kitada, Wataru Nakano, Fumio Amano, Hiroto Shimauchi, Daiki Matsuda, Pomi Jong, and Masuo Sono.

ZeLo (Foreign Law Joint Enterprise)

ZeLo is a Tokyo-based law firm founded in 2017. It now has over two dozen lawyers, as well as other professionals, with experience in many cutting-edge practice areas. ZeLo has represented over 100 start-up companies and actively participates in discussions with administrative authorities regarding legislative and regulatory proposals for innovative business models and ideas to support start-ups. In addition to space law and policy, ZeLo advises on FinTech matters, blockchain and crypto assets, cybersecurity, and data protection. ZeLo also advises on more traditional practice areas, such as M&A, corporate restructuring, and dispute resolution. In 2020, ZeLo received the Asian Legal Business Japan Award for "Rising Law Firm of the Year."

Acknowledgements

The editors and co-authors are most grateful to Dr. Marietta Benkö for her enthusiastic encouragement of this project, as well as for her review of earlier drafts and judicious editorial advice. They also are very appreciative of the patience and assistance of Ms. Selma Soetenhorst-Hoedt and her colleagues at Eleven International Publishing.

Essential Air and Space Law (Series Editor: Marietta Benkö)

Volume 1: Natalino Ronzitti & Gabriella Venturini (eds.), The Law of Air Warfare – Contemporary Issues, ISBN 978-90-77596-14-2

Volume 2: Marietta Benkö & Kai-Uwe Schrogl (eds.), Space Law: Current Problems and Perspectives for Future Regulations, ISBN 978-90-77596-11-1

Volume 3: Tare Brisibe, Aeronautical Public Correspondence by Satellite, ISBN 978-90-77596-10-4

Volume 4: Michael Milde, International Air Law and ICAO, ISBN 978-90-77596-54-8

Volume 5: Markus Geisler & Marius Boewe, The German Civil Aviation Act, ISBN 978-90-77596-72-2

Volume 6: Ulrich Steppler & Angela Klingmüller, EU Emissions Trading Scheme and Aviation, ISBN 978-90-77596-79-1

Volume 7: Heiko van Schyndel (ed.), Aviation Code of the Russian Federation, ISBN 978-90-77596-80-7

Volume 8: Zang Hongliang & Meng Qingfen, Civil Aviation Law in the People's Republic of China, ISBN 978-90-77596-91-3

Volume 9: Ronald M. Schnitker & Dick van het Kaar, Aviation Accident and Incident Investigation. Concurrence of Technical, ISBN 978-94-90947-01-9

Volume 10: Michael Milde, International Air Law and ICAO, second edition, ISBN 978-90-90947-35-4

Volume 11: Ronald Schnitker & Dick van het Kaar, Safety Assessment of Foreign Aircraft Programme. A European Approach to Enhance Global Aviation Safety, ISBN 978-94-9094-793-4

Volume 12: Marietta Benkö & Engelbert Plescher, Space Law: Reconsidering the Definition/Delimitation Question and the Passage of Spacecraft through Foreign Airspace, ISBN 978-94-6236-076-1

Volume 13: Heiko van Schyndel (ed.), Aviation Code of the Russian Federation, second edition, ISBN 978-94-6236-433-2

Volume 14: Alejandro Piera Valdés, Greenhouse Gas Emissions from International Aviation: Legal and Policy Challenges, ISBN 978-94-6236-467-7

Volume 15: Peter Paul Fitzgerald, A Level Playing Field for "Open Skies": The Need for Consistent Aviation Regulation, ISBN 978-94-6236-625-1

Volume 16: Jae Woon Lee, Regional Liberalization in International Air Transport: Towards Northeast Asian Open Skies, ISBN 978-94-6236-688-6

Volume 17: Tanveer Ahmad, Climate Change Governance in International Civil Aviation: Toward Regulating Emissions Relevant to Climate Change and Global Warming, ISBN 978-94-6236-692-3

Volume 18: Michael Milde, International Air Law and ICAO, third edition, ISBN 978-94-6236-619-0

Volume 19: Nataliia Malysheva, Space Law and Policy in the Post-Soviet States, ISBN 978-94-6236-847-7

Volume 20: Philippe Clerc, Space Law in the European Context, ISBN 978-94-6236-797-5

Volume 21: Benjamyn Scott, Aviation Cybersecurity: Regulatory Approach in the European Union, ISBN 978-94-6236-961-0

Volume 22: Dick van het Kaar, International Civil Aviation: Treaties, Institutions and Programmes, ISBN 978-94-6236-972-6

Volume 23: Lasantha Hettiariachchi, International Air Transportation Association, ISBN 978-94-9094-758-3

Volume 24: Masataka Ogasawara & Joel Greer (eds.), Japan in Space, ISBN 978-94-6236-203-1